Climate
through the Ages

Climate
through the Ages

A STUDY OF THE CLIMATIC FACTORS
AND THEIR VARIATIONS

C. E. P. BROOKS

SECOND REVISED EDITION

DOVER PUBLICATIONS, INC.
NEW YORK

Published in Canada by General Publishing Company, Ltd., 30 Lesmill Road, Don Mills, Toronto, Ontario.

Published in the United Kingdom by Constable and Company, Ltd., 10 Orange Street, London WC 2.

This Dover edition, first published in 1970, is an unabridged and unaltered republication of the revised (1949) edition of the work originally published in 1926. It is reprinted by special arrangement with the original publisher, Ernest Benn, Limited.

Standard Book Number: 486-22245-4
Library of Congress Catalog Card Number: 74-100543

Manufactured in the United States of America
Dover Publications, Inc.
180 Varick Street
New York, N.Y. 10014

PREFACE TO THE FIRST EDITION

SOME years ago I had the privilege of laying before the public " The Evolution of Climate." As its title implies, that book mainly consists of an account of the variations of climate during geological times, and it deals especially with the Quaternary Ice-Age and the Post-glacial period. The causes of the events described were touched on only very briefly, and chiefly for the purpose of emphasising the climatic importance of the period of continental emergence which intervened between the mild climates of the Mesozoic and Tertiary periods and the beginnings of the Ice-Age. It was my intention then to write a companion volume dealing with the mechanism of climatic changes and drawing on the first volume for illustrations. But there were many problems requiring further investigation and much reading to be done before such a work could be approached with any confidence, and so for several years desire outran performance. Then came the work of Wegener on the theory of continental drift, and of Köppen and Wegener on the interpretation of the climatic record in terms of the travels of the continents across the parallels of latitude, and this stimulated me to complete the investigation in order to examine Wegener's theory from the climatic side.

" Climate through the Ages " has not been altogether easy writing ; the present is the key to the past, but there have been many strange meteorological situations for which we have no present parallel to guide us. Moreover, the meteorology of the present has still to solve many problems of its own, and I am even encouraged to hope that the meteorology of the past may at times help in the study of the present. The theory of the circulation of the earth's atmosphere, for instance, is not yet complete, and it may be that the modifications of the circulation during the varied climatic history of the globe, as deduced from the distribution of rainfall and temperature, will provide just the additional material required for a solution.

The point of view being meteorological, it has not been considered necessary to keep to a strict chronological order in discussing the climates of different geological periods, as was done in " The Evolution of Climate." For convenience of reference, a list of the geological periods, with some idea of their duration and a brief note on the general type of climate, has been given in Appendix I. For the purposes of the discussion, geological climates are classed as "warm" or "glacial." We are living now in a "glacial" period, though fortunately not at its maximum, and the meteorological conditions during the Quaternary Ice-Age are not really strange to us. The "warm" periods, in which a genial climate extended almost or quite to the poles, are meteorologically much more remote. In the best developed warm periods it is probable that ice was unknown. This approach to uniformity of temperature was inevitably associated with very great changes in the winds, the ocean currents, and the distribution of rainfall. The way in which this situation came about was not simple, and I have thought it best to open the book by setting out, in an Introduction, the geological evidence as to the existence and nature of these warm periods, and especially the extent to which climatic zones were developed. This greatly simplifies the subsequent meteorological discussion by allowing us to take the main facts for granted and to consider the variation of the individual factors of climate one by one, until the ground has been prepared for a more complete and logical discussion in Chapter X.

"Warm" periods have been the rule in the history of the earth, but from time to time the revolutionary forces of the underworld have succeeded in overthrowing this quiet and genial existence and have brought about crises in the earth's history, and the greatest of these crises have been marked by ice-ages. There have been at least four major ice-ages, in the early Proterozoic, in the late Proterozoic or Algonkian, in the Upper Carboniferous, and in the Pleistocene-Recent periods. Of the first two of these we know little, but that little suggests that they were entirely analogous to the Quaternary. The Upper Carboniferous glaciation, on the other hand, was highly abnormal, in that the greatest ice-sheets developed in regions which are now not far from the equator. It seems probable that there was another great period of mountain-building at the close of the Cretaceous, which developed in such a way that the distribution of land and sea, and of mountain ranges, was not favourable for extensive glaciation.

The book is divided into three sections. In the first, the various factors of climate are discussed, and the scope which they offer for the introduction of climatic changes is considered, quantitatively when possible. This part of the work is essentially a text-book of meteorology, in which, however, some of the constants of the ordinary meteorological text-book are treated as variables. Various theories of climatic change are discussed in successive chapters as they arise, but no attempt has been made to include all the theories which have been put forward from time to time. No useful purpose would be served by solemnly refuting, for instance, the theory that the mild polar climates of the Tertiary were due to radiation from a warm moon, while the author of another theory of the warm periods, who attributes them to the heat set free by decomposing animal remains in the stratified deposits, can only be described as a " Prophet of the Utterly Absurd." A fairly full list of palæoclimatic theories, possible and impossible, is given in Appendix II.

Even when we leave out of consideration the " Patently Impossible and Vain," however, there still remain theories innumerable. Most of the earlier adventurers in this subject had a single cause—oceanic circulation, carbon dioxide, eccentricity of the orbit, etc.—and saw in it complete explanation of the whole of climatic history. This spirit is not yet entirely dead, but of recent years a broader outlook has become manifest :—

> There are nine and sixty ways
> Of constructing tribal lays,
> And every single one of them is right.

There are at least nine and sixty ways of constructing a theory of climatic change, and there is probably some truth in quite a number of them. The greatest extremes of climate are not to be attributed to the abnormal development of any one factor, but to the co-operation of a number of different factors acting in the same direction. It seems probable that the dominant factors have always been " geographical," but we must give this term a very wide meaning, and include not only the distribution of land and sea and the systems of ocean currents which result from it, but also the vertical circulation of the sea, the general elevation of the land, and the amount of explosive volcanic activity. It will be seen in Chapter XII. that all these " geographical " factors are so closely inter-related that no discussion which takes account of only one of them can be complete.

But although the results of this work seem to indicate that the " geographical " factors have dominated the climatic history of the globe, this does not mean that other non-geographical factors are without effect. In Chapter XXII. it is shown that during the past two thousand years, when the distribution and elevation of the land remained practically constant, a number of minor fluctuations of climate have been associated with changes of solar activity. No doubt similar fluctuations have occurred throughout geological time, but they have been masked by the much greater effect of the " geographical " factors, and it is only exceptionally that solar influences can be recognised, as in the " varve " clays formed during the melting of the Quaternary ice-sheets. When the " varve " clays of earlier glacial periods, the rings of the fossil trees found in various geological horizons, and other remains which depend on seasonal changes of temperature or rainfall come to be examined in detail, it is to be expected that further evidence will be discovered of the existence of minor climatic fluctuations which can be attributed to small variations of solar radiation. Any such material which can be analysed in the way that the tree-rings or the Nile flood records have been analysed will be especially welcomé.

We know that the more or less regular variations in the inclination of the earth's axis, the eccentricity of its orbit, and the precession of the equinoxes must have influenced the seasons in the past. So far no climatic variations have been found of such a nature that they can be unhesitatingly attributed entirely to these astronomical factors, though it is suggested in Chapter V. that they may have been responsible for the sequence of deposits in the coal measures and in the Tertiary of Southern Europe, and there is good reason to believe that the large annual range of temperature in Northern

Europe during the Boreal or Continental period was partly
due to a high inclination of the earth's axis. Even in the
last example, however, the main cause of the extreme climate
was probably geographical, not astronomical, while the
historical period in which geographical causes fell into the
background has not been long enough for any of these
astronomical factors to change sufficiently to make their
presence felt.

Any change, geographical or astronomical, can only become
effective through its action on the atmosphere, but it will be
seen that this action is complex, and may at times produce
unexpected results. During the warm periods, for example,
there was a large amount of water vapour in the air, yet the
cloudiness, rainfall, and evaporation were less than at present,
because the extension of the sub-tropical anticyclones into
higher latitudes led to more stable conditions. The discussion
of the effect of cloudiness in Chapter VII. shows that Marsden
Manson was on the right lines when he emphasised the great
climatic importance of cloudiness, but he erred in attempting
to dissociate this element from the pressure distribution and
the atmospheric circulation in general, and in relating it
directly to earth heat. His neglect of meteorological principles
led him into some absurd conclusions, and the same is true
of many other theorists in the domain of palæoclimatology.
The theory which attributed all the great climatic changes to
variations in the amount of carbon dioxide is a case in point ;
after a series of highly laborious and intricate discussions from
the geological side, it was laid gently to rest by the application
of simple physical experiments, and it is now generally
recognised that variations in the amount of carbon dioxide,
while not entirely without some climatic significance, are of
relatively slight importance. Chamberlin's ingenious theory
of the reversal of the oceanic circulation to give warm periods
in high latitudes, although it never reverberated so widely
through the world as did the original carbon dioxide theory,
really has a better physical basis. It is shown in Chapter III.
that a change in the type of oceanic circulation, although on
different lines to Chamberlin's, may have been of great
climatic importance, but only as an auxiliary factor in conjunction with favourable geographical and meteorological
conditions.

The second section of the book applies the principles laid down in the first section to the various problems presented by geological climates. It opens with a comparison—by the method of correlation—of the climatic history of the north temperate and polar regions with the corresponding changes of elevation, land and sea distribution, oceanic circulation, and volcanic activity. In this way numerical measures are obtained of the influence of these various geographical factors on geological climates, which are found to be in good agreement with the theoretical values deduced from recent meteorological work. The next two chapters are mainly devoted to a consideration of the theory of continental drift. In Chapter XIII. the theory is stated, with some reference to its geological basis. Chapter XIV. considers the extent to which the theory satisfies the requirements of palæoclimatology, and it is shown that even on Wegener's reconstruction of the Upper Carboniferous geography, the climate of that period still presents many difficulties. Next, in Chapter XV., the distribution of land and sea, mountain ranges and ocean currents during the Upper Carboniferous is set out on the basis of the present positions of the continents, and the probable climate to which such a distribution would give rise is deduced from the principles developed in Part I. It is found that these give at least as good an approximation to the facts as does Wegener's theory, and it is accordingly inferred that continental drift is not necessary to account for the distribution of past climates.

I may, I hope, be excused for the length to which the discussion of continental drift has run. The theory is at present on its trial before the tribunal of the world's scientists, and the verdict appears to be wavering in the balance. Geology, naturally, will be the final arbiter, but the voices of other sciences are not without some weight. Meteorologists, thinking always in terms of the well-known distinction that while weather changes from day to day climate goes on all the time, are naturally averse to considering the possibility of changes in the atmosphere circulation of sufficient magnitude to give climatic revolutions on the geological scale. To most meteorologists, therefore, Wegener's theory of continental drift presents an easy escape from the palæothermal problem. But it seems to me that this point of view is similar to that of the captain of a vessel who, crossing the North Atlantic, should attempt to determine his latitude with the aid of a thermometer and isothermal chart.

There seems also to be some confusion of thought as to Wegener's theory itself. There is some definite geological evidence in favour of that part of the theory which states that the alternate stretching and compression of the crust, in conjunction with tidal forces, has led to the gradual drifting apart of the continents in an east-west direction. This being granted, there is a tendency to assume that the remaining part of Wegener's theory is also well grounded, namely, that which attributes to the continents, in addition to the comparatively small east-west movements, enormous drifts from north to south or vice versa. The only real evidence adduced in support of this view is climatological, and, practically speaking, the climate of the Upper Carboniferous; the geological evidence is quite inadequate. It is therefore necessary to examine the climatic evidence carefully from all points of view, and especially to explore the possibility of alternative explanations to it. Wegener's explanation, though not probable, is possible; in Chapter XV. I have attempted to set out an alternative explanation which likewise, though not probable, is possible. Which of us is on the right lines, time will show. But even if the theory of continental drift should ultimately be established in all its parts, the necessity for a science of " palæometeorology " will still remain. It is now quite beyond doubt that the earth has passed through periods of mountain formation alternating with long periods of comparative rest, and that in consequence the average elevation of the land has varied considerably from time to time. Similarly, on any theory (and on Wegener's theory more than on any other), the distribution of land and sea has varied greatly from one period to another. These changes must inevitably have produced corresponding changes in the distribution of climate, which can only be arrived at by applying the principles of meteorology. Hence it is hoped that this book will be of service to geologists and palæoclimatologists no matter what basal theory of past geography they may adopt.

The third section of the book deals in considerable detail with the climates of different parts of the world during the historical period, or from about 5000 B.C. to the present day. In recent years a large amount of valuable work has been published dealing with the climatic changes during this period, and this part of the book is materially more complete and definite than would have been possible had it been written even as little as ten years ago. I have been fortunate in being able to make use of a number of detailed and entirely independent records for different parts of the world, such as the annual rings of the big trees of America, the literary and historical records of Europe and of China, the levels of the Caspian, the racial movements of Asia, and the floods and low-water stages of the Nile, and these have shown so good an agreement with each other and with such records of solar activity as we possess, that I cannot but feel that the climatic fluctuations portrayed are definitely real and demonstrate the solar control of climate in the absence of disturbing geographical changes.

Apart from its meteorological interest, I hope that this part of the work will prove of service to archæologists and historians. Mr Harold Peake, in his brilliant study of " The Bronze Age and the Celtic World," found it necessary to call in the aid of climatic changes in order to understand the migrations of the " Aryans," which laid the foundations of the present distribution of peoples in Europe and Western Asia. Similarly, Ellsworth Huntington believes that the rise and decline of the ancient Maya civilisation of Yucatan can only be explained by changes in the climate, and there are other examples. The literature of historic climates is, however, so chaotic that the historian or archæologists has little inducement to trust himself to its mazes. A co-ordination of the evidence was urgently needed, and this need I have attempted to meet.

I wish to make acknowledgment to the Council of the Royal Meteorological Society for permission to reprint parts of several papers published in the *Quarterly Journal*, including Figures 1, 8, and 16 to 19, for several of which they have kindly lent the blocks. Mr W. H. Dines, F.R.S., has kindly accorded me permission to use his illustration of the heat balance of the atmosphere (Fig. 11). Messrs G. W. Bacon & Co. have accorded permission for the use of their outline chart of the globe on Mercator's projection as a basis for several of the charts. Several friends have read parts of the manuscript and have made valuable criticisms and suggestions, or have assisted me with expert advice on different aspects of a subject which is so wide that no one man can hope to master it.

C. E. P. B.

September 1926.

PREFACE TO THE SECOND EDITION

The twenty-two years which have elapsed since the preparation of the first edition have added to the literature, among other works, Sir George Simpson's theory of variations of solar radiation and Professor Zeuner's brilliant exposition of the astronomical causes of the succession of glacial and interglacial periods ; the explanation of the glacial succession in the Quaternary now seems to rest with one or possibly both of these theories. Neither of them accounts for the occurrence of the Ice-Age as a whole or for the warm periods, and with the piling up of objections to Wegener's hypothesis of continental drift, the case for the " geographical " theory has been strengthened. There has been a general acceptance of the idea of " glacial " and " non-glacial " climates which forms the basis of that theory. In the post-glacial period the principal changes of the past twenty years have been a new conception of the climate of the Sub-boreal and the dating of the beginning of the Sub-atlantic as 500 B.C. instead of 850 B.C., both of which make the post-glacial sequence more intelligible. It is now possible to present a much more complete interpretation of the changes of climate in the historical period than could be given in 1926.

While the general plan of the book remains unchanged, several chapters have been almost entirely rewritten. It seemed unnecessary to reprint the Appendix describing the mathematical theory of correlation, which is now widely known and is available in many text-books on Statistics. My thanks are due to the many authors who have sent me copies of their publications dealing with climatic changes, which have saved me much arduous search in libraries, to Sir George Simpson and the Manchester Literary and Philosophical Society for permission to reproduce Figures 9 and 10, to the publishers for bringing out a new edition in the face of great difficulties, to Miss N. Carruthers for reading the revision and making a number of valuable suggestions, and especially to my wife for her great practical help and encouragement during the course of the revision.

C. E. P. B.

February 1948.

CONTENTS

LIST OF ILLUSTRATIONS

INTRODUCTION

The Normal Climate of Geological Time

DURING some hundreds of millions of years with which we are acquainted through the records of the rocks, the surface of the earth has passed through some strange climatic vicissitudes. At least four, probably five or more times in its history great ice-sheets have spread out from various centres, covering the plains and even filling the shallow continental seas. There have been other periods of somewhat less intense climatic stress, when perhaps only small glaciers were able to develop among the high hills. These strenuous episodes have been of great importance in the development of the earth's living beings, but always they have been brief, and always after them the earth has returned to more genial conditions, which have endured for long periods. Hence we may regard these genial conditions as the normal state of affairs on this our earth, and the glacial periods as episodes disturbing the normal climate for a brief time, as at long intervals a passing cyclone disturbs the peaceful life of a tropical island. Many references to these warm periods will occur in the next few chapters, and it is necessary to have some preliminary knowledge of the climatic conditions which characterised them. I am therefore beginning this study, not with the stir and strife of an ice-age, but with the everyday life of the genial periods. We are not yet in a position to discuss their meteorology, but we can see what the geologists have to say about their general climate.

The first point to notice about these periods, however, is not climatic, although we shall see later that it has a very great bearing on their climatology; it concerns the absence of mountain ranges. Warm periods were without exception, periods of low relief, while glacial periods occurred when the earth's crust had been thrown into folds and ridges by great internal convulsions. The processes of denudation—wave action on the coasts, running water on the slopes and in the valleys, and blown sand in the steppes and deserts—go on all the time, and are constantly tending to level up the earth's crust, and they are powerfully aided during the cold periods by the action of frost and ice on the high ground. Against them are set the internal mountain-building forces, which act with great intensity for comparatively short periods at great intervals. The forces of denudation are most powerful when the land is highest ; following a period of mountain-building the general level is reduced very rapidly at first, and the normal stage of low relief is soon reached again.

During the periods of low relief also the sea generally encroached on the land and the continental areas were greatly reduced in size. This we shall see was another factor of great climatic importance, first, because it did away with the areas of intense winter cooling in the centres of the great continents in middle to high latitudes ; and secondly, because the extension of the seas allowed the warm ocean currents to penetrate very readily into the polar regions through many broad channels. This gives us the background for our climatological study—groups of broad low islands rather than continents, with rounded hills instead of mountains, and wide oceans extending through broad channels from pole to pole.

The next point is the comparatively small difference of temperature between the tropical and the polar regions. The marine fauna, and to a less extent even the land vegetation, differ little whether we are in Greenland, in Yorkshire, or in Central America. When the rich fossil faunas and floras from high latitudes were discovered in the middle of the nineteenth century and this similarity in the life from widely different latitudes was first remarked, it was believed that during the early geological periods there were in fact no biological zones at all, and that life was really uniform over the whole extent of the seas. Radio-activity had not been heard of, and it was believed that the high temperatures of the earth's interior indicated by volcanic eruptions were mainly a legacy from the time, not long before the Cambrian, when the earth was entirely molten. It was believed that for a long time after the formation of a solid crust, the condensation of water as oceans, and the origin of life, this internal reservoir of heat continued to make itself felt at the surface, and maintained genial temperatures in all latitudes throughout the Palæozoic, the Mesozoic, and a large part of the Tertiary. The pre-Cambrian and Carboniferous glaciations were unknown, and the apparently uniform, equable temperatures of these earlier periods fitted excellently into this theory of a cooling earth. It is now recognised, however, that the earth has owed its surface warmth to the sun as far back as we can see into the past, and that even in the warmest periods there must have been zonal differentiation of climates ; the tropics were perhaps a little warmer than they are to-day, while the polar oceans and their shores had a temperature now found in temperate latitudes. Since the deposits which have come down to us were almost invariably formed in the sea or in the estuaries of wide rivers, we know comparatively little of the plants which grew in the interior of the continents in high latitudes ; what we do know suggests that they were of hardier types than those from the low coastal valleys.

Neumayr (1) was the first to bring forward definite evidence of a zonal distribution of animals in earlier geological periods ; his conclusions, as revised by V. Uhlig (2), distinguish five faunal zones during the Jurassic :—Boreal, Mediterranean-Caucasian, Himalayan, Japanese, and South Andean—but the four latter all seem to be facies of a tropical zone which is contrasted with the cooler boreal. The boundary between these zones does not run strictly parallel with the lines of latitude, but shows divagations which may reasonably be attributed to ocean currents. Similarly, during other geological periods, the faunas of different latitudes when examined critically invariably show differences between different latitudes which are best accounted for by a slow decrease of temperature towards the poles ; for example, during the Cambrian, the *Archæocyathinæ* of the Antarctic show the same species as those of Australia, but in a dwarfed and crippled condition. In the Lower Cretaceous we have a clear division of the marine faunas into northern boreal, Mediterranean-equatorial, and southern boreal, the latter containing some species identical with those in the first, but absent in the second. The principal difference between the temperate and the polar faunas throughout the Mesozoic is the absence or dwarfing of the corals in the latter.

Similarly, although a rich vegetation apparently extended as near the present poles as the land surfaces permitted, there are considerable differences between the lower Tertiary floras of the sub-Arctic regions and those of lower latitudes. Thus E. W. Berry (3) records " the total dissimilarity between the Canadian floras, which are a part of the [Upper Eocene] Arctic flora of Alaska, Greenland, Iceland, Spitsbergen, etc., and the contemporaneous flora of the [Mexican] Gulf States." The Arctic flora is found also in Northern and Eastern Asia, so that its distribution covers an irregular area surrounding the pole, the southern limit varying from 45° N. to the Arctic Circle. The plants include—poplar, willow, alder, birch, hazel, beech, oak, plane, laurel, andromeda, ash, guelder rose, cornel, magnolia, ivy, spindle-tree, buckthorn, sumach, and hawthorn. All these are now found in some parts of the cool temperate zone, and the assemblage, while presenting a very different picture from the present life of the sub-Arctic regions, cannot be described as sub-tropical. Berry considers that the identifications of palms, etc., which have given rise to the ideas as to very high polar temperatures in the early Tertiary, are erroneous. The most northerly Eocene flora at present known is from near Cape Murchison in Grinnell Land, latitude 71° 55′ N. ; Berry reduces the list of Heer's determinations of this flora to the following :—horse-tail, yew, pine, spruce, poplar, birch, hazel, sedge or grass, and apparently water-lily—but he remarks that even this flora is sufficiently remarkable when we consider the scantiness of the present vegetation of the region, and would require the present isotherms to swing 15° or 20° northward.

Not only has the northern boundary of this cool temperate flora moved southward since the Eocene, but so also has its southern boundary, the shift being about 10° of latitude. This suggests that the tropical as well as the polar regions were warmer in the Eocene than they are at present, though not to the same extent, and that definite climatic zones were in existence at that time.

The third point to notice about the warm periods is their general aridity. Deserts have apparently existed throughout geological time, but during most of the warm periods, and especially in the Mesozoic, they expanded greatly, extending from the sub-tropical regions far into the present temperate zones. Among the most remarkable of these " fossil deserts " we have the " Old Red Sandstone " of the Devonian period, consisting mainly of inland lake or lagoon deposits formed under arid or semi-arid conditions. This régime extended from Ireland across Britain to Southern Norway, Poland, Courland, and the White Sea, and similar deposits are found in Nova Scotia, New Brunswick, Canada, and the north-east of the United States. This region was not altogether rainless, but the rain came in occasional heavy showers which caused brief floods in the water-courses, carrying large quantities of coarse detritus into the lakes and depressions. The desert character of the land was accentuated because the specialisation of the plants for life on dry land had scarcely begun, and it is quite likely that regions with a similar climate at the present day would have a moderately rich vegetation. They have therefore been described as biological rather than climatic deserts.

Desert deposits are again well developed in the Permian, in Britain, France, Germany, and the Tyrolese Alps. During this period a great inland sea was formed over a large part of Europe ; since evaporation exceeded rainfall this sea became highly saline and deposited thick layers of salts. These show in part regular annual layers, gypsum, which is more soluble in cold than in warm water, being formed in summer, while in winter its place was taken by rock salt. Kubierschky, quoted by Köppen and Wegener, on the basis of similar phenomena in the Sahara, estimates the annual range of temperature as between 60° and 95° F. The number of annual layers indicates that the salt lake existed for some 10,000 years, after which the salt deposits were covered by a layer of desert sand.

In the Triassic we have a very wide development of desert formations, especially in Western and Central Europe and in the east of the United States, Texas, Colorado, and Idaho. Many of these may have been biological rather than climatic deserts, but the occurrence of numerous salt deposits, especially in Central Europe, shows that the rainfall was less than the evaporation over wide areas. In the Jurassic and Cretaceous there is less evidence of extensive deserts, partly because the climate was not so arid as that of the Triassic, but partly because the land vegetation had evolved more specialised types which were able to exist without the constant presence of water. Occasional salt deposits show, however, that the climate of Europe was still rather dry. During the Tertiary, the Eocene period shows a return of moister conditions, but the Oligocene and, to a less extent, the Miocene, were again dry over much of Europe. The Mediterranean regions had for a time a true arid climate, while Egypt was for long an absolute desert, but in Central Europe, although the summer was hot and dry and was prolonged into autumn, there was heavy rainfall of the thunderstorm type in spring, which tore leaves and twigs from the trees and bushes and carried them to the lakes, where they were buried in mud. The winter was generally mild, and there were many evergreens, but some of the Miocene leaves from Central Europe show traces of frost action, indicating that there were occasional cold nights. Generally speaking, the climatic zones in Europe lay 10 or 15 degrees north of their present position.

In Western U.S.A., where the succession of events has been worked out in great detail, the Eocene and early Oligocene enjoyed a sub-tropical climate with ample rainfall, becoming temperate farther north in Alaska. In middle Tertiary the climate became increasingly cooler and more arid, especially to the east of the Rocky Mountains, this process culminating in the cool semi-arid climate of the Pliocene.

It is probable that one of the most momentous steps in the history of life was taken during a period of drought, namely, the origin of air-breathing land vertebrates. A similar development on a smaller scale was the evolution of the Lung-fish (*Ceratodus*), adapted to breathe both air and water, which to-day inhabits regions subject to alternating floods and droughts. Ancestral forms appeared in the Devonian, and the true lung-fish as early as the Permian ; it was widely distributed in the Triassic of Central Europe.

Thus it will be seen that the predominant features of the normal geological climate were warmth and dryness. Broadly speaking, the polar regions had the climate of the present temperate belts, while the latter had the climate of the sub-tropics. In the latitude of the British Isles the rainfall came almost entirely in the form of brief, heavy showers, of the type associated with thunderstorms, and the steady but gentle rains which are characteristic of our winter months did not occur. These rains are associated with the passage of extensive barometric depressions or cyclones, and from their absence we can infer that during the warm periods such depressions either did not occur, or were limited to high latitudes. Thus the meteorology of these warm periods was very different from that of to-day, and it is these differences which we have to analyse and, if possible, account for, in the following pages.

REFERENCES

(1) NEUMAYR, M. " Über klimatische Zonen während der Jura- und Kreide-zeit." Wien, *Denkschr. K. Akad. Wiss., Math. nat. Kl.*, 47, 1883, p. 277.

(2) UHLIG, V. " Die marinen Reiche des Jura und der Unterkreide." Wien, *Mitt. Geol. Ges.*, 4, 1911, p. 329.

(3) BERRY, EDWARD W. " A possible explanation of Upper Eocene climates." *Proc. Amer. Phil. Soc.*, 61, 1922, p. 1.

PART I

THE CLIMATIC FACTORS AND THEIR VARIATION

CHAPTER I

" Glacial " and " Non-Glacial " Periods

IN the preceding Introduction we saw that one of the
characteristics of the " warm " periods which have formed
the major part of geological time was the small temperature
difference between equatorial and polar regions, small, that
is, relative to the present difference. This at once introduces
a difficulty which the meteorologist feels very acutely in
attempting to account for geological climates, for the basis
on which the existence of the major climatic zones depends
is not to be found in any surface features of the earth, which
might have been different in some former geological period,
but in the fact that the earth is very nearly spherical. So
long as the axis of rotation remains in nearly its present position
relative to the plane of the earth's orbit round the sun, the
outer limit of the atmosphere in tropical regions must receive
more of the sun's heat than the middle latitudes, and the
middle latitudes more than the polar regions ; this is an
inviolable law. It is not difficult to think of causes, such as
a change in the heat of the sun, or the formation of a veil of
volcanic dust in the atmosphere, which will slightly raise or
lower the mean temperature of the earth as a whole, or change
the temperature of equatorial regions more than that of the
polar regions ; it is much more difficult to think of a cause
which will raise the temperature of polar regions by some
30° F. or more, while leaving that of equatorial regions almost
unchanged, and so bring about an approach to the distribution
of climatic zones during the warm periods.

Let us consider what is the essential difference between
polar and equatorial regions to-day. We can say that the
mean temperature of the polar regions is 20° F., and that
of the equatorial regions 80° F., and the difference between
20° and 80° is very striking. But the zero on the Fahrenheit
scale is purely arbitrary ; the true zero of temperature is at
−459° F., or to adopt the units employed by physicists, −273°
C., and if we express these temperatures on the absolute

scale the figures become—polar, 266° A, equatorial, 299° A, and the difference does not look so alarming. The essential difference is not between 20° and 80° F., but between the fact that the former is below the freezing point of water and the latter is above it. The real difference between the centre of Africa and the centre of Antarctica is that in the former, water is water and in the latter, water is ice. Similarly, the essential way in which the polar regions during the warm periods differed from the polar regions at present was that they had no great ice-sheets and no floating ice on the sea, that water was water, and they were " non-glacial."

The winter cold of Siberia is well known, and is explained in the geographical text-books by the statement that the climate is " continental." With the coming of the short days, in which the sun has little heating power, and the long nights, the surface of the ground loses its heat very rapidly, and soon falls below freezing point. At a depth of a foot or two the soil may be much warmer, but heat passes very slowly through the ground, and this underground store of heat does not help appreciably to keep up the temperature of the surface. Soon snow falls on the frozen ground ; snow, too, is a very poor conductor of heat, and moreover its white surface reflects four-fifths of the sun's heat which falls on it ; it absorbs heat very slowly and loses heat very readily, hence the temperature falls still more rapidly, until life becomes barely supportable. Compare this with the conditions in the centre of the Pacific Ocean in the same latitude. As the surface layer of water cools, the winds and waves, aided by convection, mix it with the underlying water, and in this way the cooling which was practically limited to a few inches of soil or snow over the land, becomes spread through many feet of water in the oceans. This means that the sea surface cools very much more slowly than the land, quite apart from the fact that water has a higher heat capacity than soil or rock. But when water freezes and loses its mobility, it takes on some of the properties of land, and when snow falls on the ice, we have a surface which is indistinguishable from the surface of the snow-covered land. Thus an Arctic Ocean covered by a snow-encrusted layer of floating ice acts almost as a northward extension of the continents of Asia and America. The temperature does not fall quite so low as in Siberia, partly because, the surface being more level, the cold air can escape more readily, and partly because a little more heat finds its way from below through ice than through the subsoil, but a temperature of −62° F. was recorded over the ice in March 1894 during Nansen's " Fram " expedition.

But the cooling power of a snow-covered surface is not limited to the surface itself; the winds spread its influence over the surrounding land or ocean. The winds blowing off the snow of Canada lower the mean temperature of the whole of North America except the Pacific coast, and this cooling allows the snow-cover to extend much farther south than it would do if, for instance, these cold winds were held off by a range of mountains extending from east to west. Investigations to be described in detail later (Chapter VIII.) have shown that if in the middle of a large open polar ocean a circular cap of ice were formed, with its centre at the pole and its southern edge in latitude 80° N. (that is, having a radius of ten degrees of arc or about 690 miles), the temperature would be lowered by about 45° F. at the pole and about 20° F. in latitude 80° on the edge of the ice, while the cooling would still be quite appreciable at a distance of 500 miles south of the ice-edge, across the open ocean.

Now let us suppose (after Brooks (1)) that the earth is entirely covered by water for a distance of more than 2,000 miles from the North Pole, and further suppose that the surface of the water at the pole itself is just warm enough to prevent freezing, while the temperature rises outwards from the centre at a uniform rate. Now suppose that a small uniform decrease of temperature occurs over the whole ocean. This results in the formation of a mass of floating ice in the centre, which will extend to the limit of the area reduced below the freezing point by the initial fall of temperature. This ice itself exerts an additional cooling effect on the air, not only above the ice, but for some distance round it, so that a further area of the sea surface has its temperature reduced below freezing point and becomes converted into ice. The ice-cap accordingly extends beyond the limits of the area of freezing due directly to the initial fall of temperature. Equilibrium is reached only when the initial decrease of temperature, plus the cooling effect exerted by the ice-cap at a point on its edge, is balanced by the rise of temperature due to increasing distance from the pole.

Now we come to the consideration that while the rise of temperature due to the horizontal temperature gradient is proportional to the distance from the pole, the area of the ice-cap increases with the square of its radius. Consequently, if the cooling effect of an ice-cap at a point on its edge varies directly with its area, the cooling effect is small at first, but increases more and more rapidly as the ice-cap grows in size, while the counteracting effect of the normal horizontal gradient increases at a uniform rate. Sooner or later a point is reached at which a further growth of the ice-sheet causes a greater lowering of temperature on its edge than can be neutralised by the horizontal temperature gradient, and beyond that point, in the conditions postulated, the ice-cap must continue to grow indefinitely. The critical point at which the ice-cap becomes unstable depends on the numerical values assigned to the cooling power of ice and to the horizontal gradient.

This conclusion is so important, underlying as it does the whole theory of climatic changes as set out below, that I may be excused for dwelling on it at some length. First, is the cooling power of an ice-cap on its edge proportional to its area ? The winter cooling by floating ice is quite comparable to the winter cooling due to the presence of land in high latitudes, since in winter the land is almost invariably snow covered. It is shown in Chapter VIII. that with increasing land area, the cooling effect exerted by each square kilometre of land in winter actually increases at first with the increase in total area, so that by doubling the area we more than double the cooling. In the centre of a round island the maximum cooling effect per square mile of total area occurs when the island has a radius of about ten degrees of arc ; beyond this the more distant parts cease to exert their full effect, and while the total cooling effect continues to increase with increasing area, the average effect per square mile begins to decrease. On the edge of a large island the maximum effect is exerted only by that portion which lies within a circle of ten degrees radius centred at a point on the edge ; when the radius of the island exceeds eight degrees, any further increase of radius makes a comparatively small difference in the area within this ten-degree circle. We shall not be far out if we assume that the fall of temperature due to the ice at a point on its edge is directly proportional to the area of ice included in a circle of ten degrees radius, drawn round that point, everything outside that circle being ignored. As the average effect per square mile of land between ten and twenty degrees away is only one-fifth of the average effect of land inside the circle of ten degree radius, this simplification is justified.

The next point is to determine the critical temperature given by definite values of horizontal gradient and cooling power of ice. We still start with a simple numerical example. Let us put the cooling power of a floating ice-cap as $0 \cdot 5°$ F. for each one per cent. of a circle of ten degrees radius which is occupied by ice. Then on the edge of the ice-cap the lowering of temperature due to the ice will be $0 \cdot 5°$ F. when it has a radius of one degree ; four times $0 \cdot 5$ or $2 \cdot 0°$ F. when it has a radius of two degrees ; nine times $0 \cdot 5$ or $4 \cdot 5°$ F. when it has a radius of three degrees, and so on. We will suppose that before the ice-cap developed the temperature increased uniformly outwards from the pole at the rate of $1°$ F. for each degree of latitude. Then, taking the freezing point of sea water as $28°$ F. and supposing that initially the sea at the pole was just on the point of freezing, we have the following distribution of temperatures before the formation of the ice-cap and at different stages of its growth :—

Distance from pole, degrees of latitude	0	1	2	3	4	5
Initial temperature, °F. .	28	29	30	31	32	33
Cooling due to ice-cap, °F.	0	0·5	2	4·5	8	12·5
Temperature on edge of ice-cap, °F. . . .	28	28·5	28	26·5	24	20·5

Table 1.—Temperature at edge of floating polar ice-cap.

This table shows that the floating ice-cap will experience most difficulty in establishing itself ; once it has reached a certain size the temperature on its edge will be below the freezing point of sea water, and it will continue to expand owing to the lowering of temperature which the ice itself introduces. Suppose that, starting with a temperature of $28°$ F. at the pole, we have a general fall of temperature by $0 \cdot 5°$ F. An ice-cap will form and will grow until it has a radius of one degree of latitude. At this stage the temperature at its edge, at a distance of one degree from the pole, which was originally 29° F., will have been lowered 0·5° F. by the initial cooling and a further 0·5° F. by the ice, a total decrease of 1° F. The water on the edge of the ice will therefore be exactly at freezing point ; lower the general temperature a little more, and the ice-cap will continue to grow until it reaches very large dimensions ; lower it a little less, and the ice will extend less than one degree from the pole. This result is not dependent on the special numerical values which have been taken. Suppose for instance, that the cooling power of ice is only 0·25° F. for each one per cent. of a ten-degree circle, instead of 0·5° F. as before. Then the critical radius of the ice-cap is four degrees of latitude instead of one degree, and for it to grow to large dimensions the initial lowering of temperature must exceed 1° F. instead of 0·5° F.

Let us now put the matter quite generally. Let the horizontal gradient of temperature be h degrees per degree of latitude and let the cooling power of ice be k degrees for each one per cent. of a circle with a radius of ten degrees. An initial small fall of temperature by t degrees will cause the formation of an ice-cap with a final radius greater than t/h ; call this radius R, where R is measured in degrees of latitude. The cooling power of this ice-cap is $k\mathrm{R}^2$, and in a condition of equilibrium the total cooling $t+k\mathrm{R}^2$ is balanced by the horizontal increase of temperature $h\mathrm{R}$. Thus we have the equation :—

$$t + k\mathrm{R}^2 = h\mathrm{R}$$

the solution of which is

$$\mathrm{R} = \{\, h \pm \sqrt{h^2 - 4kt}\,\} / 2k.$$

When $t = 0$, $\mathrm{R} = 0$, so that only the negative value of the root term is required here. The critical point occurs when $4kt = h^2$, or $t = h^2/4k$; for values of t above this amount the equation has no solution, and on the assumptions made the ice-cap has no finite limit.

The average value of the horizontal gradient of air temperature in January (h) which would prevail over an ice-free ocean between 80° and 90° N. latitude is difficult to estimate, but would probably be of the order of 0·5° C. (0·9° F.) per degree of latitude. The cooling power of land in winter in high latitudes is approximately 0·5° C. (0·9° F.)

for one per cent. of a circle with a radius of ten degrees ; the cooling power of a continuous surface of thick ice would be nearly the same, but where the ice is interrupted by lanes of open water caused by ocean currents the cooling power is less, and we may take k as $0 \cdot 25°$ C. ($0 \cdot 45°$ F.) for one per cent. Hence the critical value of t for which the ice-cap becomes unstable is

$$\frac{h^2}{4k} = \frac{0 \cdot 81}{4 \times 0 \cdot 45} = 0 \cdot 45° \text{ F.}$$

at which stage the radius R is one degree of latitude. We can now calculate the total cooling on the assumption that the initial fall of temperature is $0 \cdot 6°$ F. and that ice distant from any point more than ten degrees of arc has no influence on the temperature of that point.

The results are as follows :—

Radius of ice-cap R, in ° .	0	1	2	3	4	5	6	8	10	15	20	25
Total cooling on edge, °F.	0·6	1·05	2·4	4·6	7·8	11·8	14·5	17·2	18·2	20·4	21·3	22·2
Rise due to normal horizontal gradient h, °F. .	0·0	0·9	1·8	2·7	3·6	4·5	5·4	7·2	9·0	13·5	18·0	22·5

Table 2.—Cooling at edge of ice-cap.

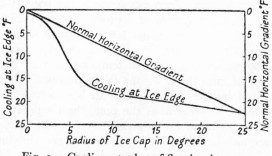

Fig. 1.—Cooling at edge of floating ice-cap.

These results are shown graphically in Fig. 1, in which the curve represents the cooling on the edge of the ice, while the inclined straight line represents the normal horizontal gradient. Over almost the whole of the diagram the curved line representing cooling is below the straight line representing the normal rise of temperature with increasing distance from the pole, and the two do not meet until the ice-cap attains a radius of nearly twenty-five degrees of latitude. Hence an initial winter cooling to 0·6° F. below the freezing point will result in the formation of a floating ice-cap with a radius of nearly twenty-five degrees.

We have now to consider to what extent this ice-cap will melt in summer. Let us suppose that the floating ice-cap has been formed in winter owing to a depression of the temperature to 24° F. or 4° F. below the freezing point of sea water. When summer comes, there is a general warming up ; let us suppose (in accordance with a calculation by F. Kerner (2)) that under normal or " non-glacial " conditions the summer temperature would be 13·5° F. above that of winter, or 9·5° F. above the freezing point. The edge of the ice-cap, which was just at freezing point in winter, will now rise above freezing point, and the ice will begin to melt. But although the " non-glacial " temperature, even at the pole, is now above the freezing point of water, this does not mean that the ice-cap will necessarily entirely vanish. As the ice melts back towards the pole, its cooling effect at a point on its edge decreases slowly—very slowly at first—and on the edge of an ice-cap of ten degrees radius it is only 4° F. less than on the edge of an ice-cap of 25 degrees radius. But the natural or " non-glacial " temperature decreases towards the pole at the rate of 0·9° F. for each degree of latitude, and in a distance of 15 degrees or 1,035 miles the temperature will have fallen from this cause by 13·5° F. This is exactly equal to the 9·5° F. by which the summer temperature rises above the freezing point, plus the 4° F. due to the decrease in size of the ice-cap, and by the time the ice-cap has melted back to a distance of ten degrees from the pole, the temperature at its edge will again be at freezing point and it will be unable to melt back any farther. Hence, so long as the amount of the summer warming does not exceed the greatest difference between the curved line and the straight line in Fig. 1, the ice-cap will not entirely disappear even in summer ; instead a core of ice will remain from which the ice-cap will spread out again in the following winter. It is evident that this central core will in time come to consist of very thick old ice, while the outer region in which melting takes place in summer will consist only of thinner ice of one season's growth.

In order that the ice-cap may entirely disappear in summer, the summer temperature must rise so far above the freezing point that the curved line in Fig. 1, representing the cooling at the ice-edge, lies entirely above the straight line representing the normal horizontal gradient of temperature. From Table 1, we see that these lines are 10° F. apart when the ice-cap has a radius of eight degrees, and in order that the ice may disappear in summer, the summer temperature must rise above the freezing point by at least 10° F. Allowing an annual range of 13·5° F., this means that the winter temperature must be above 24·5° F.

Fortunately this hypothesis as to the conditions under which an extensive floating ice-cap develops does not rest entirely on theoretical arguments. The assumptions made represent very fairly a generalisation of the Arctic Ocean, and we can confirm the conclusions reached by reference to the existing ice-conditions in that ocean. In winter, with the exception of an embayment west of Norway kept open by the Gulf Stream Drift, the whole ocean north of the Arctic Circle is covered by floating ice, much of which breaks up in summer. Inside the outer or winter limit is another or summer limit representing the area over which there exists a solid mass of very thick ice (the " Palæocrystic Ice ") which never melts, winter or summer. The limit of this inner ice-cap has an average latitude of about 78 degrees, or 12 degrees from the pole, and this may be taken as representing the limit of the region of maximum depression of temperature below freezing point.

It may be remarked in parenthesis that this theory of a critical temperature applies equally well to the case of an ice-sheet over land, and affords a satisfactory explanation of the rapid growth and decay of the Quaternary ice-sheets at certain stages of their existence.

The critical point for an ice-sheet on a conical continent with a surface slope of 1 in 1,000 (e.g., height 2,000 metres, radius 20 degrees of latitude) occurs when the radius of the ice-sheet is about one hundred miles. Once it extends beyond this limit, it will grow rapidly until it attains a radius of more than 500 miles, and then more slowly to a much greater radius.

Conversely, during a period of amelioration of climate, a large ice-sheet will decrease slowly in size until it attains a radius of rather less than ten degrees (600 miles), after which it will shrink rapidly, even without any further change of climate, until its radius is less than one hundred miles.   This is exactly what happened in Scandinavia during the closing stages of the Quaternary Ice-Age.   The ice-sheet withdrew slowly at first, until the distance from the centre to the southern edge was about nine degrees ; after this the retreat became more and more rapid until the ice-sheet had sunk to inconsiderable dimensions.   Here again the concordance of the theory with the observations is as good as could be wished.

From this discussion we can see that if over an open polar ocean the winter temperature at the pole falls only as much as $0 \cdot 6°$ F. below the freezing point of sea water, an ice-cap will develop, which will extend rapidly until it reaches a latitude of about 78°.   From this stage it will grow more slowly to about 65°, unless in the meantime its growth is arrested by land, by the seasonal change of temperature, or by a warm ocean current.   The ultimate lowering of the winter temperature brought about by the initial small fall of temperature of $0 \cdot 6°$ F. will amount to about 45° F.

This estimate of the cooling effect may appear large, but consider the difference of conditions a little way below and a little way above the critical point.   Below the critical point all the cooled water remains on the surface in the form of ice, spreading freezing temperatures all round it, and even when the ice melts, the cold thaw water, owing to its comparative freshness, is lighter than the warmer more saline water and remains on the surface, freezing again very readily. Fresh snow falls on the ice and increases its thickness, until the surface of the central parts of the Arctic Ocean resembles that of the Antarctic Ice Barrier, with correspondingly low temperature.   On the other hand, above the critical point, the water as it cools becomes denser and sinks to the bottom, where it drains away as a ground current, to be replaced by warm surface currents from lower latitudes.   Every inlet is flooded with warm water, no ice can form, and the temperature never sinks below the freezing point of sea water.   The difference between these two pictures shows that they may easily have a temperature difference of 45° F.

This good agreement between the conditions found in the Arctic Ocean and the conditions deduced from theory suggests that the " non-glacial " winter temperature of the Arctic Ocean may not be far below the freezing point of sea water, and that if the Arctic ice could once be swept away, it might find some difficulty in re-establishing itself. It therefore becomes an interesting speculation to try to determine what the " non-glacial " temperature of the Arctic would be. The problem may be stated thus : to determine what would be the distribution of temperature if sea water, with all its other properties unchanged, reached its freezing point and point of maximum density at a very much lower temperature than 28° F. This problem may be attacked in two ways. One solution, which is due to F. Kerner (2), starts with an attempt to determine the temperature distribution over an entirely open ocean, and then superposes on this the effects of the present land and sea distribution. Kerner considers that there are only two regions in which the oceans are free from the effects of floating ice, namely, the seas of the East Indies and the centre of the North Pacific anticyclonic area, and he takes the temperatures of the sea surface in these two regions as the real oceanic temperatures which would obtain if there were no transference of heat across the parallels of latitude by ocean currents. From these temperatures, taking account of the influence of solar and terrestrial radiation, he calculates by various methods the " akryogenous " (*i.e.*, " non-glacial ") temperature of every tenth parallel of latitude, and obtains for the North Pole an average annual temperature of 28° F. He computes the annual range of temperature as 13·5° F., and this gives him for the January temperature 21·3° F., which he regards as the lowest limit for the temperature of an open ocean near the pole in the absence of ocean currents crossing the lines of latitude. (This, of course, is well below the freezing point of sea water, but we have guarded against this contretemps by a suppositious change of the freezing point.)

Kerner next analyses the present distribution of temperature in January along the 75th parallel of latitude, and obtains a geographical formula which expresses the temperature of any point in this latitude in terms of the land and sea distribution. The geographical effects are twofold ; an elevation of temperature by the Gulf Stream Drift, and a depression of temperature owing to the cooling effect in winter of the large land-masses which nearly surround the Arctic Ocean in about 70° N. From Kerner's discussion it appears probable that the Gulf Stream effect is slightly the larger, and that the " non-glacial " temperature near the pole in an Arctic ocean with the present configuration would be a degree or two above the figure of 21·3° F. calculated for the open ocean.

The second way in which a rough calculation of the " non-glacial " temperature of the North Pole may be made depends on my investigations of " continentality and temperature " described in Chapter VIII., and referred to earlier in this chapter. By comparing the distribution of temperature with the distribution of land and of ice, I obtained measures of the thermal effects of land and of ice, in January and in July in different latitudes. These enabled me to eliminate the land and ice effects, but not the effect of ocean currents crossing the parallels of latitude, and consequently gave me the temperature of an open ocean under " non-glacial " conditions, with a steady interchange of water between equator and pole. The calculation was carried up to 70° N., but the figures calculated for January are very erratic. The results for July are more regular, since in that month the interference by ice is much less than in January, and the July figures gave :—

Latitude, N.	40°	50°	60°	70°
Mean July temperature, °F.	64·6	55·0	45·9	43·3

By extrapolation the July water temperature of 90° N. would be about 40° F., and since there is very little land within 20° of the North Pole, we may take the figure of 40° F. to represent the " non-glacial " July temperature of the North Pole with the present land and sea distribution. The annual range over an open ice-free polar ocean being 13·5° F. according to Kerner's calculation, the January " non-glacial " temperature would be about 26·5° F. There are, however, many objections to this method of calculation, and it cannot give us more than a rough approximation.

Taking account of all three sources of information, the ice-conditions in the Arctic which suggest a January " non-glacial " temperature of not more than 24° F. (p. 39), Kerner's calculation of not less than 21·3° F., and my own rough calculation of 26·5° F., we find that the first of these figures, 24° F., is probably not far from the truth. This gives us some very interesting information about the stability of the present climatic régime. Quite a small rise in the general temperature of the earth, say, 2° F., would suffice to make the whole ice-sheet unstable in summer. This does not necessarily mean that it would completely break up and disappear during a single summer. The Palæocrystic ice is so thick and extensive that in order to melt it a very large amount of heat would be required. Hence the most that could happen during a single abnormally warm summer would be that, after the fringe of thin ice of one winter's growth had been broken up, a border a few miles wide round the Palæocrystic ice might be attacked before the winter cold came again to reverse the process. But suppose that there occurred a real and permanent change in the general climatic conditions, so that the warm summers came year after year, alternating with mild winters in which the outer fringe of ice did not form again with quite its old thickness or extent. Under these conditions, aided by the gradual drift of the whole ice-mass across the Arctic Ocean, the Palæocrystic ice would gradually break up and be replaced by much thinner and looser ice. After this stage had been reached it might be possible for the ice to disappear completely during favourable summers, forming again in the following winter.

Such a " semi-glacial " condition would require a nice adjustment of temperatures, for it seems that it would be possible only if the temperature in winter fell so little below the critical figure of 27·5° F. that the initial stages in the formation of the ice-cap took place very slowly. If the worst of the winter were over by the time the ice-cap had attained its critical radius of one degree, the subsequent extension beyond this radius would be small and weak, and the ice-cap would be readily dissipated in summer. It will be shown in Part III. that this condition in which ice formed in winter but disappeared in summer, probably existed during the fifth to seventh centuries of the Christian era.

If the general warming up went a little farther, so that the " non-glacial " winter temperature at the pole rose above the freezing point of sea water, the process would at first be very similar, though presumably more rapid. The Palæocrystic ice would melt back farther and farther each summer, but so long as it persisted with a radius exceeding one degree, an extensive ice-cap would form again in the winter. Finally, however, there would come a summer in which the ice-cap completely disappeared, and in the following winter it would not form again ; the climate would become definitely " non-glacial."

Apart from the transitional or " semi-glacial " stage, which can have occurred but rarely in geological times and then only for short intervals, only two types of oceanic climate

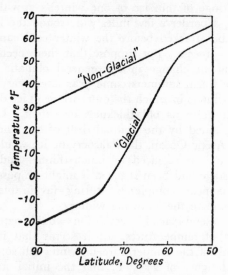

Fig. 2.—Temperature difference between " non-glacial "
and " glacial " climate.

are possible, the " glacial " and the " non-glacial," and these are very wide apart. The former is characterised by a very rapid decrease of temperature in high latitudes, the latter by a slow decrease. In order to show this, I have constructed Fig. 2, which gives the variations of temperature between latitudes 50° and 90°. The straight line represents conditions in winter over an open ocean in which the temperature falls at the rate of 0·9° F. for every degree of latitude, the water at the pole being one degree above the freezing point. The curve below it represents the temperatures over the same ocean in winter which would be the final result of a general initial fall of temperature by 5° F., causing the formation of an ice-cap in the way described above. The cooling power of ice is taken as 0·45° F. for each one per cent. of a ten-degree circle. The original fall of temperature by 5° F. has increased to 50° F. in high latitudes as the result of the formation of the ice-cap.

This result greatly simplifies the problem presented by warm polar climates. Instead of having to account for changes of temperature of the order of 50° F. we have only to account for initial changes of 5° F. or so, since we can safely leave the floating ice to make up the odd 45° F. Before we are in a position to begin a serious search for the causes of these greatly reduced changes, however, it will be necessary to carry our preliminary studies a little further, and obtain some information about the winds and ocean currents, and the way they may have been modified by the change from a " glacial " to a " non-glacial " climate.

REFERENCES

(1) BROOKS, C. E. P. "The problem of mild polar climates." London, Q. J. R. Meteor. Soc., 51, 1925, p. 83.
(2) KERNER, F. "Das akryogene Seeklima und seine Bedeutung für geologischen Probleme der Arktis." Wien, Sitzungsber. Akad. Wiss., 131, 1922, p. 153.

CHAPTER II

Pressure and Winds

THE weather at any place on any day is mainly governed by the winds which are blowing at the time ; similarly, the climate is largely determined by the winds which blow most frequently. The winds in one place are closely related to the winds at other places, the whole forming a more or less orderly system which ultimately depends on the differences of temperature between different latitudes. This system of winds over the earth as a whole is known as the circulation of the earth's atmosphere. Since the winds are bound up with the distribution of pressure, their discussion must also include the distribution of pressure in the various seasons.

Figs. 3 and 4 show the average pressure distribution, reduced to mean sea-level, prevailing at present in January and July respectively. In January there is a belt of relatively low pressure (below 1,012 millibars) extending the whole length of the equatorial regions, with local minima over the continents south of the equator (South Africa, South America, and Australasia). On either side of this low-pressure belt are the sub-tropical high-pressure belts between 20° and 40° latitude. The high-pressure belt in the Northern Hemisphere is especially well developed ; in addition to the maxima over the oceans there are marked anticyclones near the centres of the continents. In the Southern Hemisphere the anticyclones lie entirely over the oceans, with their centres rather nearer the eastern than the western shores.

On the poleward sides of these high-pressure belts lie marked areas of low pressure. In the Northern Hemisphere these are in the form of isolated depressions over the oceans. One area, known as the Aleutian low, is centred near the Aleutian Islands in latitude 52° N. ; the other, the Icelandic low, lies west-south-west of Iceland in 60° N., and extends in a long tongue towards the Arctic Circle between Norway and

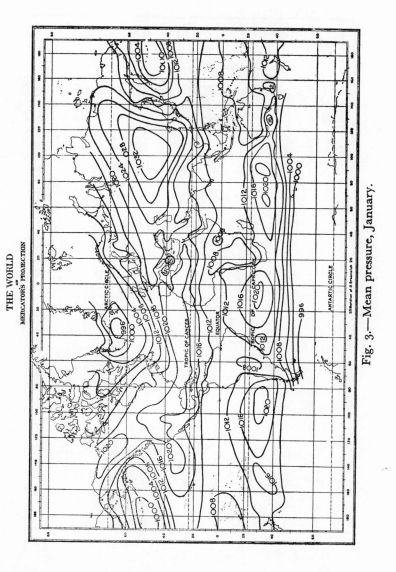

Fig. 3.—Mean pressure, January.

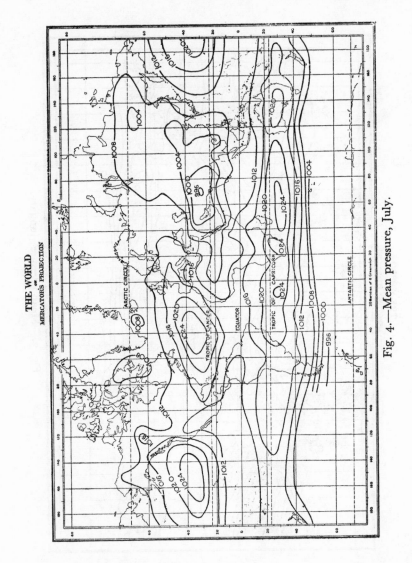

Fig. 4.—Mean pressure, July.

Spitsbergen. In its centre the average pressure is below 996 mb. In the Southern Hemisphere, where the Southern Ocean is not broken up by land-masses, the area of minimum pressure is remarkably deep, and extends in a continuous belt completely round the globe in about 60° S.; it is, however, accentuated in the Ross Sea and Weddell Sea. On the poleward sides of these minima, pressure rises again, especially towards the Antarctic continent; in the Arctic Ocean the polar increase is masked by the Siberian and American anticyclones.

In the chart for July (Fig. 4) the relative distribution over the two hemispheres is partly reversed. The equatorial low-pressure belt is still shown, but the lowest pressure is now found north of the equator, near Jacobabad in North-western India, about latitude 30° N. In the Northern Hemisphere the sub-tropical high-pressure belt is limited to two areas over the Atlantic and Pacific Oceans, but in the Southern Hemisphere it is nearly continuous in about 25° S. The Icelandic minimum still persists with greatly reduced intensity, but the Aleutian low has disappeared; on the other hand, the low-pressure belt in the Southern Ocean is very sharply defined. As in January, pressure rises again near the poles.

If the earth were not rotating about its axis, air would simply flow directly from areas of high pressure to areas of low pressure. Owing to the rotation the air is deflected, to the right in the Northern Hemisphere and to the left in the Southern Hemisphere, and in the free air it blows parallel with the isobars. Close to the earth's surface, however, the wind blows obliquely inwards towards the low pressure, and in the Northern Hemisphere a system of straight isobars with low pressure to the north and high pressure to the south gives surface winds from between west-south-west and south-west. Thus in accordance with the pressure distribution shown in the figures, on the eastern and south-eastern sides of the sub-tropical high-pressure centres there are great systems of winds, blowing towards the equator, known as the Trade Winds, north-east in the Northern and south-east in the Southern Hemisphere. Near the equator these two wind systems unite in a slow drift which on the whole is from east to west, though it is diversified by frequent calms and variable winds; this region is known as the Doldrums. On the poleward side of the high-pressure belts occur westerly winds, which blow round the globe in temperate latitudes and are varied by frequent " cyclonic depressions." Finally, on the edge of the polar high-pressure areas easterly winds again increase in frequency.

This generalised system is modified by the continents, which tend to be occupied by high-pressure areas in winter and by low-pressure areas in summer. These are associated with continental systems of winds known as monsoons, blowing outwards from the continents in winter and inwards towards the continents in summer. The classic example of a monsoon area is Asia ; the winter pressure over Siberia, when corrected to sea-level, is the highest known on the earth, on the average exceeding 1,040 mb. south of Lake Baikal, while the summer low pressure in North-western India rivals in intensity the great barometric minimum in the Southern Ocean. There is a correspondingly great reversal of wind direction, for instance on the China coast the winds blow from the north or north-east with remarkable persistence for several months in winter, and from the south with almost equal steadiness in summer. The neighbouring continent of Australia south of the equator is undergoing the same alternation, though not so marked, in the reverse sense, so that when pressure is high in Asia it is low in Australia and vice versa ; consequently there takes place an immense ebb and flow of air between these two regions.

It is evident that we have to deal with two factors which combine to form the main features of the distribution of pressure and winds over the globe. The first is known as the planetary circulation, the second is the influence of the continents and oceans and also of the distribution of ice ; it must have varied greatly from one geological epoch to another in accordance with the varying land and sea distribution, and may be termed the geographical circulation. The planetary circulation also cannot be regarded as a constant ; it is governed by the contrast of temperature between low and high latitudes, and before we can estimate the part which the atmospheric circulation has played in geological changes of climate, it will be necessary to discuss the planetary circulation somewhat more fully, especially in relation to the vertical and horizontal distribution of temperatures.

Near the surface of the earth, temperature decreases upwards at the average rate of about 19° F. per mile, but this decrease does not continue indefinitely. It is found that at a height of several miles temperature ceases to fall, and above that height it remains constant or even rises a little. The lower part of the atmosphere, in which temperature decreases upwards, is termed the troposphere, the succeeding layer in which there is no change with height is termed the stratosphere, the junction between them being the " tropopause." Now the tropopause is not always found at the same height ; it is slightly higher in summer than in winter, it is higher over anticyclones than over cyclones, and it is very much higher near the equator than near the poles. Over the equator its average height is about 18 kilometres (11 miles), over the British Isles about 11 km. (7 miles), near the North Pole only about 7½ km. (less than 5 miles). This curious structure of the atmosphere has some very important consequences. In the troposphere the temperature at any given height is roughly proportional to that at sea-level ; the air five miles above the equator is warmer than the air five miles above the pole. But at a height of ten miles the conditions are very different. Let us work it out. At sea-level the temperatures are, say, 80° F. at the equator and 20° F. near the North Pole. Allowing a decrease of 19° F. for each mile of height, at five miles we have :—equator, −15° F., pole, −75° F. But as we go on to a height of ten miles the temperature above the equator goes on falling, while that above the pole stays at −75° F., and at ten miles we have :—equator, −110° F., pole, −75° F., that is, the air ten miles above the pole is 35° F. *warmer* than the air at the same height above the equator. In fact, the lowest temperatures naturally existing anywhere on earth are found at a height of about 10½ miles above the equator. Cold air is heavier than warm air, and therefore the colder the air above any locality, the greater will be the pressure there.

The barometric pressure on any part of the earth's surface is the result of the whole column of air above it, but we may follow Sir Napier Shaw (1) in dividing the atmosphere into two parts, one above the level of 8 kilometres (5 miles) and the other below that level, and we may further suppose that at a height of about 20 kilometres (12½ miles) the pressure differences between different parts of the earth are small. The pressure at eight kilometres is inversely proportional to the temperature (on the absolute scale) of the higher layers of the atmosphere, and since in the upper air temperature increases from low to high latitudes, at eight kilometres the pressure decreases from low to high latitudes. If the surface of the globe were fairly homogeneous this would result in the formation of two great systems of westerly winds in the upper air (the " polar whirls ") blowing completely round the earth with their centres near the poles. Such a polar whirl may in fact exist in the Southern Hemisphere, but in the Northern Hemisphere, with its geographical and meteorological complexities, the circulation is greatly distorted, so that it is more correct to speak of a " zone of sub-polar whirls." This condition, however, must be an exception from the geological point of view, and during the warm geological periods the polar whirls were probably in existence in the upper air in both hemispheres. The reason for the exception in the Northern Hemisphere at present is not the low temperature over the Greenland ice sheet, since the layer of abnormally cold air above the ice is comparatively thin ; the circulation apparently takes place round the low-pressure centre and is due to the fact referred to previously, that, above the low pressure the stratosphere extends down to a low level and is therefore abnormally warm.

In the stratum between the surface and eight kilometres the average temperature decreases from lower to higher latitudes, and the pressure at the surface due to this layer of air alone is therefore greater in high than in low latitudes. The effect of this layer would be to cause easterly winds at the surface in all latitudes. Thus we have two apposing tendencies ; the temperature distribution above eight kilometres tends to produce westerly winds at all levels from the surface up to nearly twenty kilometres, while the temperature distribution below eight kilometres tends to produce easterly winds at the surface. Which of these two directions, east or west, predominates in the resultant surface wind of any latitude, depends on the vertical temperature distribution. If the decrease of temperature towards the poles is rapid, as in the " glacial " periods, the weight of the cold air in high latitudes in the lowest eight kilometres may be more than sufficient to counterbalance the pressure difference between the equator and the poles at eight kilometres, and there will be a polar cap of east winds as at present. But if the decrease of temperature towards the poles is slow, as it was during the " non-glacial " periods, the weight of the lowest eight kilometres of air may not be enough to counterbalance the pressure difference between the equator and the poles at eight kilometres, and in that event westerly winds will prevail at the surface up to the immediate neighbourhood of the poles. This must not be interpreted to mean that there were no depressions, or only one stationary depression concentric with the pole. Some theoretical work of H. Jeffreys (2) leads him to the conclusion that the friction of the wind against the earth's surface must inevitably introduce a system of moving cyclones surrounding the pole. In the absence of glacial anticyclones, however, it appears that cyclones would be fewer and less intense than at present, and would occur as a rule in very high latitudes.

At this point we may pause to consider a possible objection. Since the atmospheric circulation depends on differences of temperature, it would seem to follow that the greater the temperature difference between equator and poles, the stronger would be the atmospheric circulation, and consequently the greater the amount of heat carried from low to high latitudes by the winds and wind-driven ocean currents. This would tend to restore the balance and keep the temperature difference between equator and poles more or less constant. This way of looking at the atmospheric circulation is very plausible, but it ignores the fact brought out in the preceding discussion that there is a critical point for the planetary atmospheric circulation just as for the distribution of temperature in a polar sea. In the glacial state, with low temperatures near the poles the excess of air density in the layer of the atmosphere below eight kilometres causes easterly winds in high latitudes ; these have an equatorward component at the surface which hinders the poleward surface component of the westerly winds from carrying warm air into high latitudes. This shutting out of the warmth-bringing equatorial air helps to keep the polar regions cold. In the " non-glacial " state, the excess of density in the lowest eight kilometres is not sufficient to counterbalance the warmer stratosphere of high latitudes and the winds are therefore westerly with a poleward surface

component, up to high latitudes. The influx of warm
equatorial air then helps to maintain the high polar temperatures. Owing to the existence of this critical point the
atmospheric circulation, so far from smoothing out the temperature contrast, maintains or even magnifies it.

The boundary between the polar east winds and the
temperate west winds (the " polar front ") is at present
the chief seat of the development of the barometric
depressions of temperate latitudes. There seems to be no
doubt that the presence of two adjacent air currents at
different temperatures which are in motion relative to one
another is an important factor in bringing these depressions
into existence (3). Owing to the frequent passage of depressions the average pressure in about 60-65° latitude is lower
than it would otherwise be. These belts of cyclonic activity
and low pressure limit the poleward extension of the sub-
tropical anticyclones and help to cause the relatively sharp
differentiation of the climatic zones at present ; in the absence
of polar east winds we should expect a gradual fall of pressure
from maxima in middle latitudes to minima near the poles.
The chief centres of storminess would be found over the oceans
in the Arctic and Antarctic regions ; in middle latitudes there
would be only occasional feeble depressions moving slowly
along irregular tracks.

The zone of westerly winds which blow round the world
in middle latitudes is broken up to some extent near the
surface by travelling depressions and anticyclones, but at a
height of a few miles the winds are much stronger and more
steady. A steady west wind blowing round the globe would,
however, be unstable; as described for example by C. G.
Rossby (4), any slight disturbance of the west-east movement
would set up wave-like disturbances, with the crests towards
the pole and the troughs towards the equator. The wave-
length depends on the strength of the winds ; the average
wind velocity in winter in the northern hemisphere at present
is such that the wave-length is about 3,000 miles, and this is
the approximate distance between the Aleutian and Icelandic
low-pressure centres (the corresponding centre in about 60° E.
is masked by the Siberian winter anticyclone). The weaker
the winds the shorter the wave-length ; hence when for any
reason the average speed of the westerly winds falls off, the

circulation tends to break up into a number of smaller cells. The deep single Aleutian and Icelandic low pressure centres are replaced by smaller and less stable double centres. The winds are less strong, but the exchange of air between high and low latitudes is maintained by larger north-south and south-north components.

Since the strength of the west-east circulation depends on the temperature gradient between low and high latitudes, in the non-glacial periods we should expect a weak circulation and hence numerous small areas of low pressure. This would tend to equalise climatic conditions along the same parallel of latitude, whereas the large semi-permanent Aleutian and Icelandic lows which predominate at present lead to great extremes, *e.g.* between Labrador and north-west Europe.

In the early stages of the Quaternary glaciation, as pointed out by R. F. Flint and H. G. Dorsey (6), the zonal circulation must have increased in strength. At present, when the circulation is especially strong the Icelandic low tends to spread eastwards ; a similar change in the Quaternary would have facilitated the eastward extension of the Scandinavian glaciation. On the other hand strong west winds crossing the Rocky Mountains would give Föhn effects on the leeward side and so limit the eastward extension of the Cordilleran ice-sheets, favouring a separate centre of glaciation in eastern Quebec and Labrador. When fair-sized ice-sheets had developed in the north, however, semi-permanent glacial anticyclones would form which would displace the tracks of depressions southward. This would weaken the zonal circulation ; depressions would tend to stagnate south of the ice-sheets, facilitating the extension of the latter southward on their western margins while farther east comparatively warm southerly winds checked the southward expansion. These ideas are interesting and warrant further investigation, but the pattern suggested by Flint and Dorsey is complicated by other factors such as isostatic changes due to the weight of the ice.

In tropical and sub-tropical regions the existing planetary circulation over the oceans (equatorial low-pressure belt, trade winds, sub-tropical anticyclones) is clearly the result of zonal temperature differences and the rotation of the earth. In warm periods thermal zones must have existed though probably somewhat less marked than now, and there is no reason to suppose that trade winds and sub-tropical anticyclones did not exist. The absence of the polar fronts, however, and the pole-ward displacement of the tracks of depressions would permit the anticyclones to extend farther poleward, just as they do now in the North Atlantic in summer, when the south-north thermal gradient is smallest, compared with winter, when the thermal gradient is greatest.

Fortunately Nature herself carries out for our inspection an experiment which suggests the way in which the general atmospheric circulation to be expected during periods with a small temperature difference between equator and poles would differ from the circulation during periods with a large temper-ature difference. In summer the air over the Arctic Ocean is now more than 30° F. warmer than in winter, while in equatorial regions there is very little change of temperature throughout the year. The temperature difference between equatorial and polar regions is therefore about 50° F. in summer and more than 80° F. in winter. As a rough approxi-mation we may say that the temperature difference between low and high latitudes during the present summer in the Northern Hemisphere is similar to the temperature difference which existed during winter in the warm periods. Nature has a habit of complicating her experiments with irrelevant details, and in all the oceans but one there are, even in summer, large quantities of ice and ice-cooled water which, coming into contact with warm ocean currents, produce great differences of temperature in short distances, but in the North Pacific Ocean this complication is reduced to a minimum, and we may illustrate the probable system of winds during the winter of a warm period by the system of winds prevailing at present over the North Pacific in summer (Fig. 6). Of course it is not to be expected that this will give us an exact picture of what happened during a warm period ; the differences in other oceans, and especially in the Southern Hemisphere, would inevitably introduce some modifications, but at least it will serve as a suggestion.

The distribution of pressure at present has been shown in Figs. 3 and 4. In January (Fig. 3) the anticyclone is small and sharply defined ; it does not extend north of 40° N., and between 40° N. and 60° N. lies a well-marked area of low pressure, the Aleutian low. In July the anticyclone is less sharply defined, but extends almost to 60° N., and *the Aleutian low has completely disappeared*. This disappearance of the cyclonic centre from the North Pacific in summer is of great importance ; the Icelandic minimum in the North Atlantic does not disappear during summer and the low-pressure belt in the Southern Hemisphere also persists throughout the year. There seems no doubt that this peculiarity of the North Pacific is directly due to the absence of ice in summer—both floating ice in the sea and large ice-sheets on the neighbouring land-masses. The surface winds over the Pacific corresponding with the pressure distribution in January and July are shown

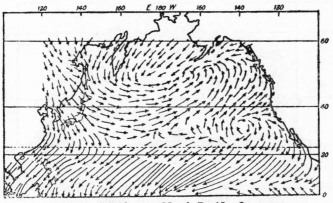

Fig. 5.—Winds over North Pacific, January.

in Figs. 5 and 6. In January (Fig. 5), representing winter conditions at present, a belt of north-easterly trade winds extends completely across the ocean south of latitude 25° N. and is almost entirely separated by a belt of calms and variable winds from another system of winds, mainly westerly, between 30° and 50° N. North of 50° N. the prevailing wind is from north-east. Thus the system of winds in January, while it is calculated to bring mild winters to the American coast south of 55° N., greatly intensifies the rigour of the winter in sub-polar latitudes and on the eastern coasts of Asia.

In July (Fig. 6), representing summer conditions at present and winter conditions during the warm periods, the system of winds is very different. The direction of the north-east trade is more easterly, and west of 180° longitude even south-easterly ; it passes without a break into a great system of southerly winds which blow from 30° N. to the Arctic Circle. The south-easterly direction of these winds north of about latitude 50° N. is due to the monsoonal inflow into the great land-mass of Asia ; during the warm periods the winds probably retained their south-westerly direction into high latitudes. These winds bring high temperatures over the whole of the North Pacific coasts (except the western coast of the United States), and are especially favourable to Alaska and North-eastern Asia, which enjoy a warm climate at this

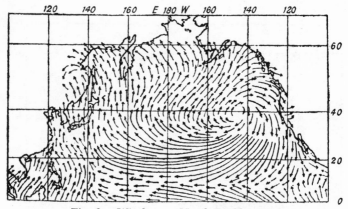

Fig. 6.—Winds over North Pacific, July.

season. The winds induce favourable oceanic currents, and in spite of the narrowness and small depth of Bering Strait some warm water succeeds in penetrating into the Arctic Ocean through that opening in summer. The Arctic Ocean retains a large amount of floating ice throughout the summer, so that the polar cyclone cannot develop properly, and the southerly winds of the North Pacific are greatly weakened ; there seems no doubt that if the Arctic Ocean were free of ice the system of southerly or south-westerly winds would attain great steadiness. If at the same time the Bering Strait were replaced by a wide and deep gap, practically the whole surface of the ocean north of 30° N. would be set in motion in a north-easterly direction, and an immense volume of warm water would be driven into the Arctic Ocean.

This picture of the pressure and winds during the geological periods characterised by widespread warmth, which we have obtained first from some theoretical considerations, and secondly from an examination of present summer conditions over the North Pacific, fits in admirably with what we know of the climates of these warm periods. The system of stable southerly winds, extending across the middle latitudes, would give them fine quiet weather in place of the present succession of storms. As we saw in the Introduction, apart from the warmth and rich vegetation of the polar regions, the most striking feature of these periods was the widespread development of semi-desert conditions in temperate regions. True deserts were probably less extensive than they are to-day, owing to the smoothing out of the zonal contrasts, but very large regions had a " Mediterranean " type of climate, with a small rainfall during the mild winter and a long dry hot summer. This explains, for example, the widespread distribution of the characteristic " sub-tropical " (*i.e.*, Mediterranean) vegetation of the first half of the Tertiary period.

Let us look at the reverse of this picture, and see what would happen during an ice-age. An extension of the cold areas over a large part of the present temperate regions, such as occurred during the Quaternary, would bring the " polar front " between the polar east winds and the temperate west winds nearer to the equator, and by increasing the temperature contrast between low and high latitudes would increase the storminess. This means that the sub-tropical high-pressure belts would be sharply limited on their poleward sides. The result would probably be an intensification of the present winter conditions—a small but intense anticyclonic belt in about 20° to 25° latitude, a narrow belt of powerful trade winds, and a deepened equatorial trough of low pressure. The circulation would be much less stable than that of the warm periods, and the anticyclonic belts would be subject to great and rapid displacements. Hence the rainfall would be increased over all the tropical and sub-tropical regions ; outside the new storm tracks the increase would be greatest and most regular near the equator, while towards the tropics the rainfall would be less in amount and very variable from year to year. Here again the theoretical conclusions are supported by geological results ; outside the great ice-sheets there is evidence of a much greater rainfall (snowfall on the mountains) in two belts, one along the new storm tracks a short distance equatorward of the ice-edges, and the other along the equator. The former gave rise to the great lakes of the Great Basin of America and of the interior of Asia, and to a large number of mountain glaciers, the latter to the greatly increased lakes of Central Africa and to the glaciers of Kenya, Kilimanjaro, Ruwenzori, and parts of the Andes. Between these two belts the evidence of greater precipitation is more indefinite and irregular.

We must now return to the geographical circulation. This we have seen is characterised by the presence over the continents of high pressure in winter and low pressure in summer, resulting in monsoonal winds. The intensity of the monsoons over a continent depends on a number of factors—the limits of latitude, the size, the presence of large arid basins surrounded by mountain ranges, and the strength and direction of the winds over the neighbouring seas. The intensity of the Siberian winter anticyclone is due partly to the great size of Eurasia, and very largely to the way in which its surface is broken up by mountain ranges. The highest pressure occurs over the great enclosed basin south of Lake Baikal, from which the cold air finds difficulty in escaping. The winter cooling is of course essential to the development of the anticyclone, but the existence of the anticyclone in turn intensifies the cold. In North America, where owing to the absence of transverse mountain ranges the cold air finds less difficulty in escaping, the winter anticyclone is comparatively feeble. A good example of the effect of favourable orographical conditions on the pressure distribution is the Iberian Peninsula, which is occupied by a well-marked anticyclone in winter. In North-west Europe, on the other hand, the influence of the Icelandic minimum and the barometric depressions which originate in the Atlantic make the maintenance of a winter anticyclone very difficult.

The establishment of continental low pressure in summer depends on similar factors, but the centres occur nearer the equator than do those of the winter anticyclones. In Asia the area of low pressure in July extends over a large part of the continent, but the actual minimum occurs comparatively near the sea in North-west India, and is very intense ; this position of the minimum is due largely to the position of the mountain ranges which interfere with the free circulation of the air (6) and entirely inhibit the North-east Trade, which is the natural wind of that latitude. Directly to the westward over the Sahara, where the air is as hot or hotter, pressure is much higher, and in fact throughout the year the Sahara is practically a continuation of the belt of the North-east Trades, probably because the mountain ranges are not high enough to shut out these winds entirely.

These scattered instances show how difficult it is to analyse the geographical circulation exactly, but they do give us some basis for estimating the results of various geological changes on the local wind circulations. Consider, for instance, a warm period in which there are extensive oceans and some flat continents of moderate size, little diversified by mountain ranges. It seems that these continents would not greatly modify the planetary circulation described above. In summer, when the winds are generally weakest, the continents in low and middle latitudes would be hot, and there being no mountain ranges to cause the ascent of air, the only rainfall would occur in sporadic thunderstorms and squalls. In winter they would be cooler than the oceans, but in the absence of a polar reservoir of cold air, they would not be intensely cold, especially since the prevailing winds in temperate zones would be from the equator. The centres of the anticyclones, as now, would probably lie over the oceans, but there would be a tendency for the high-pressure areas to extend nearer the poles over the continents, giving southerly winds on the west coasts and westerly winds on the east coasts. A diagrammatic reconstruction of the pressure and winds during winter and summer in a warm period is shown in Fig. 7.

There are two special parts of the geographical circulation which demand further notice. One of these is the south-west monsoon of the Asiatic continent ; the other is the circulation over ice-sheets. The south-west monsoon (6) is remarkable because it involves a large transference of air from one hemisphere to the other ; there is a continuous pressure gradient from the sub-tropical anticyclone in the South Indian Ocean to the minimum over Asia, and the surface winds are south-easterly south of the equator and south-westerly north of the equator. The air-flow actually in places surmounts the mighty barrier of the Himalayas, nowhere less than 12,000 feet in height, and arrives in Tibet as a dry descending current. Now it is possible that under certain conditions there may be a quite considerable difference between the mean annual temperatures of the two hemispheres, leading to a more or less permanent circulation of the type of the south-west monsoon. A circulation of this type affords a possible

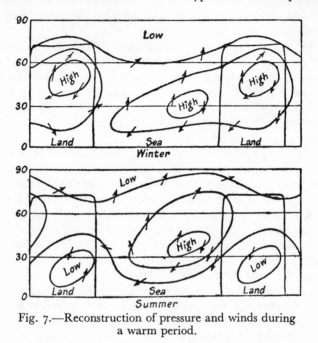

Fig. 7.—Reconstruction of pressure and winds during
a warm period.

explanation of the peculiar climate of the Upper Carboniferous period (see Chapter XV.).

The circulation over a large ice-sheet is of great importance for the study of the meteorology of glacial epochs. A surface of snow and ice reflects a large part of the solar radiation which falls on it, and, owing to the dryness of the air above it, also radiates freely and receives little return radiation from the air. Hence during most of the year the surface is very cold, and this cools the lower layer of the air in contact with it. Since cold air is relatively heavy, barometric pressure must be higher over a large ice-sheet than at the same level over neighbouring seas. This is interpreted by W. H. Hobbs (7) to imply that an ice-sheet is the site of a nearly permanent anticyclone, with outflowing winds on all sides, supplied by air which flows in at higher levels and descends over the ice-sheet. This raises the difficult question of the supply of moisture to maintain the flow of ice, for descending air is warmed by compression and is normally dry, but Hobbs pointed out that the surface of the ice is intensely cold, and may be much colder than the air at a height of several thousand feet. The moisture of the upper winds reaches ground level in the form of vapour, but a large part of it is immediately condensed as ice-mist or deposited directly as hoar-frost. The outflowing winds sweep these crystals before them and so maintain the marginal parts of the ice-sheet.

There is undoubtedly a good deal of truth in this theory. Owing to difficulties in reduction to sea-level, pressure on the surface of a large ice-sheet cannot be compared directly with that over the surrounding ocean, but the winds do on the whole tend to blow outwards and must be made good by descending currents, while the vertical distribution of temperature is such that there must be some condensation as ice-crystals or hoar-frost. On the other hand, the supposed high pressure may be entirely due to a quite thin layer of air near the ground, above which the structure ceases to bear any relation to an anticyclone, while the outflowing winds may be purely katabatic—rivers of cold air such as flow down any slope on a cold clear night—and have no relation to the general pressure distribution. In such a case the normal processes of precipitation could go on unchanged above the surface layers of air, the only difference being that precipitation would fall entirely as snow. It is very unlikely that the deposit of hoar-frost could suffice to supply the enormous quantities of ice which issue from an ice-sheet each year as glaciers, icebergs and glacial streams.

F. E. Matthes (8) discussed the light thrown on this question by observations at Eismitte near the centre of Greenland during the Wegener expeditions of 1929 and 1930-31, the latter covering more than a year of continuous instrumental readings. These observations show clearly that quiet fine conditions are the exception and stormy cloudy conditions the rule, that the katabatic winds are feeble and easily overpowered by storm winds, and that by far the greatestfactor in the nourishment of the ice-sheet is ordinary snow. The great oscillations of pressure and temperature closely resemble though much lower in the scale, those in typical cyclonic, regions such as New England. The winds are strong, but blow mainly from east, with a secondary maximum from S.S.E. True precipitation is actually more frequent than on the west coast of Greenland and not much less common than on the rain-swept coast of Norway, but owing to the low temperatures the total amount is not large, the average being estimated from the firn layers as 12·4 inches of water a year.

The frequency of easterly winds is not due entirely or even mainly to katabatic winds from the ice divide to the east. It means that depressions pass mainly to the south, but extend their influence right to the centre of the ice-cap. There is now no difficulty about the supply of moisture, which is derived from the Atlantic Ocean and carried inland, over-riding the shallow katabatic winds of the coast, to be condensed into snow over the ice-shed and swept on to be gradually deposited on the leeward slopes. Atmospheric pressure is undoubtedly higher than it would be if the surface, at the same level, were unglaciated, but it appears that the sub-continent is not big enough to form the basis for a self-supporting glacial anticyclone.

The Antarctic presents a more difficult problem. W. Meinardus (14) considered that the Antarctic anticyclone is limited to the lowest 2,000 metres, above which the circulation is cyclonic, and since the greater part of the Antarctic continent is above this level, he supposed that the greater part of the land is subjected to cyclonic air motion. Sir George Simpson pointed out (9) that an extensive ice-covered plateau must be occupied by a glacial anticyclone just as if it were at sea-level ; he accordingly divides the Antarctic continent into two parts, a plateau at about 3,000 metres (10,000 feet) and a plain near sea-level. The latter is occupied by an anti-cyclone at sea-level, but owing to the rapid decrease of pressure with height, conditions in the free air are cyclonic at a height of about 3,000 metres. Thus at the latter height there is an anticyclone at the surface of the plateau and a cyclone in the free air above the plain.

It appears that an ice-sheet does not develop a stable

glacial anticyclone until it attains a certain size. The Greenland ice-sheet is not broad enough to prevent the influence of large depressions from extending even to its centre, but these depressions are deflected southwards by the ice-sheet so that their centres rarely pass directly over the central regions of Greenland. The greater part of the Antarctic continent is immune from travelling depressions. On the other hand, the smaller ice-masses of Iceland and Spitsbergen appear to have little effect on the pressure distribution. Thus the critical point comes at a diameter somewhat larger than the width of Greenland. In 75° N. Greenland is about 650 miles across. From studies of the January temperature distribution over land areas in latitude 50 to 70° N. (10), which are normally snow covered in winter and are therefore similar to ice-sheets in their effect, I think the diameter which a circular ice-sheet must reach before it begins to dominate the pressure distribution is between seven hundred and a thousand miles. When the diameter is less than 700 miles, the normal winds of the region sweep over the ice-sheet without much hindrance, and the only effect of the ice is to cool the air slightly by conduction. When the diameter reaches, say, 1,000 miles, a glacial anticyclone develops, with clear skies and intense cooling by radiation. The outwardly-directed winds spread Arctic conditions in a broad zone round the margin of the ice, and may even result in the " sympathetic " glaciation of a neighbouring mountain range. It is probable that the great development of Alpine glaciers during the Quaternary was partly due to cooling by the winds blowing off the Scandinavian ice-sheet. It will also be remembered that during the earlier stages in the final retreat of the Scandinavian ice-sheet the ground vacated by the ice was occupied by a dwarf flora of Arctic plants, but later, when the ice-sheet was smaller, the retreating edge was immediately followed by a temperate flora. This probably indicates the stage at which the glacial anticyclone broke down.

Every cause or factor which is put forward to explain climatic changes has to take into account the modifications which would be introduced into the atmospheric circulation by its operation. Even the possible occurrence of modifications of the circulation sufficient by themselves to give rise to great climatic changes, without the intervention of any other factor, has been discussed. Thus W. H. Dines remarks (11) :— " There seems to be no particular reason why the winds known as the ' trades ' should not be westerly and the winds of temperate latitudes easterly. Perhaps such a system is possible and might be stable if once established. It would explain the glaciation of North-western Europe, for it would very greatly lower the temperature of that region, but it is not feasible as an explanation of the glacial epoch, because it would raise the winter temperature of North America." A restoration of the meteorological conditions of the Quaternary Ice-Age was attempted by the late F. W. Harmer (12) on the assumption that the glaciations of North America alternated with those of Europe. It is hard to conceive of great changes such as these without some ulterior reason, such as a change in the land and sea distribution or in the solar radiation, and we must regard the part which the atmospheric circulation plays as that of a regulator, at times perhaps an amplifier, but probably not an originator of major climatic oscillations.

It must be admitted, however, that the part played by the circulation of the atmosphere in climatic changes is not yet fully understood. In the past hundred years, for example, there has been a marked recession of glaciers in all parts of the world, accompanied by a large rise of temperature in the Arctic and a rise of winter temperature over a much wider region. R. Scherhag (13) attributes these phenomena to a strengthening of the atmospheric circulation and consequently of the Gulf Stream, but this only pushes the problem one stage further back, i.e., to the cause of the stronger circulation. The answer certainly does not lie in a change of land and sea distribution, and to the best of our knowledge there has been no appreciable change of solar radiation. It is not unlikely that the cause lies in the atmosphere itself, or in its interactions with the oceans, owing to some process initiated almost by " accident " in the constant turmoil of depressions and anticyclones, but which, once begun, will automatically increase in intensity until it becomes unstable or is reversed by some other " accident." The atmosphere and hydrosphere are so vast that such self-reinforcing actions may well persist for many decades. It is possible that the majority of temporary swings of climate are of this nature. If so, they cannot be said to have a " cause," any more than can a run of luck in a game of pure chance.

REFERENCES

(1) SHAW, SIR NAPIER. " The air and its ways." London, 1924.
(2) JEFFREYS, H. " On the dynamics of geostrophic winds." London, Q. J. R. Meteor. Soc., 52, 1926, p. 85.
(3) BJERKNES, J., and H. SOLBERG. " Life-cycle of cyclones and the polar front theory of atmospheric circulation." Kristiania, Geofysiske Publ., 3, No. 1, 1922.
(4) ROSSBY, C. G. " The scientific basis of modern meteorology." Washington, Yearb. Agric., 1941, p. 599.
(5) FLINT, R. F., and H. G. DORSEY. " Iowan and Tazewell drifts and the North American ice-sheet." New Haven, Amer. J. Sci., 243, 1945, p. 627.
(6) SIMPSON, G. C. " The south-west monsoon." London, Q. J. R. Meteor. Soc., 47, 1921, p. 151.
(7) HOBBS, W. H. " The glacial anticyclones. The poles of atmospheric circulation." New York (Macmillan), 1926.
(8) MATTHES, F. E. " The glacial anticyclone theory examined in the light of recent meteorological data from Greenland." Trans. Amer. Geoph. Union, 27, 1946, Pt. I, p. 324.
(9) BRITISH ANTARCTIC EXPEDITION, 1910-1913. Meteorology, vol. i. Discussion, by G. C. SIMPSON. Calcutta, 1919.
(10) BROOKS, C. E. P. " Continentality and temperature." London, Q. J. R. Meteor. Soc., 43, 1917, p. 164.
(11) DINES, W. H. " Circulation and temperature of the atmosphere." Washington, D.C., Monthly Weather Review, 43, 1915, p. 551.
(12) HARMER, F. W. " The influence of the winds upon climate during the Pleistocene epoch : a palæometeorological explanation of some geological problems." London, Q. J. Geol. Soc., 57, 1901, p. 405.
(13) SCHERHAG, R. " Die Erwärmung der Arktis." Copenhague, J. Cons. int. Explor. Mer., 12, 1937, p. 263. See also Meteor. Mag., London, 73, 1938, p. 29.
(14) DEUTSCHE-SÜDPOLAR-EXPEDITION, 1901-1903. III Bd., Meteorologie. Berlin, 1911.

CHAPTER III

The Circulation of the Oceans

OWING to the high specific heat of water, the great oceanic currents and the variations in the surface temperature of the sea to which they give rise are of very great climatic importance. The classic example of this is the Gulf Stream Drift and the high winter temperatures of North-west Europe, but there have been still more notable instances in the geological past, in the highly favourable climates of the polar regions during the warm periods. It is estimated that at present about half the transfer of heat from low to high latitudes is due to ocean currents, the remaining half being due to interchange of air. Ocean currents are due chiefly to two causes, differences of density, and the winds, which drive before them the surface layers from which motion is imparted by friction to the underlying layers. If two masses of water of different densities lie side by side a circulation will be set up between them, resulting in a surface flow from the lighter to the heavier mass, and a flow at a greater depth from the heavier to the lighter, and if the earth were at rest this would continue until the horizontal differences of density had been removed and all the heavy water lay at the bottom, with all the light water on top. Owing to the earth's rotation, currents of water are deflected to the right in the Northern Hemisphere and to the left in the Southern Hemisphere just like currents of air, and the ultimate flow is at right angles to the gradient of density, giving two currents moving side by side in opposite directions.

Density depends almost entirely on two factors, the temperature and the salinity. With falling temperature, water increases in density until it reaches a temperature of 4° C. (39° F.) if it is fresh, but about −2° C. (28° F.) if it is average sea water. Since 28° F. is also the freezing point of sea water, the latter on being cooled will always sink through water of equal salinity, while fresh water at a temperature of 39° F. will sink through fresh water having either a higher or lower temperature. The temperature of the oceans almost everywhere falls with increasing depth, and the lower parts of the oceans are occupied by a stratum of water at about 39° F.

Density also increases with salinity ; the average salinity is greatest in the sub-tropical open oceans where evaporation is great and rainfall slight. Since these are also in general the areas where the temperature is high, the effect of salinity on density partly balances that of temperature. Density becomes especially great where an ocean current from the tropics penetrates into high latitudes, losing much of its heat but retaining a high salinity. Thus the relatively warm salt water which originates in the Gulf Stream and penetrates into the Arctic Ocean is heavier than the colder but much fresher water of local origin ; the latter remains on the surface and freezes readily, giving rise to great quantities of floating ice. Fresh water is always lighter than sea water of average salinity (35 parts per thousand) at temperatures which are met with in nature.

In the oceanic circulation as it is developed at present (Fig. 8), the winds apparently play a much greater part than differences of density, especially in tropical and temperate latitudes, where the direction of the ocean currents is almost everywhere the same as that of the air currents. On a non-rotating globe this agreement would be easy to undertsand, but as it is, the matter is more complex. The wind drives the surface layers before it, the movement being communicated by friction and vertical interchange of water to the underlying layers, but owing to the rotation of the globe, the surface current caused by a steady wind is inclined to the wind direction at an angle of 45 degrees, to the right in the Northern Hemisphere and to the left in the Southern Hemisphere. As we go below the surface the currents deviate more and more from the winds, and the main mass of the water moves at right angles to the direction of the wind. Consider now the case of a centre of high pressure in the Northern Hemisphere round which the winds circulate in a clockwise direction. All these winds will be driving the water to the right, that is, towards the centre of the system. The result will be that water is piled up in the centre ; we shall have an oceanic " hill "

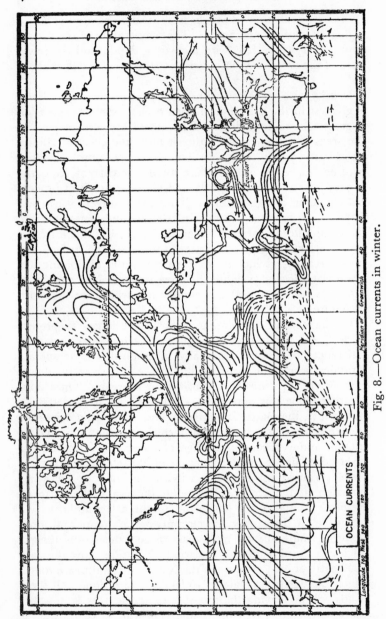

Fig. 8.—Ocean currents in winter.

down the slopes of which the surface water will commence to run. The rotation of the earth will deviate this water to the right, and when a steady state is reached it will be flowing round the central " hill " in a clockwise direction. The final result will be in fact as we now find it, a system of oceanic currents surrounding a central area of stagnant ocean, closely resembling the system of winds blowing round a central area of calm. The tendency of the winds to drive water towards the centre is just balanced by the tendency of the accumulation of water in the centre to flow outwards. Winds irregular in direction and velocity are less effective than steady winds in causing ocean currents, and with variable winds the angle between the resultant wind and the surface current is generally less than 45 degrees (1).

The best known system of currents is that in the North Atlantic (2, 3), (Fig. 8). Commencing with the tropical part of the Atlantic Ocean, we find that the North-east and South-east Trades give rise to currents which turn more and more to the eastward and increase in volume as they approach the equator, until they unite to form the Equatorial Current. For convenience, the northern and southern portions of this current are called the North and South Equatorial Currents, but there is no definite dividing line except that from May or June to November a very shallow current, the Equatorial Counter Current, sets eastward where the winds are weakest between about 3° and 10° N., ultimately entering the Gulf of Guinea. The greater part of the North Equatorial Current turns north-westward as the Antilles Current, which passes between Cuba and the Bahamas, and unites with the Gulf Stream flowing through the Strait of Florida. The Antilles Current is estimated to convey nearly forty cubic miles of water an hour past Porto Rico.

Owing to the greater strength of the South-east Trades, the South Equatorial Current is stronger and steadier than the North Equatorial. It is directed slightly north of west ; striking Cape San Roque on the Brazilian coast in 5° S., it divides into two branches, of which the southern turns south-westwards as the Brazilian Current, while the northern and more extensive passes along the coast of Guiana and unites with the western branch of the North Equatorial Current. The combined current flows towards the coasts of Honduras and Yucatan, and thence mainly through the Yucatan Channel into the Gulf of Mexico. Here it spreads out and turns eastward, passing between Florida and Cuba as the Florida Current.

In Florida Strait the Gulf Stream moves with a speed of 80 nautical miles a day in the centre, and conveys about 22 cubic miles of water an hour. The combined Florida and Antilles Currents move northward to Cape Hatteras with an average velocity of 70 miles a day in the centre, and half this amount on the edges and convey about 47 cubic miles of water an hour. Off Cape Hatteras is the so-called " Delta of the Gulf Stream," where it begins to break up into several branches. South of Nova Scotia the velocity is about 38 miles a day. At the south-eastern and southern edge of the Grand Banks of Newfoundland the Gulf Stream comes into conflict with the cold southward-flowing Labrador Current, which greatly lowers its temperature. The average temperature of the water flowing towards Europe after passing the Grand Banks is 10-15° F. lower than the temperature of the Gulf Stream off Cape Hatteras, and the greater part of this cooling must be due to the Labrador Current, either directly by intermingling of the warm and cold waters and the melting of icebergs which enter the warm current, or indirectly by the winds which blow from the cold to the warm water. The salinity of the water is also lowered somewhat.

Off the Newfoundland Banks the Gulf Stream divides into several branches ; the most northerly flows towards West Greenland, another flows towards Iceland, and a third, containing the main body of the water, flows towards Europe and the Mediterranean. The West Greenland Current is felt as far as 66° N. ; it is this current which causes the decay of the ice brought round Cape Farewell by the East Greenland Current. North of 66° it appears to curve round and join the Labrador Current, but part of it may continue northward beneath the surface and cause the " North Water," the wide sheet of navigable water found in the upper end of Baffin Bay in summer and autumn. The branch of the Gulf Stream which passes directly towards Iceland usually reaches the south-west coast, where it ameliorates the climate somewhat, after which it is lost.

The third or main branch of the Gulf Stream passes directly

eastward, again dividing in about 45° N., 40° W. into two branches. The southern branch turns south-eastward, skirting the coasts of South-west Europe and Africa as a cold current, and ultimately re-entering the North-east Trade Current. Between Cape Verde and Gibraltar, and even farther north in summer, it sets off the coast, and is separated from the land by a belt of cold upwelling water, to which we will refer again later. The northern branch, gaining renewed velocity from the prevailing south-west winds, crosses the Atlantic with an average speed of 12 miles a day, and bathes the shores of Western and North-western Europe from the Bay of Biscay to the North Sea. This current is banked up against the coast, greatly ameliorating the climate. The bulk of the water passes north of Ireland and Scotland to the North Sea, from which one arm passes west of the Faroes to Iceland, turning east again north of Iceland and mingling with a south-easterly branch of the East Greenland Current in a series of great whirls.

The larger arm, under the influence of the prevailing southerly winds, drifts along the Norwegian and North European coasts to Novaya Zemlya, where it is largely overlain by colder but fresher water and loses its identity. From North Cape an arm goes northward to Spitsbergen, where it mingles with the westward-flowing Arctic Drift in another series of whirls ; this branch gives rise to the relatively favourable climate of Spitsbergen, and keeps the western coasts of this archipelago almost free of ice. In the Faroe-Shetland channel the volume of the warm current is about 2 cubic miles an hour, and it has decreased to less than 1 cubic mile off the Lofoten Islands. Even this amount, small compared with the volume of the Gulf Stream off the Atlantic coast of Florida, is of very great climatic importance, and its variations from year to year have important effects on the Norwegian harvests. An increased volume and high temperature (the two usually go together) of the Atlantic Current off Norway in May gives good harvests in the autumn of the same year, and diminishes the amount of drift ice in the Barents Sea one or two years later.

We have seen that in the Arctic Ocean the last remnants of the Gulf Stream are finally lost beneath a layer of colder, fresher water. The latter originates chiefly in the great rivers of Eurasia and North America which discharge into the Arctic Ocean, and this surface stratum, in which ice-formation is very active, forms the mainspring of the return cold circulation. Passing north of Spitsbergen it continues towards the east coast of Greenland ; the main mass of the water follows this coast southwards as the East Greenland Current, bearing great quantities of ice. Owing to the earth's rotation, this current is banked up against the coast ; it rounds Cape Farewell and passes up the west coast of Greenland as far as Disco Island. Here it turns westward under the influence of the prevailing easterly winds, and finally, mingling with the West Greenland Current, it flows southward as the Labrador Current, gaining important accessions from Smith Sound and other channels in the Northern Archipelago. Off the Newfoundland Banks the Labrador Current meets the Gulf Stream, as we have seen, and helps to lower its temperature. When the Labrador Current and Gulf Stream meet, their densities are approximately the same, and they mix along the junction. The mixture is, however, slightly heavier than either of the original currents, and this produces a " density wall," on either side of which the currents are opposed. The maximum density is from 20 to 30 miles inside the cold wall, so that there is a cold current flowing alongside the Gulf Stream in the same direction. There are also continual eddies breaking off from the cold wall and drifting eastwards.

Before leaving the subject of the North Atlantic and Arctic circulation it is necessary to emphasise the part played by floating ice. It has been pointed out that cold water can only remain at the surface above warmer water by virtue of being lighter, because it is less saline. This relative freshness can be brought about in three ways : by heavy rainfall, as in the doldrums, by great rivers, or by the addition of ice. Ice, even when formed in the sea, is fresh, and though some salt water is usually mixed up with the ice at first, this tends to drain out. Hence, when the ice melts it decreases the salinity of the neighbouring sea water. A thin layer of relatively fresh water is constantly gaining salt from below by mixing and diffusion, and unless it continually received accessions of fresh water, by the time it had travelled a thousand miles or so it would differ little in salinity from the underlying water. Hence it could no longer exist as a cold surface current. There are not likely to be great differences of rainfall over adjacent parts of the ocean (if there were, the heavier rainfall would most likely be over the warmer water), and the accession of river water is only possible near the coast, so that the only way in which a current can remain relatively fresh while traversing the open ocean is by bearing with it large quantities of floating ice, the melting of which continually renews the surface layer. In the absence of ice, the current would either lose its identity, or become heavy and sink below the surface. Thus, for instance, in the Arctic Ocean the fresh water from the great rivers is conserved by being frozen, instead of mixing with the more saline underlying water, and helps to form the floating ice-cap or Palæocrystic ice. The East Greenland Current is initiated by this floating ice-cap and maintained by the ice which it carries with it, and which gradually melts. In the same way the Labrador Current is supplied partly by the remains of the East Greenland Current and partly by the ice from the innumerable channels of the Arctic Archipelago of America. It seems highly probable that if there were no floating ice in the Arctic Ocean the East Greenland and Labrador Currents would not exist ; the water as it cooled would sink, and the return to lower latitudes of the water brought by the warm currents would take place not at the surface, but below the surface, if not actually at the bottom.

I have described the North Atlantic circulation, with its extension into the Arctic, in some detail, because a good grasp of it is necessary in order to understand the way in which changes of the oceanic currents controlled the warm periods. The other oceans may be dismissed more briefly. The western halves of the North Pacific, South Atlantic, and, to a less extent, the South Pacific all have warm currents resembling the Gulf Stream. In the North Pacific the warm current is unable to penetrate the Bering Strait, and therefore turns south-eastward and washes the western coast of North America. In the South Atlantic and South Pacific the warm currents enter a great stream of water which circumnavigates the globe in the Southern Ocean, picking up ice from the Antarctic and sending branches northward along the western coasts of all the continents. In the tropical

Indian Ocean the currents are largely controlled by the monsoons.

I have referred to the importance of upwelling cold water. Cold water, owing to its greater density, tends to sink towards the bottom, so that, except in the presence of ice, there is generally a steady decrease of temperature as we go deeper below the surface. Since the maximum density of sea water occurs at its freezing point of about 28° F., so long as there is a plentiful supply of water at this temperature the bottoms of the great oceanic basins will be occupied by water not much above the freezing point. The warm surface layer may be likened to a skin, and wherever this skin is broken, the colder underlying layers will be exposed. This will happen whenever there are winds over adjacent areas blowing away from each other (divergent winds), or whenever the winds blow off the coast. It will also happen whenever a current flowing along a coast-line is deflected away from it by the earth's rotation. The latter happens with the currents on the west coast of South America and South Africa—the Humboldt and Benguela Currents—which turn away from the coast and cause belts of upwelling cold water to form between them and the shore. The low temperature of these currents is due quite as much to this upwelling of cold water as to the original low temperature of the surface water. The temperature of the surface layers is also lowered slightly by breaking waves, which mix up the surface " skin " with the underlying colder layers and so cause a diffusion of heat through the whole depth affected by the waves.

Along the edges of an ocean current travelling across the open sea there is usually a certain amount of eddy motion. This is especially noticeable off the Newfoundland Banks, where the Gulf Stream meets the Labrador Current. This must result in a certain amount of mixing, and a decrease in the volume or temperature of the warm current. It also lowers the average velocity, and therefore increases the loss of heat by radiation and conduction while the current is travelling a given distance. The loss both of volume and of heat is greater in proportion from a weak current than from a strong one. Thus we may sum up the causes which lead to the decrease of temperature or volume in a warm ocean current as follows :—

1. Mixing with colder surface water by eddy motion along the edges.

2. Melting of floating ice which drifts on to the warm current.

3. Mixing with colder underlying water by
   (a) upwelling due to divergent winds or motion directed away from a coast ;
   (b) breaking waves.

4. Cooling by conduction to the air, especially to cold winds.

5. Cooling by radiation.

We have next to consider the variations which these factors may have undergone in the geological past, and especially during the warm periods. In Chapter II. we found that during the warm periods, on the poleward sides of the sub-tropical high-pressure maximum, the winds tend to blow directly towards the poles over the whole surface of the ocean. These winds would act on the water in the way described in the first paragraph of this chapter ; that is, they would drive a body of water towards the right (in the Northern Hemisphere) or towards the eastern shores of the ocean. There would be a piling up of the water in the east, so that the surface of the ocean would slope downwards towards the west. In the steady state this would give rise to a wide oceanic surface current directed from south to north, in the same direction as the winds. There is no reason to suppose that the inter-tropical circulation during the warm periods differed from that found at present, and the warm currents due to the winds of middle latitudes would be reinforced by the warm inter-tropical water driven westward in the Equatorial Currents and rounding the western ends of the sub-tropical highs. Under these conditions, and with the complete or almost complete absence of floating ice, the occurrence of adjacent areas of water at different surface temperatures would be reduced to a minimum. Thus, the cooling under headings 1, 2, and 4 would be much less than at present.

The temperature of the water at the bottom of the deep oceanic basins cannot be lower than that of the coldest part of the sea surface, in fact, owing to earth heat, it must be a few degrees higher. During the warm periods, therefore, when the surface waters of the polar oceans were well above freezing point, there must have been a corresponding rise in the temperature of the bottom layers. This implies a marked decrease in the vertical temperature gradient, and while there must always have been upwellings of underlying water, due to off-shore winds and currents leaning away from the coasts, their cooling effect must have been much less than at present. Further, with the decrease of storminess consequent on the absence of the polar fronts, steady light or moderate winds would prevail in middle latitudes, and there would be a great diminution of divergent winds and of wave motion. Thus the cooling of surface ocean currents under headings 3(a) and 3(b) would also be less than now.

Finally, we come to 5, cooling by radiation. As will be seen in Chapter VI., the higher temperature of the air implies a greater amount of water vapour, especially above the oceans, but probably not an increase in the cloudiness, and this means that a larger part of the earth's radiation would be absorbed by the air, part of it being returned to the surface of the sea and helping to maintain the temperature. Thus we see that during the warm periods all the circumstances worked together to maintain the temperature of the warm ocean currents into high latitudes. Since these warm currents were also accompanied by warm winds, it will be seen that with large, open oceans and low, level continents, the extension of warm temperate oceanic climates into the immediate neighbourhood of the poles does not involve any insuperable difficulty.

An investigation of the probable systems of ocean currents in the northern hemisphere during the various geological epochs was made by P. Lasareff (4). He placed plaster models of continents in a circular plane basin filled with water, and directed streams of air obliquely towards the circumference to represent the trade winds. When the model reproduced the present land- and sea-distribution, the currents produced resembled existing currents even in detail. The horizontal temperature gradient was simulated by passing a heating coil round the edge, which represented the equator. The results are of great interest ; in the models representing the warm periods ocean currents passed across the pole, whereas in the cold periods no current crossed the pole.

The results were especially effective in reproducing the variations of climate in Europe. The currents shown by the models agree with the directions of migration of marine animals.

The picture we have drawn of the oceanic circulation during the warm periods—warm currents extending from shore to shore of the oceans and steadily drifting poleward, to return to low latitudes beneath the surface—is not the only one which has been presented. T. C. Chamberlin (5) has arrived at very different conclusions ; he supposes that during the warm periods there was very great evaporation in low latitudes, so great in fact that the increased salinity of the water caused it to become heavy enough to sink to the bottom. Here it travelled slowly north and south towards the poles, retaining its heat, and rising to the surface in high latitudes, where it caused highly favourable climates. An illustration of this type of circulation on a small scale has been described earlier in this chapter (the " North Water " in Baffin Bay). But I do not think that this " reversal of the oceanic circulation " is a practicable explanation of climatic changes, for at least two reasons. In the first place, it will be seen in Chapter VI. that greater warmth does not necessarily mean greater evaporation ; once the air is saturated, it cannot take up any more moisture unless either the temperature rises still further or some of the water vapour it already contains is first condensed as rain. The temperature cannot go on rising indefinitely, and the conditions during the warm periods were less favourable to rainfall in low and middle latitudes than at present ; hence, evaporation was probably less active rather than more active than now. Secondly, there does not seem to be any obvious reason why the saline water should rise at the poles when it got there. In the absence of a wind-driven circulation, we should expect the heavy water to remain at the bottom and accumulate there until it occupied all the oceans except a thin surface layer. Here there would be a slow drift of water from the regions where precipitation exceeded evaporation, and from the mouths of the great rivers, to the regions where evaporation exceeded precipitation, the drift being just enough to maintain equilibrium. The example of the " North Water " is not to the point ; the warm water here comes to the surface either in an eddy or because the prevailing off-shore winds drive away the surface layer of colder but fresher water.

During the Quaternary Ice-Age the warm currents stood less chance than now of carrying an appreciable portion of their original warmth to high latitudes. Owing to the enormous quantities of floating ice which existed in the oceans, and which have left traces of their existence in the submarine accumulations of glacial material which have been dropped by icebergs, the surface waters must have been very cold. We have evidence of great icebergs in the English Channel, which dropped boulders weighing many tons on to the sea-floor at Selsey, and there are glacial accumulations off the west coast of Ireland and in many other localities. The winds from the glacial anticyclones must have driven these icebergs and their cold thaw water far across the oceans. This water, being light because of its freshness, spread over the warmer but more saline water of the warm currents in mid-Atlantic, just as it does to-day in the Arctic, so that the Gulf Stream, for instance, must have lost its identity in relatively low latitudes.

In addition to Chamberlin's hypothesis referred to above, changes in the oceanic circulation induced by alterations of the land and sea distribution have often been suggested to account for climatic changes. F. Kerner has been especially active in explaining the warm periods in this way, but I am deferring a consideration of his work until Chapter VIII., since his method is to analyse the distribution of temperature resulting from the present land and sea distribution, and to apply the results to the geography of former geological epochs. We may refer here to a very old idea that the Quaternary Ice-Age was brought about by the omission of the Gulf Stream from the economy of the North Atlantic. Four ways have been suggested in which this may happen ; the opening of a wide gap between North and South America by the submergence of the Isthmus of Panama, allowing the Gulf Stream proper to pass into the Pacific instead of being bent back into the North Atlantic ; the northward extension of the eastern shore-line of South America in such a way as to deflect the greater part of the Equatorial Current to the south instead of to the north ; an increase in the velocity of the North-east Trades in the Atlantic relatively to the South-east Trades, shifting the whole system of Equatorial Currents southward with the same result ; and the formation of an extensive " Antillean Continent " across the path of the Gulf Stream and Antilles Current, forcing them to pass eastward much farther south than at present and form a closed circulation in tropical and sub-tropical regions. Any one of these changes might modify the surface circulation in the North Atlantic and introduce corresponding climatic changes in eastern North America, and especially in Europe, and we must examine them.

The separation of North and South America by a strait across the Isthmus of Panama occurred during the greater part of the Tertiary period, and was responsible for the great difference in the faunas of these continents, but it does not appear to have persisted into the Quaternary. Hence from this factor alone we should have expected a cold climate in North-west Europe during the Tertiary, becoming warmer in the Quaternary, which is the reverse of what actually happened. Evidently the opening and closing of this gap did not greatly affect the Gulf Stream. South America stands in the course of the South Equatorial Current like a mighty wedge, with its apex at Cape San Roque in 5° S., and all that portion of the current which lies between the equator and 5° S., and which would normally be deflected southwards by the earth's rotation, is turned to the northward, and enters the North Atlantic as the Guiana Current. It is the Guiana Current which mainly supplies the warm water in the Gulf of Mexico. If, owing to geographical changes, the apex were shifted two degrees farther north, the amount of water which it deflects from the Southern to the Northern Hemisphere would be decreased by about forty per cent., an event which would appreciably affect the warmth of the North Atlantic. The north-eastern part of South America appears to have been slightly lower during the Quaternary than at present, but the configuration both of the land surface and of the ocean floor is such that a change of a thousand feet or more, whether elevation or depression, would make very little difference in the latitude of the apex of the wedge. There is, therefore, no reason to suppose that the geographical changes in this region during and since the Quaternary have been sufficiently great to introduce any important modifications in the volume of the Guiana Current.

There is a considerable amount of evidence that during at least the early part of the Quaternary period the Gulf of Mexico was largely dry land ; this would merely turn the waters of the Guiana Current north into the Antilles Current, and would not greatly affect the temperature of the Gulf Stream off the east coast of the United States. In fact, so far as the geography of the Quaternary can be reconstructed, it was as favourable as the present for the existence of a powerful warm current in the North Atlantic.

This leads us to ask if the effect of minor geographical changes on the great oceanic circulations has been over-estimated. When the project of a Panama Canal was first mooted, there was some popular outcry that it would allow the Gulf Stream to pass through into the Pacific and so interfere with the climate of Europe. This fear was quite unnecessary, but it illustrates the importance which the " man in the street " attaches to the slender barrier of the Isthmus of Panama. Actually, as we have seen, only about one-third of the water which forms the Gulf Stream off the east of Florida passes through the Gulf of Mexico at all ; two-thirds of it is derived from the Antilles Current, which takes the whole of the water from the North Equatorial Current. But the real reason for the existence of the Antilles Current is not the chain of islands known as the Antilles, it is the limitation of the sub-tropical anticyclonic centres to the eastern halves of the oceans, combined with the rotation of the earth, which deflects the North Equatorial Current to the right, i.e., northwards.

In the North Pacific there is a gap between the Philippine Islands and China which is wide open to the waters of the North Pacific Equatorial Current, but the latter ignores the invitation, and instead turns northward in the open ocean to form the warm current which gives its favourable climate to Japan. In fact, under normal conditions, water which is travelling westward north of the equator must turn north, and water which is travelling westward south of the equator must turn south, unless hindered from doing so by some geographical obstacle. This flight from the equator will take place most readily where the isobars also trend away from the equator at the western ends of the sub-tropical oceanic anticyclones.

Finally, there remains the fourth consideration, the unequal strength of the Trade winds in the two hemispheres. At present the atmospheric circulation over the whole Southern Hemisphere is stronger than that over the Northern Hemisphere, partly because the greater area of the oceans leads to a smaller loss of energy through friction, and partly because, owing to the very low temperatures over Antarctica, the temperature gradient between equator and pole is greater in the Southern Hemisphere. Hence the South-east Trades are the stronger, and owing to their momentum are able to blow right across the equator into the Northern Hemisphere. The difference is greatest in June to July and least in December to January, but the doldrums lie north of the equator throughout the year. Hence the greater part of the Equatorial Current, both in the Atlantic and Pacific Oceans, lies north of the equator, and is deflected northward by the earth's rotation.

Now we know that the glaciation of the Antarctic Continent began during the Tertiary, while the Northern Hemisphere was still enjoying genial climates in high latitudes. Hence we may suppose that at this period the South-east Trade crossed the equator to an even greater extent than at present, and that this helped to maintain the temperature of the Northern Hemisphere and to depress that of the Southern Hemisphere. Then the Northern Hemisphere also became glaciated, and, owing to the greater land area in the glaciated regions, these northern ice-sheets outweighed the southern, and caused the North-east Trades to become as strong as or stronger than the South-east Trades. This caused the Northern Hemisphere to receive a smaller share of the equatorial warm water, intensifying the glacial conditions in that hemisphere still further, while the glaciation of the Southern Hemisphere remained relatively slight. Hence we see that during the Quaternary period the variations of the oceanic circulation must have tended to exaggerate the climatic oscillations in the Northern Hemisphere and moderate them in the Southern Hemisphere.

The partial closing of the gap between Greenland and Europe by the elevation of the submarine ridge which passes through Iceland and the Faroes to Scotland, which occurred during the Quaternary Ice-Age, must have deflected the

Gulf Stream Drift into lower latitudes and displaced the Icelandic minimum southwards, altering its alignment to west-east or even north-west-south-east, instead of south-west-north-east as at present. This must have profoundly modified the climate of the countries bordering on the North Atlantic, and probably increased the severity of the glaciation in these regions. These changes have been discussed by the late F. W. Harmer (6, 7), who attempted to reconstruct the pressure distribution and storm tracks which would prevail under these conditions. His papers were written before the work of G. de Geer on annual clay varves had demonstrated the contemporaneity of at least the Wurmian glaciation in North America and Europe, and he supposed that the glaciations of these two continents alternated, but I think the pressure distribution which he deduces would have favoured increased winter snowfall over the north-eastern parts of North America, and hence brought about glaciation rather than deglaciation.

H. J. E. Peake and H. J. Fleure (8) in some comments on Harmer's papers suggest a complete explanation of the Quaternary glacial sequence in Europe in terms of the elevation and depression of a Labrador-Greenland-Iceland-Scotland land-bridge. They point out that with this bridge *complete* there would be little or no ice in the North Atlantic in summer, and the climate of the British Isles would be dry and sunny, similar to that of the coast of British Columbia, and unfavourable for glaciation. A somewhat smaller elevation, however, which left some gaps in the land-bridge, " would probably increase the amount of ice in the North Atlantic so long as the northern lands remained much higher than at present," and would cause a deterioration of the climate of Western Europe. They point out that the first or Gunzian glaciation was limited to Scandinavia and the Alps, and did not extend to France, the North Sea, or the English plain, so that it might well have been the result of elevation only, and they suggest that during this glaciation the land-bridge was complete and the climate of England favourable. The cold period in Eastern England, represented by the Weybourne Crag and Chillesford Beds, which appears to correspond with the Gunzian glaciation, would then fall either just before or more probably just after the time of maximum elevation.

The later glaciations occurred during periods of lesser elevation, when the land-bridge was not complete and there was much ice in the North Atlantic ; the interglacial periods occurred during the intervals of subsidence in which the land fell below its present level.

There is quite a lot to be said for this interpretation of the Quaternary sequence in Europe. The suggestion that a large amount of cold water was accumulated in a closed Arctic basin, and that the level of the Arctic Ocean may even have risen well above the general level of the remaining oceans, this mass of cold water being subsequently released by a depression of the land and flooding southwards into the Atlantic, may be especially fruitful. It gives a plausible explanation of the sudden appearance of the Arctic fauna in middle latitudes, for example, in the Sicilian (300-foot) raised beaches of the Mediterranean, which contain a fauna now found only in the northernmost parts of Europe. But we must not forget that the Quaternary sequence in Europe was paralleled by that in other parts of the world, such as the Himalayas, and that any explanation of the phenomena found in Europe must fit into place in a larger scheme which takes account of the whole world.

The final rôle of the oceans in climatic changes to which we have to refer is that of regulator. It is well known that owing to the high specific heat of water and to the fact that the changes of temperature penetrate to a greater depth, a large sea or ocean takes much longer to warm or to cool than does a land surface in the same latitude. Hence the annual range of temperature on islands or windward coasts is much less than that in the interior of great continents. But this conservation of heat is not limited to periods of a year or even a few years. During a period of cooling climate, the cooled sea water sinks to the bottom of the oceans and the warmest water remains at the top.

If the ocean covered the whole surface of the earth it would have an average depth of 2,600 metres (8,500 feet); if we take the amount of heat reaching the outer limit of the earth's atmosphere from the sun as 720 calories per square centimetre per day (see next chapter), we find that the whole of this solar heat for a year would have to be absorbed and retained by this universal ocean, to raise the temperature by one centigrade degree. Again, suppose that at the end of a long warm period the mean temperature of the oceans is 10° C. (18° F.) higher than at present. When we remember that a large portion of the oceanic water is now little above freezing point, it is seen that this is not an exaggerated assumption. Then if for some reason the equilibrium temperature sank to its present level, the heat conserved in the oceans would suffice to maintain the average temperature 2° C. (3·6° F.) above the present for a period of nearly 250 years. The retardation of warming up after a cold period would be less effective, since the cold water would remain at the bottom of the ocean without greatly affecting the higher layers.

R. Spitaler (9) goes even further than this. In his theory of the astronomical cause of ice-ages (Chapter V.), he attempts to get over the difficulty that increased eccentricity of the earth's orbit would act oppositely in the Northern and Southern Hemispheres by supposing that the regulating effect of the oceans could maintain glacial conditions during a period of 10,000 years while conditions were otherwise not specially favourable for glaciation. He does not give any calculations in support of this figure, and it seems to be excessive. Spitaler's claim would perhaps have been based more soundly on the consideration that owing to the creep of cold water along the ocean floor, severe glaciation in one hemisphere would suffice to maintain low temperatures throughout the bottom layers of the whole ocean, and so to some extent lower the temperature in the other hemisphere also, though the process would not be very effective.

Huntington and Visher (10) suggest that the growing salinity of the oceans during the course of geological time may have had some climatic effect. This may be so, but I doubt if the effect can have been noticeable during the greater part of geological time. The accession of salt to the ocean is at present derived almost entirely from the sedimentary rocks, that is, it has previously been withdrawn from the oceans. The very great estimates of the duration of the pre-Cambrian period now current—nearly a thousand million years— suggest that even at the beginning of the Palæozoic the ocean had a long history behind it, and was almost as salt as it is now. This is borne out by the relatively advanced organisation of the earliest fossils, which also suggest life in salt water rather than in fresh. It is possible, however, that a smaller salt content may have been a contributory cause in the pre-Cambrian glaciations. The fact that fresh water reaches its greatest density at a temperature above its freezing point, while ordinary sea water freezes before it cools to its maximum density, is the reason why fresh-water lakes freeze more readily than ocean inlets of the same depth. In very early geological ages it is possible that the sea was less salt than now, and if the difference was sufficient to bring the temperature of maximum density above the freezing point, which would happen if the salinity were less than 24·7 parts per thousand compared with the present value of about 35 per thousand, the surface of the ocean would have frozen more readily than at present. Other conditions being equal, therefore, glaciation of coastal mountains would have been easier in very early geological ages than at present.

The fluctuations of salinity from one geological epoch to another may also have affected the capacity of the air to absorb moisture from the oceans. The withdrawal of a great volume of fresh water during the glacial periods, to be locked up in the form of ice, must have increased the average salinity slightly. Moreover, there are variations in the amount of soluble matter locked up in salt and gypsum beds, etc. ; at the close of a long warm period with shallow seas and numerous lagoons this amount must have been appreciably greater than at present. Thus we may suppose that there have been small fluctuations of salinity, with minima at the end of the long warm periods and maxima during the ice-ages, superposed on a very slow secular increase. Decreased salinity would increase the vapour pressure over the oceans, and it will be seen in Chapter VI. that an increase of water vapour in the atmosphere tends to raise the mean temperature. In this way there would be a tendency for both the warm periods and the ice-ages to be intensified with also a slight secular fall of temperature. I think, however, that at least since the middle of the Palæozoic period the variations of temperature due to this cause alone must have been so small as to be negligible compared with the other causes of variation. It seems highly improbable that at any stage in the known geological record, with the possible exception of the early pre-Cambrian, was the main mass of the sea water sufficiently fresh for its temperature of maximum density to be above its freezing point.

REFERENCES

(1) DURST, C. S. "The relationship between current and wind." London, Q. J. R. Meteor. Soc., 50, 1924, p. 113.

(2) HEPWORTH, M. W. CAMPBELL. "The Gulf Stream." London, Geogr. J., 1914, p. 431.

(3) SVERDRUP, H. U. "Oceanography for meteorologists." New York, 1942.

(4) LASAREFF, P. "Sur un méthode permettant de démontrer la dépendance des courants océaniques des vents alizés et sur le rôle des courants océaniques dans le changement du climat aux époques géologiques." Beitr. Geoph., 21, 1929, p. 215.

(5) CHAMBERLIN, T. C. "An attempt to frame a working hypothesis of the cause of glacial periods on an atmospheric basis." J. Geol., Chicago, 7, 1899, pp. 545 and 667.

(6) HARMER, F. W. "The influence of the winds upon climate during the Pleistocene epoch." London, Q. J. Geol. Soc., 47, 1901, p. 405.

(7) HARMER, the late F. W. "Further remarks on the influence of the winds upon climate during the Pleistocene epoch." London, Q. J. R. Meteor. Soc., 51, 1925, p. 247.

(8) PEAKE, HAROLD J. E., and H. J. FLEURE. "The Ice-Age." Man, 1926, p. [4].

(9) SPITALER, R. "Das Klima des Eiszeitalters." Prag, 1921. (Lithographed.)

(10) HUNTINGTON, E., and S. S. VISHER. "Climatic changes, their nature and cause." New Haven, 1922.

CHAPTER IV

RADIATION FROM THE SUN

IT has been shown in the preceding chapters that a comparatively small initial change in the mean temperature of the polar regions might be so magnified by secondary effects, especially in connexion with the polar ice-caps, that the final result would be a very great change of climate, sufficient to account for the genial polar climates of the " warm " periods. We have now to begin our search for such possible initial causes, and the most obvious place to look is the great source of all warmth and life on the earth—the sun. Careful studies are being made, especially by the Astrophysical Observatory of the Smithsonian Institution (1), of the amount of heat radiated by the sun, and of the variations to which it is subject. The measurement aimed at in the first place is the amount of heat which would reach a unit area of the earth's surface exposed to the solar beam at right angles, if none of it were intercepted by the earth's atmosphere. As it is impossible to get outside the atmosphere to obtain these measurements, the result has to be arrived at indirectly. Observatories have been established at several points on high mountains, where the air is normally very dry and the sky clear ; the chief of these observatories are at Montezuma in Chile (8,895 feet), Table Mountain, California (7,500 feet) and Mount St. Katherine, Egypt (8,500 feet). Elevated stations are chosen because the mass of air through which the sun's rays have to penetrate is less ; dry regions because part of the solar radiation is absorbed by water vapour and also because a clear sky is essential for regular observations. At these places the direct heating power of the sun is therefore greater than it is on low, humid and cloudy plains, but there is still a large amount of absorption. This is calculated in two ways. When the sun is nearly vertical, the thickness of air through which its rays have to penetrate is much less than when it is near the horizon, and the heat received is

consequently greater.   Making a number of observations at
intervals during a morning when the meteorological conditions
remain practically uniform is therefore almost equivalent to
making observations at the same moment at different heights
above the ground.   It is found that with radiations of certain
wave-lengths which are only absorbed slowly by the
atmosphere, the logarithm of the amount lost is proportional
to the air mass through which the rays have passed, and it is
possible to calculate the amount of heat which these rays would
deliver if there were no atmosphere.   By means of a spectro-
scope these measurements are taken for a number of different
wave-lengths, and the results are plotted.   In general, they
show a smooth curve which is, however, interrupted by a few
marked depressions corresponding with the wave-lengths of
radiations which fail to penetrate the air at all.   These rays
are almost or quite absorbed in the high levels of the
atmosphere, and in order to include them in the estimate of the
sun's total radiation an allowance has to be calculated from
the values of neighbouring wave-lengths.   As a result of a series
of measurements and calculations carried out by the Astrophysical Observatory at various stations from 1920 to 1939, the
mean value of the intensity of the sun's radiation outside the
limits of the earth's atmosphere is given as 1·945 calories per
square centimetre per minute.   This means that a layer of
cold water a centimetre deep exposed to a vertical sun,
absorbing all the solar radiation and giving out none in
exchange, would warm up at the rate of nearly 2° C. a
minute.   This value is termed the " solar constant," but the
name is somewhat misleading, as it seems probable that the
mean value of 1·945 is subject to small variations on either
side of the mean.

The range of these variations is still subject to doubt, owing
to uncertainty in the daily values.   The annual means for
the 20 years from 1920 to 1939 range from 1·928 in 1922 to
1·950 in 1921, or just over one per cent. of the mean, but both
of these are early values and somewhat uncertain.   Abbot
finds a number of short periodicities in the data, the longest
being 23 years ;   he estimates the amplitude of this cycle
as less than 0·5 per cent. of the average value of the solar
constant.   It is clear that so small a change, even if maintained,
would not suffice to bring about great changes of climate.

The possibility of much greater changes of solar radiation in geological time cannot, however, be ruled out. Sir George Simpson (2, 3) has discussed the probable effects of a large oscillation of solar radiation, and has shown that they agree with the observed climatic changes during the Quaternary.

The first effect of an increase in the solar radiation would be to raise the temperature everywhere on the surface of the earth, but more in low than in high latitudes. This would immediately increase the amount of evaporation from water surfaces, and also the strength of the atmospheric circulation. More evaporation means more cloud and more precipitation. But as will be seen in Chapter VII., cloud reflects back to space a large part of the solar radiation which falls on it. Hence an increase of cloudiness lowers the temperature. The final result of a large increase of solar radiation would therefore be a slight rise of temperature and a great increase in cloudiness and precipitation. The increase in total precipitation with increasing radiation is shown in curve I of Fig. 9, reproduced by permission of Sir George Simpson and the Manchester Literary and Philosophical Society from (2).

In high latitudes and especially on high ground, a large part of the total precipitation falls as snow. As the radiation increased, the proportion of precipitation falling as snow would decrease, but for a time this decrease would be slower than the increase of total precipitation. Hence the snowfall would increase at first, but with a further increase of radiation the general rise of temperature would cause so much less of the precipitation to fall as snow that the total snowfall would begin to decrease. The curve of snowfall is shown as II in Fig. 9.

The accumulation of snow to form ice-sheets results from the excess of snowfall over melting. The melting is shown in curve III ; so long as the summer temperature does not rise above freezing point it is inappreciable (point A on the curve). With rising temperature it grows slowly at first, so long as the snow cover is complete, but as soon as the melting is sufficient to expose part of the underlying surface it proceeds rapidly. At point B the curves of snowfall and melting intersect and above that point any accumulation of snow or ice will disappear. Curve V shows the accumulation of snow during a period of increasing radiation. With decreasing radiation the changes would proceed in the reverse order.

In low latitudes where there is no snow on low ground the result of increasing radiation would be a progressive increase of rainfall, but on equatorial mountains of sufficient height there would be a glacial cycle resembling that in high latitudes.

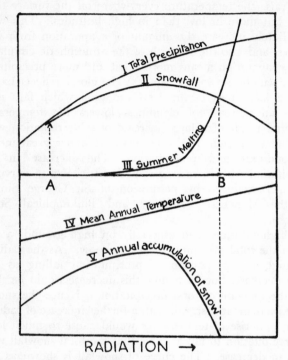

RADIATION →

Fig. 9.—Effect of increasing radiation on precipitation and accumulation of snow.

Fig. 10, also reproduced by permission from (2), shows the effect of a double oscillation of solar radiation. The first period of increasing radiation causes an ice-age in high latitudes and increasing rainfall in low latitudes. At the maximum radiation the ice melts and we have a warm moist interglacial period in high latitudes coinciding with the peak of a pluvial period in low latitudes. As radiation decreases, there is a return of glaciation at first, but as radiation and precipitation decrease still further, the latter, even though it falls entirely as snow, is insufficient to balance the loss of ice by outflow and the ice-sheets finally disappear. We now enter a long cool arid interglacial period, which lasts until a new increase of solar radiation brings about a renewal of the glacial and pluvial cycle.

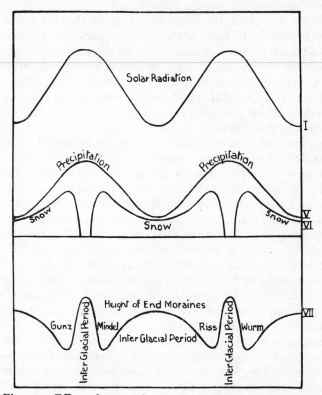

Fig. 10.—Effect of two cycles of solar radiation on glaciation.

As shown in the lowest curve of Fig. 10, this reconstruction fits in admirably with the classical sequence of glacial and interglacial periods worked out by Penck and Brückner (4). The first crest of solar radiation represents the Gunz and Mindel glaciations and the relatively short Gunz-Mindel inter-glacial. The trough of low solar radiation represents the long

Mindel-Riss interglacial. Finally the second crest of radiation represents the Riss and Wurm glaciations and the short Riss-Wurm interglacial.

Sir George Simpson worked out the geological and climatic implications of his theory in greater detail in another paper (3), and showed that they are consistent with the general climatic and biotic history of Europe during the Quaternary. There are, however, certain difficulties.

In the first place, the assumption is made that the sun is a variable star. If such variations are periodic, glaciation should have recurred at short intervals throughout geological time ; the theory cannot account for the long genial periods. Sir George Simpson tends to meet this difficulty by continental drift (see Chapter XIII.), but even so, such marked changes of precipitation as the theory requires should have left their records in the character of the sedimentary rocks, but there is no trace of such a regular cycle in the geological record. This difficulty was to some extent met by F. Hoyle and R. A. Lyttleton (5), who supposed that at intervals of time of the order of 100 million years the sun passes through clouds of interstellar matter. The particles on and near the track fall into the sun, their kinetic energy being converted into heat and giving rise to increased solar radiation. Since the cloud would in general be densest near the centre, radiation would rise to a maximum and then decrease again, the time of the passage being of the order of 100,000 years. Further, since many such clouds are irregular, the one causing the Quaternary ice-age may have had two centres, so giving two maxima of radiation and four glaciations.

Secondly, the glacial sequence was far more complex than is represented in the simple fourfold scheme of Penck and Brückner. The Riss glaciation had two or three distinct maxima and the Wurm three, in addition to retreat stadia. This objection is not very serious, however, because there is no difficulty in postulating minor but still large fluctuations of solar radiation superposed on the smooth curve of Fig. 10.

The third difficulty concerns the climates of the interglacial periods. By the solar radiation theory the Gunz-Mindel and Riss-Wurm should have been mild and very wet, the Mindel-Riss cool and very dry. The available evidence suggests, however, that the general succession of vegetation was very similar in both Mindel-Riss and Riss-Wurm interglacials, indicating that the recession of the ice in Europe was followed in each case by a rise of temperature to a level somewhat higher than the present. This rise in turn gave place to a fall and finally to the onset of another glaciation. The post-glacial period has so far followed a similar course, a rise to a maximum in the " Climatic Optimum " followed by a fall, but it is too early yet to say that we are now in the second half of an interglacial period.

Finally, the available data do not support the hypothesis that in low latitudes one pluvial period represents two glaciations. The pluvial sequence has been worked out in detail in tropical Africa, where the rises and falls of the great lakes are believed by E. Nilsson (6) to have been contem-poraneous with the advances and retreats of the mountain glaciers. For example, in the last interpluvial period, when the prehistoric Lake Kamasia dried out completely, the mountain glaciers also melted completely. Subsequent oscilla-tions of climate also run closely parallel in both lakes and glaciers.

The evaporation from the tropical oceans, and hence the total rainfall over the globe, depend on the wind velocity as well as on the temperature. Large ice-sheets in high latitudes increase the temperature gradient and hence the strength of the atmospheric circulation, while at the same time, owing to the existence of large semi-permanent glacial anticyclones and the equatorward deflection of the storm tracks, they decrease the area over which most of the rain falls. Hence, other things being equal, each glacial period must have been a pluvial period in tropical and sub-tropical regions and each interglacial an interpluvial period. Whether this oscillation was superposed on a longer oscillation in which one crest covered two glaciations is not yet clear.

We may sum up by saying that if the radiation of the sun has varied greatly in the past the variations would have had the effects postulated by Sir George Simpson, but that there is no *direct* evidence of such variation, and the indirect evidence only partially supports the theory.

F. Kerner-Marilaun in his book on Palæoclimatology (7) takes the standpoint that variations of solar radiation should only be brought in as a last resource, *i.e.*, if it is impossible to account for the climate of a period of geological time by the known factors of land and sea distribution, volcanic dust, etc. As an example he considers the temperature of Messel, near Darmstadt, Germany, during the Eocene. From geographical and astronomical factors he calculates the probable temperatures as : January, 50 to 56° F., July, 70 to 74° F. From the plant remains and the nature of the deposit the temperature is estimated as 56° in January and 68° in July. The differences are within the limits of probable error and it is not necessary to bring in unknown factors such as a change of solar radiation. In other examples the agreement is less good but it is not yet possible to say whether the discrepancies are due to errors in the calculations, in the interpretation of the geological data or to changes of solar radiation. Nevertheless the author foreshadows the time when, from a large number of such calculations in different geological periods, it may be possible to reconstruct the variations of solar radiation. Till then the question must be left open.

As another example of such a calculation I may mention a reconstruction of the probable climatic conditions in the middle Eocene of south-east England which I made for Mrs. E. M. Reid and Miss Chandler (8). I estimated the probable upper limit of the mean annual temperature as 65° F. (a remarkable agreement with Kerner-Marilaun's quite independent figure) whereas Mrs. Reid and Miss Chandler inferred from the vegetation that the mean temperature was probably not below 70° F., and I suggested that the discrepancy might possibly be due to increased solar radiation. In all other respects the climatic reconstruction agreed closely with the inferences from the vegetation.

Brief reference may be made here to a suggestion by K. Himpel (9) that the sun suffered Nova-outbreaks in Mid-Algonkian (pre-Cambrian), late Upper Carboniferous and late-Pliocene, the average interval between outbreaks being 250 to 350 million years, which he claims is about right for Novæ. Each outbreak caused a thousand-fold increase in radiation which resulted in a corresponding increase of precipitation, followed by rapid cooling to a level below the present. At one stage during the cooling there was a maximum snowfall and consequently glaciation. Further, each outbreak diminished the mass of the sun and consequently its normal radiation, leading to a general level of climate somewhat cooler than before. It seems impossible, however, that even a short-lived thousand-fold increase of solar radiation could have occurred without devastating animal and plant life, and there is certainly no trace of such a catastrophe in late Pliocene.

Total radiation is not the only solar characteristic which probably affects terrestrial climate ; we have also to take account of the nature of the radiation. An index to the latter is found in the spotted area of the sun. The spottedness is generally expressed as the " relative number " (10), which is obtained by an arbitrary formula, but which has been found by photographic comparisons to be closely proportional to the spotted area. A relative number of 100 corresponds with about one five-hundredth of the sun's visible disc covered by spots, including both umbra and penumbra. Since 1749 the monthly mean relative sunspot numbers have varied between 0 and 206.

The sunspot " relative number " does not bear any simple relation with the solar constant. Sunspots show a very marked and persistent oscillation of approximately 11·2 years, falling to an annual mean of 10 or less at sunspot minimum and rising to a maximum which has varied since 1749 from annual means of 46 to 154. This 11-year oscillation is not recognisable in the values of solar radiation, so it must represent a variation of some other solar characteristic. It seems to be an almost permanent characteristic of the sun, for a cycle of ten or eleven years has been found in the thickness of annual layers in laminated deposits of various geological ages, including the Upper Carboniferous glacial clays of Australia.

This eleven-year oscillation is not the only one shown by sunspots ; more important for our purpose is the fact that both the eleven-year means and the heights of the maxima undergo large variations over decades. Thus the annual maxima varied from 154 in 1778 to only 46 in 1817, and back again to 139 in 1870. Previous to 1749 we have no systematic observations, but there are a number of scattered references in the Chinese archives dating back nearly to the beginning of the Christian era and a few records from Europe going back to mediæval times. From these there appears to have been an important maximum of solar activity towards the close of the eleventh century, and another, which R. Wolf believed to be the absolute maximum over the whole of the Christian era, about 1372.

There is little doubt that some relation exists between the sunspot cycle and terrestrial conditions, but it is very obscure. In tropical regions temperature averages about 2° F. higher at spot minimum than at spot maximum ; this relation disappears in temperate regions but reappears in the Arctic, where it is clearly shown both in the temperature of Spitsbergen and in the amount of ice in the Barents Sea, indicating that at sunspot maximum the temperature falls in the Arctic and the area of the floating ice-cap increases. Over the world as a whole rainfall seems to have a tendency to be greatest at spot maximum, but there are many exceptions. One of the most interesting relations is between sunspot relative numbers and the frequency of thunderstorms, which are most frequent at sunspot maxima. The agreement is close in inland and tropical regions, especially Siberia, the West Indies and the tropical Pacific, but almost or quite vanishes in the stormy regions of western Europe (11). For the world as a whole the variation from sunspot minimum to sunspot maximum amounts to more than 20 per cent. of the mean number of thunderstorms.

These various relationships suggest that at periods of great solar activity as marked by large sunspot numbers, the earth as a whole should be somewhat cooler, rainier and more stormy than at times of little solar activity. Ellsworth Huntington and S. S. Visher (12) have based a complete theory of climatic change on these relationships. According to their view, the climate of the earth is largely governed by the relations between sunspots and storminess. Increased solar activity is considered to result in increased storminess, together with some displacement of the storm tracks. The greater vertical movement of the air associated with the increased storminess carries greater quantities of heat from the earth's surface to the higher levels of the atmosphere, where nearly half of it is lost by radiation to space, and when such a period of increased activity occurs with extensive and high continents, and perhaps with other favourable conditions, such as a paucity of carbon dioxide, a glaciation results. This combination of circumstances is considered to account for the Quaternary glaciation, and perhaps also for that of the

Permo-Carboniferous period. In the latter, the storm tracks are placed very far south, and higher latitudes are supposed to have remained unglaciated because they were occupied by deserts, a conclusion which ignores the extensive development of coal measures. Periods of slight solar activity and few sunspots had slight storminess and steady winds from low to high latitudes, hence they were periods of mild and equable climate over the whole earth. The variations of solar activity in the geological past are connected by Huntington with changes in the distance of the nearest fixed stars, especially double stars, a conclusion which few, if any, astronomers endorse.

It will be seen in Chapter XXII. that the variations of rainfall in the temperate zone during the Christian era have run fairly parallel with the variations of solar activity shown by the records of sunspots and auroræ. But the phenomena of the Quaternary Ice-Age were on a scale many times greater, and would require enormous and prolonged outbursts of sunspots, which seem quite improbable. Moreover it has been shown (14) that the terrestrial changes are not proportional to the sunspot relative number ; the effect falls off rapidly as the relative number increases. Hence, while variations of sunspot activity may account for some of the minor glacial oscillations, it is unlikely that they played any appreciable part in the four main glacial advances and retreats of the Quaternary Ice-Age, and still less likely that they caused that Ice-Age as a whole. The hypothesis also breaks down completely over the Permo-Carboniferous glaciation.

One other speculative work may be briefly referred to, because it is typical of a number of early hypotheses which rely on a *diminution* of the sun's radiation to explain the occurrence of ice-ages. E. Dubois (13) assumed that the sun has passed through a series of stages represented now by various fixed stars. At first, when the sun was in the stage represented by blue-white stars, radiation was more intense and the terrestrial climate was warm. The sun then changed to yellow, its radiation diminished, and the earth cooled. This was during the Tertiary period, and at intervals the sun passed into a third stage, intermediate between the yellow and red stars, during which its radiation was still feebler. These intervals formed the various glacial epochs of the Quaternary. The only " evidence " adduced in support of the theory concerns the frequency of blindness to certain colours, which is regarded as a reversion to previous geological epochs, the spectroscopic analysis of the light of phosphorescent animals, and similar peculiarities. This theory is quite untenable. In the first place the postulated changes in radiation would not have had the effects attributed to them. Secondly, the age of the sun is now estimated as between a hundred thousand million and a million million years, while the age of the oldest known rocks is not much more than a thousand million years (Appendix I.), so that it is highly improbable that the later stages of the geological record should have seen any appreciable change in the sun's constitution of the type supposed by Dubois. Such a conclusion is in fact borne out by the record of the rocks. The warm climates of the early Palæozoic period might have lent colour to a belief in a hotter sun at that time (some 400 to 500 million years ago), but it happens that the earliest reliable climatic record we possess shows us an ice-age.

Variations in the distance of the nearest " fixed " stars have been suggested as a possible source of climatic changes, through variations in the radiation received from this source, but this suggestion must be ruled out for two reasons. At present the heat received from all the fixed stars together is not one ten-millionth of that received from the sun, and there is thus no possibility of explaining colder periods than the present on these lines. The close approach of a large star might account for a warmer period, but such an event would have left its traces in a derangement of the solar system, and astronomers say that it could not have happened at any time in the geological past.

We may end this chapter with a reference to a curious hypothesis put forward by R. L. Ives (15) to explain the Permo-Carboniferous glaciation. He suggests that a small satellite of the earth, either original or captured, was disrupted by tidal and other forces in the late Palæozoic to form a ring of fragments round the equator, similar to Saturn's rings. This caused low equatorial temperatures, with storms and heavy snowfall along the boundaries of the shadow. Multiple glaciation is accounted for by the occurrence of a series of stages in the break-up. Such an occurrence seems intrinsically improbable, however, and in any case it can be readily calculated that a ring of this nature, if it cast an effective shadow, would lower the temperature of the whole earth to such an extent as to make life impossible.

REFERENCES

(1) WASHINGTON, SMITHSONIAN INSTITUTION. "Annals of the Astrophysical Observatory," vol. vi, 1942. By C. G. Abbot, F. E. Fowle, and W. H. Hoover.

(2) SIMPSON, G. C. "Past climates." *Mem. Manchr. Lit. Phil. Soc.*, 74, 1929-30, no. 1, pp. 34.

(3) SIMPSON, G. C. "The climate during the Pleistocene period." *Proc. Roy. Soc.*, Edinburgh, 50, 1929-30, p. 262.

(4) PENCK, A. and E. BRÜCKNER. "Die Alpen im Eiszeitalter." Leipzig 3 Vols., 1901-9.

(5) HOYLE, F. and R. A. LYTTLETON. "The effect of interstellar matter on climatic variations." *Proc. Camb. Phil. Soc.*, Cambridge, 35, 1919, p. 405.

(6) NILSSON, E. "Quaternary glaciations and pluvial lakes in British East Africa." *Geogr. Ann.*, Stockholm, 13, 1931, p. 249.

(7) KERNER-MARILAUN, F. "Paläoklimatologie." Berlin (Gebr. Borntraeger), 1930.

(8) REID, E. M. and M. E. J. CHANDLER. "The London Clay flora." London, 1933.

(9) HIMPEL, K. "Die Klimate der geologischen Vorzeit." *Veröff. Astron. Ges. Urania*, Wiesbaden, nr. 4, 1937.

(10) ABBOT, C. G. "The sun." London and New York, 1912.

(11) BROOKS, C. E. P. "The variation of the annual frequency of thunderstorms in relation to sunspots." London, *Q. J. R. Meteor. Soc.*, 60, 1934, p. 153.

(12) HUNTINGTON, E., and S. S. VISHER. "Climatic changes, their nature and cause." New Haven, 1922.

(13) DUBOIS, E. "The climates of the geological past." London, 1895.

(14) BROOKS, C. E. P. "Non-linear relations with sunspots." London, *Q. J. R. Meteor. Soc.*, 53, 1927, p. 68.

(15) IVES, R. L. "An astronomical hypothesis to explain Permian glaciation." Philadelphia, *J. Frankl. Inst.*, 230, 1940, p. 45.

CHAPTER V

ASTRONOMICAL FACTORS OF CLIMATE

THE radiation emitted from the sun is not the only factor in determining the solar climate of the earth. Whether or not the total heat received by the earth in the course of a year has remained constant, its distribution among the belts of latitude during the different months has certainly varied from time to time, and this distribution can be calculated. There are three variables to be considered in this respect. The first is the obliquity of the ecliptic, or the angle which the plane of the equator makes with the plane of the earth's orbit round the sun. It is this which causes the seasons ; the greater the obliquity of the ecliptic the greater is the contrast between the heat received in summer and that received in winter. In 1910 the obliquity was $23° 27' 3·58''$, and it was decreasing at the rate of $0·47''$ a year, but the limits of its variation are difficult to calculate. Lagrange found a maximum of $27° 48'$ in 29,958 B.C., a minimum of $20° 44'$ in 14,917 B.C., and a maximum of $23° 53'$ in 2167 B.C. J. N. Stockwell gives much narrower limits, ranging from $24° 36'$ to $21° 59'$, with a maximum of $24° 17'$ in 8150 B.C., since when there has been a steady decrease. Drayson, on the basis of a theory not accepted by the majority of astronomers, supposed the obliquity to range from $35°$ to $11°$, the period being 31,680 years. The latest work by M. Milankovitch (1) assumes a variation between $22°$ and $24\frac{1}{2}°$ in a period of 40,400 years.

The second variable is the eccentricity of the earth's orbit. This orbit is elliptical, with the sun at one of the foci, and the distance between the centre of the ellipse and this focus, expressed in terms of the major axis of the ellipse, is termed the eccentricity. It varies in a period of about 100,000 years from zero to a value of about $0·07$. When the earth is nearest to the sun it is in *perihelion*, when most distant, in *aphelion*. At perihelion the earth travels along its orbit more rapidly than at aphelion. Thus the season which coincides with perihelion will be short and relatively warm, that which coincides with aphelion will be long and relatively cold, but the total amount of heat received on each hemisphere in the course of a year will be the same. At present, the Northern Hemisphere has its winter in perihelion and its summer in aphelion ; with the Southern Hemisphere, of course, the reverse is the case. Hence the *solar* climate of the Northern Hemisphere is less extreme than that of the Southern Hemisphere ; the fact that, actually, the climate is much more extreme in the Northern Hemisphere is due to the preponderance of land there. The season in which perihelion falls is not constant but undergoes a cyclic variation with a period of 21,000 years. Thus 10,500 years ago the Northern Hemisphere had its winter in aphelion and its summer in perihelion, consequently a more extreme solar climate. This regular variation is termed the *precession of the equinoxes* ; it is the third of our astronomical variables.

The variation of the eccentricity and the precession of the equinoxes form the basis of Croll's famous astronomical theory of the Quaternary Ice-Age (2). He supposed that at periods of great eccentricity the hemisphere with its winter in aphelion had a climate so severe that, if geographical conditions were favourable, the snowfall during the long cold winter was heavy enough to persist through the short hot summer and thus develop ice-sheets. At the same time the opposite hemisphere was enjoying a genial or interglacial period. Croll justly points out that the power of a snow surface in reflecting the sun's rays back to space without their having any warming effect on the earth is of great importance in the heat economy of ice-ages, but, apart from this, his discussion of the meteorological changes associated with periods of maximum eccentricity is probably unsound. Moreover, recent geological investigations have shown that the glacial periods and other climatic changes are practically synchronous in the two hemispheres, while de Geer's absolute dating shows that the periods at which glaciations occurred do not fit in with those required by Croll's astronomical theory.

Rudolf Spitaler (3) re-investigated the astronomical theory of climatic changes. His first step was to relate the mean temperature of any latitude in any month to the amount of heat received from the sun in that latitude. For this purpose he analysed the existing mean temperatures of the different latitudes between 60° N. and 60° S. in January, July, and the year, and obtained an expression for the mean temperature in any latitude in any month, which may be rewritten as follows, to give Fahrenheit degrees :—

$$t = (-17 + 156 \, S_o + 29 \, S_m) \, (W) + (-36 + 264 \, S_m) \, (L).$$

Here S_o is the average daily heat received on a horizontal surface at the limit of the earth's atmosphere in the latitude in question during the year, S_m is the average daily heat received under the same conditions during the month m (the units being so chosen that the value of S at the equator during an equinoctial day is approximately $0 \cdot 5$), and L is the fraction of land and W the fraction of water covered by the line of latitude. By making the appropriate changes in the values of S_o and S_m the temperatures under various astronomical conditions can be calculated from this formula, and by varying L and W the effect of varying land and sea distribution can be introduced.

Spitaler rejects Croll's theory that the conjunction of a long cold winter and a short hot summer provides the most favourable conditions for glaciation, and adopts the opposite view, first put forward by Murphy (4) and now generally adopted, that a long cool summer and short mild winter are the most favourable. Spitaler's attempt was rational, but fails to fit in with the actual sequence of events ; in particular his time-scale is hopelessly impossible. For example he ends the Wurm glaciation at 89,680 B.C. whereas geological evidence puts it at a mere 18-20,000 B.C.

W. Köppen and A. Wegener (5), while accounting for the Quaternary Ice-Age as a whole by the latter's theory of continental displacements (Chapter XIV.), bring in astronomical causes to account for the succession of glacial and interglacial periods. Like Spitaler, they emphasise the importance of a low summer temperature, but they employ a different method, based on work by M. Milankovitch (1), in which the amount of radiation received at any point during the summer half-year is expressed as the equivalent latitude of that point with respect to present astronomical conditions. With perihelion in June and a great obliquity of the ecliptic, the radiation received by any point in the north temperate or Arctic zones will be greater than the present, so that the equivalent latitude will be lower. Thus, in 65° N., the summer radiation in 9,500 B.C. was as great as that now received in 60° 20′ N., while in 20,400 B.C. the summer radiation was as low as that now received in 68° N. It is pointed out that a decreased obliquity of the ecliptic must increase the temperature contrast between pole and equator in summer without making much difference in winter. Hence the atmospheric circulation in summer would be strengthened, giving an increased frequency and intensity of cyclones in the temperate belt, and the summer rainfall of Southern Europe would be greater.

The plane of the ecliptic varies in a period of 40,400 years, while perihelion makes the circuit of the seasons in 20,700 years. In each hemisphere the coldest summers occur when the smallest obliquity corresponds with summer in aphelion during a period of maximum eccentricity. Hence the authors suppose that the glacial periods occur at maximum eccentricity, and consist of two periods of cold summers 40,000 years apart, which are united in one glaciation through the power of persistence of an ice-sheet. Any retreat stadia formed during the intervening period of warmer summers are masked by the readvance of the second phase, which starts from the remains of the first ice-sheet and therefore attains a greater development. The glaciations in the two hemispheres are therefore roughly synchronous, but a maximum in the Southern Hemisphere occurs about 10,000 years before or after the corresponding maximum in the Northern Hemisphere. On these lines the authors give in Table 3 a chronology of the Quaternary Ice-Age.

This theory marks a great advance on Croll and Spitaler but the chronology is still a long way from fitting the geological evidence, ending the Wurm glaciation, for example, about 66,000 B.C. and dating the post-glacial climatic optimum at 9,100 B.C. instead of 5-3,000 B.C. when the astronomical climate differed from the present by an insignificant amount.

The latest and most elaborate reconstruction of glacial history from astronomical data, also based on Milankovitch, is by F. E. Zeuner (6). Zeuner studies the glacial history of Europe in minute detail, including also the periglacial regions, from which he makes inferences as to the character of the

Northern Hemisphere.	Southern Hemisphere.

(The figures are in thousands of years before the present.)

Climatic Optimum, 9·1.

Baltic Stadium, 25.	Pre-Baltic Stadium, 33-30.
Wurm II., 74-66. Wurm I., 118-110.	Post-Wurm I., 110-103.
Riss II., 193-183. Riss I., 236-225.	Post-Riss II., 200-195. Post-Riss I., 226-218.
Nameless,[1] 305-302.	About (442), 389, 350, 312, 270.
Mindel II., 434-429. Mindel I., 478-470.	Post-Mindel I., 468-462.
Gunz II., 550-543. Gunz I., 592-585.	Pre-Gunz II., 560-554.

Table 3.—Köppen's interpretation of Quaternary sequence.

climate. He follows Köppen and Wegener in expressing variation of summer radiation as equivalent latitude. He shows that the variation of the present snow-line with latitude closely follows the variation of radiation received in the summer half-year. He argues that with changing astronomical conditions a rise of the winter temperature increases the snowfall and the corresponding fall of summer temperature enables the snow to persist through the year. He also follows up various secondary effects of glaciation such as the reflecting power of the snow surface, shift of tracks of depressions, and the periglacial belt of east winds, as well as the effect of changes of sea level caused by the locking up of great quantities of water in the ice-sheets, and the delayed effect of the weight of the ice in depressing the land.

The result is a very detailed scheme of changes of both climate and sea-level which fits in well with the most recent geological interpretations, as exemplified by the curve of

[1] H. Gams and R. Nordhagen consider that between the Mindel and Riss glaciations in the Alps, but nearer the latter than the former, there was an additional glaciation, which they term the Mühlbergian.

equivalent latitude of 65° N. His dating is briefly as follows (dates in thousands of years before present day) :—

Zeuner.			Penck and Brückner.	
Late Glaciation	III.	25	Wurm	40-18.
	II.	72		
	I.	115		
Penultimate	II.	187	Riss	130-100
Glaciation	I.	230		
Antepenultimate	II.	435	Mindel	430-370
Glaciation	I.	476		
Early Glaciation	II.	550	Gunz	520-490
	I.	590		

Table 4.—Zeuner's interpretation of Quaternary sequence.

On Zeuner's view the long interglacial between the Antepenultimate and Penultimate Glaciations (or Mindel-Riss) was on the whole about 4° F. warmer than the present, with a rather oceanic climate, but it was interrupted by a minor cold period (see footnote to Table 3).

Zeuner recognises that the astronomical theory does not account for the Quaternary Ice-Age as a whole, only for the details within the period. His scheme comes much nearer to the geological dating than do those of Spitaler and Köppen and Wegener. The astronomical dates still tend to be earlier than those favoured by the geologists, but this may reasonably be attributed to the natural lag in the accumulation and disappearance of great ice-sheets, which Zeuner considers may have amounted to some thousands of years.

Whatever we may think of the parallelism between Milankovitch's curve and the succession of glacial stages, it is clear that astronomical changes, and especially changes in the eccentricity of the earth's orbit, must have had quite appreciable effects on the climates of the past. At periods of maximum eccentricity the hemisphere with winter in aphelion must, other things being equal, have more pronounced seasons than that with winter in perihelion, and these pronounced seasons must have increased the strength of monsoons and other seasonal changes.

According to W. H. Bradley (7) the Eocene of Colorado, Utah and Wyoming includes beds with annual layers covering a duration of five to eight million years, which show periodicities of $11\frac{1}{2}$ years (sunspot cycle), 23 years, 50 years and about 21,000 years, the latter presumably representing the cyclic changes of eccentricity of the earth's orbit and the precession of the equinoxes. Alternations of layers in the Cretaceous of U.S.A. also suggest a cycle which is estimated as about 21,000 years, but there are no annual layers.

It is possible that the coal seams in the Upper Carboniferous represent periods of great eccentricity with winter in perihelion in the Northern Hemisphere and a small obliquity of the ecliptic giving a climate in the Northern Hemisphere with little annual range of temperature, and consequently no annual rings of growth in the woody stems. In that case the intervening beds of sandstone and clay would represent the other halves of the periods in which winter in the Northern Hemisphere was in aphelion and the climate consequently more extreme, and the whole succession from the base of one coal seam to the base of the next would represent a period of 21,000 years. Nearer the present, the alternation of brown coal and bauxite beds in the Tertiary of South-eastern Europe may be due, as suggested by Köppen and Wegener, to a similar alternation of equable and extreme astronomical climates. Similarly, as suggested by Dacqué (8), the extreme climate of the Old Red Sandstone may be due to the deposits having been formed during a period of great eccentricity, but in the absence of absolute dating of these deposits, both these suggestions must remain speculative. There is more support for the supposition that during the closing stages of the retreat of the Quaternary ice-sheet of Scandinavia, the extreme " continentality " of the climate was due partly to the greater obliquity. The maximum obliquity of 8500 to 7500 B.C. (according to the work of Stockwell) would have had the effect of lowering the winter temperature and raising the summer temperature by about $1°$ F. Even in this example, however, the changes due to purely geographical causes were probably much greater than those due to the increased obliquity, and the astronomical effect by itself would have been hard to distinguish.

REFERENCES

(1) MILANKOVITCH, M. " Théorie mathématique des phénomènes thermiques produits par la radiation solaire." Paris, 1920.

(2) CROLL, J. " Climate and time in their geological relations." London, 1875.

(3) SPITALER, R. " Das Klima des Eiszeitalters." Prag, 1921 (Lithographed).

(4) MURPHY, J. J. " Glacial climate and polar ice-cap." London, Q. J. Geol. Soc., 32, 1876, p. 400.

(5) KÖPPEN, W., and A. WEGENER. " Die Klimate der geologischen Vorzeit." Berlin, 1924.

(6) ZEUNER, F. E. " The Pleistocene period ; its climate, chronology and faunal successions." London, Ray Soc., 1945.

(7) BRADLEY, W. H. " The varves and climate of the Green River epoch." U.S. Geol. Surv., Prof. Paper, 158, 1929, p. 87.

(8) DACQUÉ, E. " Grundlagen und Methoden der Palæogeographie." Jena, 1915.

CHAPTER VI

The Absorption of Radiation by the Atmosphere

THE mean temperature of the earth is determined by the balance between the radiation received from the sun and that given out by the earth. The solar radiation entering the earth's atmosphere undergoes various transformations before it finally leaves the atmosphere again as radiation to space. These changes have been set out very clearly in a diagram by W. H. Dines (1) (Fig. 11, reproduced by permission of the Royal Meteorological Society). In this diagram, A is the radiation reaching the limit of the earth's atmosphere from the sun. The numerical value of A (measured in calories per square centimetre per day) is about 720. Part of this radiation is reflected from the surfaces of clouds, snowfields, and to a lesser degree from all parts of the earth's surface, both land and water, without undergoing any change. Another part of the solar beam is scattered by the molecules of the gases and the particles of dust in the atmosphere ; some of this scattered radiation ultimately reaches the surface of the earth, but some of it is entirely lost to the earth. The amount lost by reflection and scattering is represented by D, which has a value of about 320. [The numerical values are taken partly from a later source (2).] Another part of the radiation, C, is absorbed by the air, and the remainder, B, is absorbed by the earth. The average value of B is about 350, leaving only 50 for C.

The surface of the earth is losing heat in three ways. An amount, G, which depends on the temperature, is radiated outwards ; of this a small part, M, is reflected back to the earth, mainly from the under surfaces of clouds, without undergoing any change ; another part, H, is absorbed by the air, and the remainder, K, is lost to space. Another part of the surface heat is transferred to the atmosphere by evaporation of moisture from the surface, the heat of evaporation being given up to the air when the moisture is again condensed, and a third part is communicated from the earth to the air by conduction and carried upwards by convection, but this is partly balanced by the reverse process, conduction from the air to the earth, and the net result is probably small. The heat transferred by evaporation and by the net conduction together is indicated by L.

The air is constantly receiving heat, C, from the sun, H and L from the earth. This heat is in turn radiated by the air, part E going back to the earth and part F to space. Thus the incoming radiation A is finally given out again in three forms, D by direct reflection, K by radiation from the earth, and F by radiation from the air, and since, practically speaking,

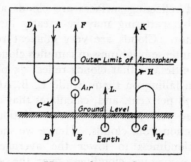

Fig. 11.—Heat exchange of atmosphere.

there is no gain or loss of heat from one year to the next, A=D+K+F.

The value of G, the radiation from the earth's surface, is related to the temperature in accordance with the well-known Stefan-Boltzmann law, which states that the radiation from a black body is proportional to the fourth power of its absolute temperature. The earth's surface is not quite a black body, but the radiation must be greater the higher the temperature. G and L are together equal to B+E+M, hence the temperature of the earth's surface must be determined by the sum of these three quantities (radiation from the sun reaching the surface, radiation from the air to the earth, and terrestrial radiation reflected back to the earth), less the heat L lost by evaporation and conduction. A change in any of these quantities will therefore bring about a change in the mean temperature of the earth. W. H. Dines (1) and

(2) assigns numerical values to all the quantities A to M ; some of these values are only rough estimates, but they will serve for a discussion of the possibility of appreciable climatic changes being brought about by variations in their amount.

The amount of solar radiation absorbed by the air is probably small ; Dines gives it a value of 50 calories, or one-fourteenth of the whole solar radiation, so that for the present we can ignore it. Variations in the value of A have already been discussed in Chapter IV., and here A is considered as a constant. Variations of B, the solar radiation reaching the earth's surface unchanged, therefore depend chiefly on variations of D, the solar radiation reflected back to space or lost by scattering ; Dines gives a value of 320 for D. The chief reflecting surfaces are clouds and snowfields, and the loss by scattering may be greatly increased by the presence of dust. Clouds are very efficient reflectors, as is shown by the intense brightness of cumulus clouds in sunlight ; A. Angström estimates that clouds reflect 75 per cent. of the solar radiation falling on them, while L. B. Aldrich (3) puts the figure at 78 per cent. and, calculates that on the average slightly over 40 per cent. of the sun's radiation is lost to the earth by reflection from clouds. Hence we should expect an increase of cloudiness to lower the average temperature. At present the average cloudiness over the whole world is just over five-tenths ; if it increased to six-tenths without any corresponding increase in the amount of water held in the atmosphere as vapour (clouds are water droplets or ice crystals, not water vapour, so that the supposition is conceivable), the mean temperature would be considerably lower than it is now. This effect of cloudiness is of great importance ; it is discussed in greater detail in the next chapter. If the increase of cloudiness were accompanied by an increase in the amount of water vapour, the fall of temperature would be less, as will be seen.

An increase in the area covered by snow and ice would increase the heat lost by reflection, and in this case there are no compensating circumstances. Large areas of snow and ice usually have clear skies and dry air above them, and there is nothing to check the wastage of solar heat. If we assume on the basis of some work by A. Angström (4) that over an ice-sheet about four-fifths of the solar radiation is reflected back to space, compared with one-fifth in an unglaciated region, the lost solar radiation in temperate latitudes would be sufficient to melt more than 30 feet of ice in the course of a year. This must have been a powerful factor in increasing the rigour of the great ice-ages in the glaciated regions, and in facilitating the extension of the ice-sheets. According to E. Antevs (5), the ice-covered area at the maximum of glaciation was about 13 million square miles, compared with 6 million square miles at present. The increase of 7 million square miles represents 3·5 per cent. of the earth's surface. Allowing for the fact that glaciated regions are not entirely cloudless, we may estimate the increased loss of solar radiation by reflection from the ice-sheets as at least $2\frac{1}{2}$ per cent. of that reaching the whole earth. W. Wundt (6), allowing for the snowfall and drift ice of the peripheral regions, arrives at a figure of 3 per cent. The effect would be to lower the temperature of the whole earth by at least 4° F. (Wundt gives 7° F.).

A variable which may be more generally effective is the volcanic dust in the atmosphere. In the years following great volcanic eruptions of the explosive type, such as those of Krakatoa (1883), Santa Maria and Pelée (1902), Colima (1903), and Katmai, Alaska (1912), the solar radiation reaching the earth's surface (Dines' quantity B) may be 15 or 20 per cent. (45 to 60 calories) below the normal value. This radiation is not all lost to the earth ; part of it goes to warming the upper air and therefore increases the value of E, but we shall see later that the net effect is an appreciable lowering of temperature, and that volcanic dust has possibly played a part in causing ice-ages.

The quantity M, terrestrial radiation reflected back to the earth, is relatively small (60 calories) on Dines' calculation, and as its variations depend on some of the factors, cloudiness and dust, which control D, it need not be considered further here. That leaves for discussion E, the radiation from the air to the earth, and this quantity is not only large (540 calories) but probably very variable. It depends on several factors, but by far the most important is H, the terrestrial radiation absorbed by the atmosphere, to which Dines assigns a value of 600 calories per square centimetre per day. The action of the atmosphere in raising the temperature of the earth by absorbing its radiation is similar to the action of the glass roof in a greenhouse. In the atmosphere the place of the glass is taken by certain gases, notably water vapour, ozone, and carbon dioxide. W. H. Dines points out (2) that the present mean temperature of the earth's surface, 288° A. (59° F.), is about 4° A. (7° F.) above the mean temperature which would prevail if the atmosphere had no power of absorption and if there were no reflection or scattering. In addition to raising the mean temperature over the earth as a whole, atmospheric absorption, in conjunction with the mobility of the air, has an important effect in bringing about a greater approach to uniformity in the temperatures of different latitudes than would exist on a dry earth. The importance of this effect in controlling variations of climate has been emphasised by Sir George Simpson.

Water vapour has a very high coefficient of absorption of radiation in the wave-lengths chiefly emitted by the earth. Moist air is warmed by radiation from the earth and immediately returns to the surface part of the heat gained in this way. Hence on a clear night the surface loses heat more rapidly if the air is dry than if it is moist, and the amount of water vapour in the lowest layers is one of the most important elements in calculations for forecasting the occurrence of frosts. Except over cold dry continental regions, however, there is always enough water vapour present in the whole thickness of the atmosphere to absorb practically the whole of the radiation in most parts of the spectrum of terrestrial radiation. An increase in the amount of water vapour would therefore have little effect on the proportion of terrestrial radiation transmitted unchanged to space. It would, however, have some effect on the vertical distribution of temperature in the atmosphere. Consider the atmosphere as divided into a number of concentric shells, and suppose for a moment that each of these shells absorbs all the radiation reaching it from the shell on either side, while there is a fall of temperature from the inner to the outer shells. Then it will be seen that the earth's surface is receiving radiation only from the innermost shell, while only the outermost shell is radiating to space. Hence the radiation from the air to the earth is greater than the radiation from the air to space. Although these conditions are not fully realised in the earth's

atmosphere, they are sufficiently near the truth for the above
conclusion to hold. An increase of water vapour, by increasing
the completeness of absorption in each of our hypothetical
shells, would make this difference somewhat greater, and
so raise the temperature of the earth's surface, while it
would increase the temperature of the air at a height of
four or five miles more than that at the surface, and so lessen
the decrease of temperature with height.

More aqueous vapour in the air may be due either to
greater evaporation at the same temperature or to a higher
initial temperature. Greater evaporation may be due to a
greater expanse of sea in low latitudes, to a greater average
wind velocity, or to a greater vertical interchange of air.
In any case more heat is taken from the surface and carried
to the level of condensation, whence it is partly radiated to
space. The result is an increase of L, the heat lost to the
earth's surface otherwise than by radiation, and it therefore
causes a decrease of the earth's radiation G, *i.e.*, a fall of the
surface temperature, in spite of the greater humidity. On
the other hand, if the mean temperature of the air rises from
some cause unconnected with the water vapour content, the
latter will automatically increase, since warm air has a greater
capacity for water vapour than cold air has. If all other
conditions of land and sea distribution, relief, etc., remained
unchanged, an increase in the water-vapour content due to
increased solar radiation would be accompanied by an increase
in cloudiness and precipitation as well as by a rise of temperature ; this is the basis of Sir George Simpson's theory of
glaciation described on pages 91 to 96. If, however, an
increase in the water vapour content is brought about by a
large increase in the area of the oceans, accompanied by a
decrease in the average height of the land, the solar radiation
remaining unchanged, there will not necessarily be an increase
of cloud and precipitation.

The most important way in which cloud is formed, and
practically the only way in which an appreciable fall of rain
can be produced, is by the ascent of moist air, which is cooled
by expansion until condensation takes place. Except in
limited regions of high relief, the air must become unstable
before is can be forced to rise. The stability of a column of air
depends on the temperature gradient ; the less rapid the fall

of temperature with height, the more stable is the air. We have seen that an increase in the water vapour, other things remaining unchanged, increases the absorption at high levels in the atmosphere, and therefore decreases the fall of temperature with height and makes the air more stable. The ascent of air, both in the great cyclonic storms and " barometric depressions " and in the smaller thunderstorms and local showers, becomes less frequent and extensive, and there may be an actual decrease in the cloudiness. These conditions are illustrated in the trade wind belts. The air crosses large stretches of ocean ; it speedily becomes nearly saturated and takes up further moisture very slowly, yet owing to the stability of the conditions there is very little ascent of air, the amount of cloud is small, and rainfall is very scanty. With a generally higher temperature over the earth, these conditions would extend to higher latitudes. During the hot summer of 1921 the air over the British Isles contained more water vapour than during the cool summer of 1924, but in 1921 the air was stable and there was little cloud and rainfall, while in 1924 the air was unstable and there was much cloud and a rainfall above normal.

The development of barometric depressions and consequently the formation of cloud are facilitated by the presence of horizontal temperature differences in masses of air which are moving relatively to each other. A more uniform distribution of temperature over the earth, such as prevailed during the warm periods, would therefore largely remove one important source of cloudiness at the present day. Thus it is possible that a general elevation of temperature over the globe could increase the amount of water vapour in the air without increasing the cloudiness ; the result would be a further rise of temperature, intensifying the original increase.

In certain theories of climatic change (See Appendix II., Section IX.) extraordinary importance is attached to variations in the amount of carbon dioxide in the atmosphere. Periods during which the atmosphere was rich in this gas are considered to have been uniformly warm, those in which it was poor to have been cold or even glacial periods. Chamberlin has devoted great ingenuity to a discussion of the probable variations in the quantity of carbon dioxide in the atmosphere during different geological periods, and though his conclusions are probably somewhat exaggerated, there appears to be little doubt that the variations have been considerable. Recent physical researches have shown, however, that the part of the terrestrial radiation which is taken up by carbon dioxide is almost completely absorbed by water vapour, and no increase in the amount of the former gas could increase the total absorption appreciably. W. J. Humphreys pointed out (7, p. 567) that the only way in which an increase of carbon dioxide could affect the temperature would be by absorption at high levels in the atmosphere where water vapour is nearly absent. As explained in the case of water vapour, this would increase slightly the proportion of radiation from the air which is directed towards the earth, and decrease that which is directed towards space, and in this way, the mean temperature of the earth may have been modified to the extent of a few tenths of a degree.

In 1939, however, the question was taken up again by G. S. Callendar (8) who relates the cold of the Permian to the exhaustion of carbon dioxide by the Carboniferous forests. During the Mesozoic the relatively small development of plant life allowed the amount of CO_2, steadily replenished by the animal life of the seas, to increase again, but a great deal was locked up in Tertiary lignite formation, especially in western North America, and this may have brought about a progressive cooling which ended in the Quaternary Ice-Age. This theory cannot account for the oscillations of the individual glaciations, the time-scale of which is too short. Callendar ends by pointing out that the great coal consumption in the twentieth century has raised the amount of CO_2 in the atmosphere from ·028 per cent. about 1900 to ·030 per cent. in the 1930's, and that this increase has been accompanied by a small but steady rise in the mean temperature of the colder regions of the earth. This argument has rather broken down in the last few years, however, for the rise of temperature seems to have reached its crest and to have given place to a fall. The possibility that changes in the amount of CO_2 have been responsible for some small part of the climatic changes of geological time seems to remain open however.

The idea that volcanic dust may have important climatic effects is very old. In 1784, B. Franklin suggested that the hard winter of 1783-84 was due to the great quantities of dust in the air, and that the source of this dust might be either the destruction of meteorites or the great volcanic eruptions in Iceland. Other references to this possibility appeared from time to time, but it was not until the great diminution of radiation received at the surface after the eruption of Katmai in 1912 was noted that the subject received exhaustive discussion by Abbot and Fowle (9), and W. J. Humphreys (10, also 7, pp. 569-603). This discussion has shown that the effect of dust is due, not to absorption, but to the scattering and reflection of the solar radiation to which it gives rise. When a volcanic dust cloud is thrown to a great height in the air, as in the eruption of Krakatoa, it takes from one to three years to settle. Humphreys calculated the average diameter of the particles from the optical effects, and found that it was 1·85 microns (·00185 mm.). This is greater than the wave-length of solar radiation in the region of maximum intensity, and the particles therefore greatly interfere with the passage of the solar radiation. On the other hand, the wave-length of terrestrial radiation is six or seven times the diameter of the particles, and the terrestrial radiation passes through dusty air with little loss. The action of dust in the two cases may be compared to that of a number of wooden balls floating in a pond. Against a barrier of such balls small ripples break up and are lost, but larger waves to which the balls can rise and fall are hardly affected. Humphreys calculates that the reduction by volcanic dust of solar and of terrestrial radiation is in the ratio 30 to 1. Observations in 1912 showed that the Katmai dust reduced the solar radiation reaching the earth by about 20 per cent., which, if maintained through a long period of years, would lower the mean temperature of the earth by about 10° F., an amount quite sufficient to initiate an ice-age. The compensating processes —reflection of terrestrial radiation, and radiation from the dust itself—are negligible in comparison. During the past 160 years the average temperature of the earth has been lowered by volcanic dust possibly as much as 1° F. The most notable cold years since the beginning of the eighteenth century have all followed great volcanic eruptions, especially 1784-86, following the eruption of Asama (Japan) in 1783 ; 1816 (" the year without a summer "), following especially the eruption of Tomboro (Sumbawa) in 1815 ; 1884-86, following Krakatoa in 1883 ; and 1912-13, following Katmai in 1912. H. Arctowski (11), who has studied the temperature variations in great detail in connexion with the solar control of terrestrial temperatures, admits the great influence of the Krakatoa eruption, but considers that the effects of the eruptions of 1902 and 1912 on temperature were very small. The direct effect of volcanic dust on temperature is felt over the whole globe (according to Arctowski only over the hemisphere in which the eruption occurred) ; it is not localised in the regions which were glaciated during the Quaternary Ice-Age, but in addition to, or in consequence of, the direct effect, there may be a secondary effect on the centres of high and low pressure. A. Defant (12), from a study of the strength of the atmospheric circulation during the two years following each of the four great explosive volcanic eruptions of 1883 (Krakatoa), 1886 (Tarawera), 1888 (Bandai San) and 1902 (West Indies), considered that the result of each eruption was a strengthening of the atmospheric circulation for the next two years. C. E. P. Brooks and T. M. Hunt, however, (13), taking into account also eruptions in 1875, 1912 and 1914, found that the strengthening of the circulation was much more transient, persisting for only six months. We are left therefore with a general cooling as the only appreciable climatic effect of volcanic dust.

It may be remarked here that a large increase in the number of meteors entering the earth's atmosphere would have a similar effect to that of volcanic dust, but so far as I know, there is no evidence of such an increase during geological time.

Volcanic dust appears to be a possible explanation of climatic periods colder than the present. The variation of this element in geological time is not yet well known ; Humphreys calculates that the amount of dust required to maintain an ice-age would amount to a layer only one-fiftieth of an inch thick in 100,000 years, so that it would hardly be noticeable in the sedimentary rocks. We should expect volcanic activity to be greatest during the great periods of earth-movement and mountain-building, when the continents were highly emergent and the land and sea distribution favourable for glaciation. The explosive stage of activity generally comes later in the life-history of a volcano than the stage of fluid eruptions, so that the maximum amount of dust might be delayed for some time after the continents first became highly emergent. In the Quaternary, at least, there was in fact an appreciable lag between the first great emergence of North America and Scandinavia and the beginning of widespread glaciation, and this difficulty is met by the hypothesis that the final lowering of temperature necessary for ice-formation was given by the occurrence of widespread explosive eruptions. On the other hand, the complete absence of volcanic dust would not raise the mean temperature more than about 1° F., which is inadequate to account for the temperature of the warm geological epochs.

E. and A. Harlé (14) attributed the general warmth of the Mesozoic period to the presence of a much denser atmosphere than now exists on the earth, which helped to conserve the heat of the sun and so raised the general temperature. They base their argument on the existence of great flying reptiles in the Mesozoic, which they consider were too heavy to have flown in the present atmosphere, but as we are ignorant of the muscular development of these reptiles, this argument carries no weight. From the physical side there seems to be no reason why the earth should be losing its atmosphere, and the variations of climate during geological time are very far from suggesting the action of any irreversible force.

REFERENCES

(1) DINES, W. H. "The heat balance of the atmosphere." London, *Q. J. R. Meteor. Soc.*, 43, 1917, p. 151.
(2) "Dictionary of Applied Physics," edited by Sir RICHARD GLAZEBROOK, vol. iii., London, 1923. Article, "Radiation," by W. H. DINES.
(3) ALDRICH, L. B. "The reflecting power of clouds." Washington, D.C., *Ann. Astrophys. Obs., Smithsonian Inst.*, 4, p. 375.
(4) ÅNGSTRÖM, A. "The albedo of various surfaces of ground." Stockholm, *Geogr. Ann.*, 7, 1925, p. 323.
(5) ANTEVS, E. "The last glaciation." New York, Amer. Geogr. Soc., *Research Series*, no. 17, 1928.
(6) WUNDT, W. "Änderungen der Erdalbedo während der Eiszeit." *Met. Zs.*, Braunschweig, 50, 1933, p. 241.
(7) HUMPHREYS, W. J. "Physics of the air." 3 ed. London (McGraw-Hill), 1940.
(8) CALLENDAR, G. S. "The composition of the atmosphere through the ages." *Meteor. Mag.*, London, 74, 1939, p. 33.
(9) ABBOT, C. G., and F. E. FOWLE. "Volcanoes and climate." Washington, D.C., *Ann. Astrophys. Obs. Smithsonian Inst.*, 3, 1913, and *Smithsonian Misc. Coll.*, 60, 1913, no. 29.
(10) HUMPHREYS, W. J. "Volcanic dust and other factors in the production of climatic changes and their possible relation to ice-ages." Philadelphia, *J. Frankl. Inst.*, 176, 1913, p. 131.

(11) ARCTOWSKI, H. "Volcanic dust veils and climatic variations." New York, *Annals N. Y. Acad. Sci.*, 26, 1915, p. 149.

(12) DEFANT, A. "Die Schwankungen der atmosphärischen Zirkulation über dem nord-atlantischen Ozean im 25-jährigen Zeitraum, 1881-1905." Stockholm, *Geogr. Ann.*, 6, 1924, p. 13.

(13) BROOKS, C. E. P., and T. M. HUNT. "The influence of explosive volcanic eruptions on the subsequent pressure distribution over western Europe." *Meteor. Mag.*, London, 64, 1929, p. 226.

(14) HARLÉ, E., and A. HARLÉ. "Le vol de grands reptiles et insectes disparus semble indiquer une pression atmosphérique élevée." Paris, *Bull. Soc. Geol. France*, 11, 1911, p. 118.

CHAPTER VII

THE EFFECT OF CLOUDINESS ON TEMPERATURE

IN the last chapter we saw that one of the most potent
factors in modifying the distribution of solar radiation
over the surface of the earth was the reflection from the
upper surfaces of clouds. For this reason, cloudiness must
obviously be of great importance as a factor in the mean
temperature of any region. In Africa, for example, the highest
mean annual temperatures are found, not over the equator
where the solar radiation at the limit of the earth's atmosphere
is greatest, but over the deserts to the north and south of the
equator where the skies are clearest. The mean annual
temperature (corrected to sea-level) over the equator in the
interior of Africa is 82° F., while over the Sahara in latitude
20° N. it is 89° F. In spite of the lower mean altitude of the
sun, the small degree of cloudiness over the Sahara (2·4 tenths
of sky covered in latitude 20°, compared with 5·5 tenths at
the equator) is sufficient to make the former 6 to 7° F. warmer
than the latter. It is worth while to investigate this effect a
little further, and attempt to make some estimate of the change
of mean temperature which would result from small variations
in the mean cloudiness, all other factors being regarded as
constant.

We have seen that the mean temperature of the earth's
surface is closely related to the amount of radiation which
it gives out, and that the latter is equal to the amount of
radiation which it receives, minus the loss of heat by
evaporation and convection. If the earth were a perfect
radiator, and if its temperature were the same in all parts
of the surface, the fourth power of its temperature would
be proportional to the amount of heat which it radiates
outward. Although neither of these conditions is quite
fulfilled, it is obvious that the higher the radiation the greater
the mean temperature, and vice versa. We have also seen
that the incoming radiation from the sun undergoes a number

of changes, as a net result of which the earth's mean temperature is higher than it would be if there were no atmosphere. These changes are very complex, but we can say that an increase in the radiation received from the sun would increase all the quantities concerned. The radiation reaching the earth's surface would be greater, the return radiation from the earth to the atmosphere would be greater, consequently the amount of terrestrial radiation absorbed by the atmosphere would be increased, which would in turn increase the radiation from air to earth, and so on. An increase of 10 per cent, in the radiation received from the sun would therefore result in an increase of not exactly 10 per cent., but something of the order of 10 per cent., say, between 5 and 15 per cent., in the radiation from the earth's surface. This result would be independent of the local modifications in the distribution of temperature due to the readjustment of the wind systems, and also supposes that there is no change in the constitution of the atmosphere such as the formation of ozone in the stratosphere.

Clouds being effective reflectors of solar radiation, the presence of clouds means a loss to the earth of a certain amount of energy. It is true that clouds also reflect a part of the terrestrial radiation back to the earth, and that this partly compensates for the lost solar radiation, but the compensation is not nearly complete. For one thing, it seems probable that the percentage of the long-wave terrestrial radiation reflected from the under surfaces of clouds is smaller than the percentage of the short-wave solar radiation reflected from their upper surfaces ; the action of water droplets in this respect is probably similar in kind, though perhaps not equal in degree to the action of volcanic dust particles described on page 118. The temperatures of the surface and of the atmosphere are so intimately connected that it is impossible to conceive of a great change in one without a change in the other. Hence we may say that it is immaterial whether the radiation is lost owing to a decrease in the amount of heat radiated by the sun or a decrease in the amount which penetrates the mantle of clouds.

A. Ångström (1) estimated the reflection of solar radiation falling on clouds as 75 per cent. This figure probably applies to rather dense heavy clouds ; more recently B. Haurwitz (2) investigated the insolation received at Blue Hill, Mass., with given cloud amount, cloud density and elevation of the sun. Cloud density is expressed on a scale of 0 (very thin) to 4 (very dense cloud), and is as important as cloud amount in determining the insolation, but as we have no means of estimating variations of cloud density during geological time, it seems best to consider the results with the average density of 2·6 found by Haurwitz. He gives a table of annual insolation with various cloud amounts and cloud densities as a percentage of the insolation with cloudless skies. Interpolating for density 2·6, we have approximately :—

Cloud amount (tenths) .	0	1–3	4–7	8–9	10
Insolation, per cent. .	100	93	82	68	41

At present the average cloudiness over the earth is 5·4 tenths, giving an average of 82 per cent. of the radiation received with a cloudless sky. An increase of one tenth in the cloud amount would reduce this figure to 78 per cent., and a decrease of one tenth of cloud would raise it to 85½ per cent. Let us make the simple assumption that the radiation from the earth's surface, and therefore the fourth power of the mean temperature (absolute degrees) are proportional to the amount of insolation. Then, starting with a figure of 59° F. or 288° A. for the mean temperature of the earth's surface at present, we have the following results :—

			4·4	5·4	6·4
Cloudiness (tenths of sky) . .			4·4	5·4	6·4
Insolation (per cent. of cloudless sky)			90	86	82
Mean temperature, °A. . . .			291	288	284
Mean temperature, °F. . . .			65	59	53

These figures show that a decrease in the mean cloudiness by only one tenth of the sky would result in an increase of the mean temperature by as much as 6° F., while an increase of cloudiness by the same amount would lower the mean temperature by 6° F.

It may be noted here that large ice-sheets, by deflecting depressions into lower latitudes, would decrease the extent and permanency of the sub-tropical anticyclones, especially over the oceans. This in turn would lead to an increase in the cloudiness of these areas, and would result in an appreciable loss of heat to the world.   This may have been a contributory cause of the rapid expansion of the ice-sheets.

We saw in the Introduction that one of the salient features of the " warm " periods was their dryness.   From this we naturally infer that they were favoured by unusually clear skies, and this gives us an explanation of their warmth—or rather, it pushes the explanation one step farther back, for we still have to account for their clear skies.

Up to the present we have been assuming, for the purposes of the argument, that the cloudiness of the sky is the same in all latitudes.   That, of course, does not correctly represent the conditions prevailing at present.   We have a belt of cloudy skies near the equator (about $5\frac{1}{2}$ tenths), then two belts of clear sky along latitudes 20° to 30° in each hemisphere, with an average cloudiness of 4 to 5 tenths, decreasing to less than 2 tenths in the centres of the great deserts, and outside these clear belts cloudiness increasing again to $6\frac{1}{2}$ or 7 tenths in cold temperate and sub-polar regions.   This distribution has a considerable effect on the mean temperature of the earth, for two reasons.   First, owing to the greater average elevation of the sun, the radiation received (at the limit of the atmosphere) is much greater in low latitudes than it is near the poles.   The radiation received on a horizontal surface of one square centimetre area in the course of a year on the Arctic Circle is about half that received on the equator.   Hence a cloud on the equator reflects back to space about twice as much radiation as a similar cloud in latitude 66°.   The same average cloudiness for the whole earth would result in a much higher mean temperature if the clouds were concentrated in high latitudes than if they were concentrated in low latitudes.

Secondly, although clouds are much more effective as reflectors of solar radiation than as reflectors of the long-wave terrestrial radiation, they are not without a certain effect on the latter.   The figure given by W. H. Dines (Chapter VI., 1) is equivalent to a reflection of 8 per cent. of the terrestrial radiation compared with 60 per cent. of the solar radiation according to Haurwitz or 75 per cent. according to A. Ångström.   There is a large transference of heat from low to high latitudes by winds and ocean currents, and the temperatures of the polar and sub-polar regions are raised considerably by the heat conveyed in this way. Hence the outgoing radiation from the earth's surface in high latitudes is much greater than the incoming radiation ; let us suppose that in some particular locality it is ten times as great, and that the sky is half covered by clouds. Then it is easily seen that the gain of heat by the reflection of terrestrial radiation back to the earth exceeds the loss due to reflection of solar radiation back to space. Beyond the latitudes 67° N. and S., during the polar night, the incoming radiation from the sun is zero, and the effect is pure gain. Thus in high latitudes in winter, cloudy skies are actually effective in raising the mean temperature.

The qualification " in winter " suggests that in addition to the mean cloudiness and its distribution according to latitude, the seasonal variation is important. A locality in which the sky is generally overcast in winter and clear in summer would have a higher mean temperature than another place in the same latitude subjected to similar conditions, but with its skies clear in winter and overcast in summer. This is brought out very clearly by A. Ångström (1) in an analysis of the annual variation of temperature at Stockholm in relation to the radiation. At Stockholm the mean annual cloudiness is 6·4 tenths, and it varies from 5·1 tenths in June to 7·9 tenths in December. Ångström calculates that if the mean annual cloudiness remained at 6·4, but instead of being greater in winter than in summer was the same in all months of the year, Stockholm would be colder than it is at present in every month, the average difference being 2·2° F.

The remains of desert deposits formed during the warm periods are mainly limited to middle and low latitudes, while in high latitudes we find the remains of a rich vegetation requiring a considerable rainfall. This distribution suggests that while the cloudiness was small between about 10° and 55° latitude, it increased very rapidly beyond 55°, thus giving the most favourable conditions for a high temperature in all parts of the world.

Although the popular conception of a geological landscape is a steaming jungle rather than, as it should be, an arid plain, variations of cloudiness have played comparatively little part in theories of climatic change. Marsden Manson (3) has, however, seized on the dual rôle of clouds as reflectors alike of solar radiation and of terrestrial radiation, and has constructed an elaborate theory of geological climates on this basis, in conjunction with the gradual waning of the internal heat of the earth.   He points out that the surface of the larger planets is covered by an unbroken layer of cloud, and assumes that this must have been the condition of the earth in past times, and in fact comparatively recently.   Through this cloud canopy the sun's rays could not penetrate, and as a factor of climate the sun was almost inoperative.   The earth was radiating more than it was absorbing, and the sources of this outgoing energy were the original supply of earth-heat and radio-active minerals.   Owing to the poorly conducting crust, earth-heat was liberated, not in a steady stream, but in spasms during periods of volcanic action and crustal movement.

Let us start with one of these liberations of earth-heat. Within its protecting cloud canopy the surface, oceans and continents alike, was warm from equator to poles, but the land surfaces cooled more quickly than the heat-conserving oceans, and in due course, while the warm oceans were still supplying enough moisture to maintain the cloud canopy intact, the land surfaces began to freeze, and ice-sheets developed. Apart from some local glaciations in the centres of the larger continents this stage was first reached on a planetary scale in the Permo-Carboniferous period ; this glaciation coincided more or less with the present sub-tropical high-pressure belts, and the reason is stated to be that "cold anticyclonic winds" cooled the land most rapidly in those belts.   The cooling of the oceans continued, and with decreasing evaporation a stage was reached in which these high-pressure areas, to-day possessing the clearest skies of the world, ceased to be mantled in clouds—the sun broke through and deglaciation commenced.

Now followed a period of dual control, solar energy prevailing near the equator, earth-heat towards the poles. In spite of fluctuations, the latter gradually diminished, and just before the Quaternary glaciation the polar oceans became cold for the first time.   Then the second planetary glaciation occurred, centred in the cold temperate belts of greatest precipitation, at this time the only regions which were permanently overcast.   The cooling of the oceans continued, and evaporation ceased to supply enough moisture for even this limited cloud belt, the sun shone over the whole world, deglaciation again commenced and is still continuing.

The theory is interesting, but there are some insuperable difficulties. With warm oceans and an unbroken cloud canopy, the land surfaces, unless at a great altitude, would not be likely to freeze ; the conditions are most nearly realised at present in the equatorial rain belt, in which the land is maintained at the same temperature as the neighbouring oceans. " Cold anticyclonic winds " presuppose cooling by radiation ; even if under world-wide isothermal conditions the pressure distribution could remain unaltered, which is highly improbable, we must suppose either that the anticyclone would break down the cloud canopy, in which case the tropical sun would certainly prevent glaciation, or that the clouds would remain in spite of the anticyclone, in which case the descending air would not be cold. Finally, the moist conditions supposed by Marsden Manson to have prevailed during the warm periods are in direct opposition to the dry conditions demonstrated by the geological evidence set out in the Introduction.

L. J. Krige (4) suggested that increased cloudiness and precipitation would occur during periods of mountain building because of unusually high evaporation from ocean basins due to heat entering them from below. This type of suggestion is very difficult to discuss quantitatively but it seems that to make an appreciable difference to the amount of evaporation the quantity of earth-heat would have to be an improbably high multiple of the present supply.

REFERENCES

(1) ÅNGSTRÖM, A.   " On radiation and climate."   Stockholm, *Geogr. Ann.*, 7 1925, p. 122.
(2) HAURWITZ, B.   " Insolation in relation to cloudiness and cloud density."   *J. Met. Amer. Met. Soc.*, 2, 1945, p. 154.
(3) MANSON, MARSDEN.   " The evolution of climates."   Baltimore, Md., 1922.
(4) KRIGE, L. J.   " Magmatic cycles, continental drift and ice-ages."   *Geol. Soc. S. Africa*, 1929.

CHAPTER VIII

CONTINENTALITY AND TEMPERATURE

IN the last few chapters we have discussed the factors influencing the distribution of temperature, namely, winds and ocean currents, the heat received from the sun, the heat transmitted by the earth's atmosphere, and the reflection from cloud surfaces. We may divide these factors into those which are constant along a given line of latitude and give rise to the " solar climate," and those which vary from place to place even in the same latitude, and so give rise to the local distribution of climates. The latter may be termed the " geographical climate," since it depends mainly on the distribution of land and sea and partly also on the relief of the land surface. Hence there is an intimate relationship between temperature and land and sea distribution, and we shall expect to find that the changing outlines of the geological continents have been reflected as changes in the distribution of temperature.

If we look at a map of the mean annual temperature, we notice first of all that the isotherms run roughly parallel with the lines of latitude. There is a belt on both sides of the equator in which the temperature is above 80° F., extending from America across the Atlantic, Africa, India and the Indian Ocean, the East Indies, and part of the Western Pacific. This belt broadens out greatly over the continents and narrows over the oceans ; over the Eastern Pacific it thins out altogether. Between 30° N. and 30° S. the eastern sides of the continents are warmer than the western sides. Surrounding each pole is an irregular area over which the mean temperature is below freezing point ; the average position of the isotherm of 32° F. is north of 75° N. in the Arctic, but about 60° S. in the Antarctic. In the Arctic the isotherm of 32° F. lies farther north over the oceans and the western coasts of the continents than it does over the central and eastern parts of the continents, that is, the land is generally colder than the sea. Hence in temperate

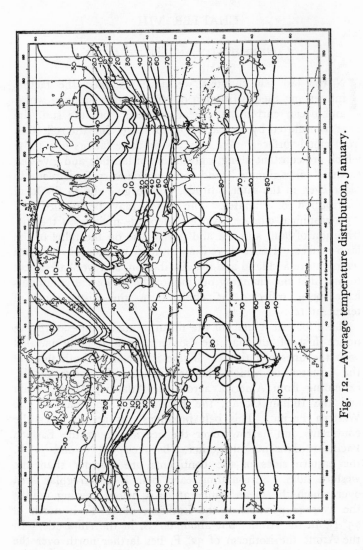

Fig. 12.—Average temperature distribution, January.

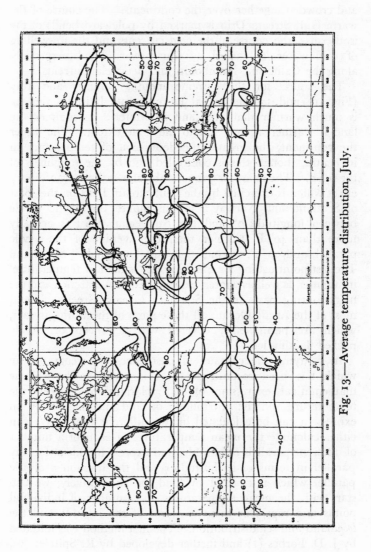

Fig. 13.—Average temperature distribution, July.

regions the isotherms are widely separated over the oceans and crowded together over the continents. The course of the warm Gulf Stream Drift is marked by poleward bends of the isotherms ; the course of the cold currents and the up-welling of cold water is marked by equatorward bends, especially along the California, Humboldt, and Benguela Currents.

The charts for the extreme months January and July (Figs. 12 and 13) show that in temperate latitudes the sea is much warmer than the land in winter ; in summer the land is warmer, though not to the same extent. The larger the continent, the greater the depression of temperature in its centre, though this is controlled also by the roughness of the relief. The chart for January also shows the extraordinary effect which is exercised by the Gulf Stream Drift on the coast of Scotland and Norway ; in this month the Arctic Circle between longitude 0° and the Norwegian coast is actually more than 40° F. warmer than the average of the whole parallel, and 90° F. warmer than the " cold pole " of Siberia. The last point calling for special notice is the low temperature, in summer as well as winter, experienced in Antarctica and the neighbouring parts of the Southern Ocean, Greenland, and most of the Arctic Ocean, all those parts of the world, in fact, which are permanently covered with ice-sheets or closely packed floating ice.

In studying the climates of past times the geography of which has been reconstructed,[1] it is useful to have some mathematical expression of this relationship between the temperature and the distribution of land and sea. Such an expression can be calculated in two different ways. We can either calculate the mean temperatures of each of a number of different parallels of latitude, and express the mean for any parallel in terms of its latitude ϕ and the fraction n of the parallel which is covered by land, or alternatively, we can start with the mean temperatures of a number of individual points and represent them in terms of the latitude and the *local* land and sea distribution. The first method was initiated by J. D. Forbes (1) and further developed by R. Spitaler (2), who obtained an expression which, converted into Fahrenheit degrees, becomes :—

$$T \ (°F.) = 27 \cdot 6 + 32 \cos\phi + 13 \cos 2\phi + 35 \ n \cos 2\phi.$$

[1] See Chapter XII.

Since cos 2ϕ is positive between 45° N. and 45° S., negative from 45° to the poles, the term $35\,n$ cos 2ϕ indicates that the effect of land in low latitudes is to raise the temperature and in high latitudes to lower the temperature. This is simply a mathematical expression of the fact that an extensive land-mass tends to give a hot desert near the equator and a cold tundra near the poles.

It is to be noted that Spitaler's formula is derived entirely from the present distribution of temperature and of land and sea, and therefore implicitly assumes the existence of the present system of winds, of ocean currents, and of ice. The fact that the formula applies reasonably well to both hemispheres, in spite of their very different configurations, shows that this is not necessarily a serious objection, but it is obvious that three conditions must be fulfilled before it can be applied to determinations of the temperature in other geological periods. First, there must be large areas of open ocean within the tropics, with free communication between low and middle latitudes, in order to allow for the great transference of heat by ocean currents. Secondly, the land and sea distribution must not be of such a nature that the system of pressure and winds is radically different from the present system ; in the language of Chapter II., the planetary circulation must not be dominated by the geographical circulation. Thirdly, there must be extensive areas of floating ice. For example, Spitaler's formula gives for the Arctic Ocean a mean temperature of about 15° F., which is well below the freezing point of sea water. But we have seen in Chapter I. that if the " non-glacial " temperature of the polar regions could be raised by some 5° F. there would be no floating ice, and the mean annual temperature would be well above 32° F. The formula in its present form, therefore, does not apply to the " non-glacial " periods ; a better representation of the zonal distribution of temperature during these periods would be given by a formula of the type :—

$$T = T_o + a \cos \phi - b \cos 2\phi + cn \cos 2\phi,$$

the negative sign of the term cos 2ϕ allowing for the decrease of the zonal contrast during these periods.

For small changes of geography, however, Spitaler's formula gives a useful means of estimating the resulting

changes of temperature. The result of a decrease of ten
per cent. in the area of land in any latitude on the mean
temperature of that latitude would be as follows :—

Latitude (degrees)	0	10	20	30	40	50	60	70	80
Change of temperature (°F.) .	−3·5	−3·3	−2·7	−1·7	−0·6	+0·6	+1·7	+2·7	+3·3

We will return to these figures later.

We must now return to the second method of calculating
geographical factors of temperature, in which the basis is
the temperature distribution at a number of individual
points instead of the mean temperature along whole parallels
of latitude. This method has been extensively employed
by F. Kerner and later by myself ; it can be used, not only for
the whole world, but also for restricted areas, though as
F. Kerner von Marilaun (Kerner-Marilaun) (3) points out
with justice, the temperature of any given point depends
not only on the geography of the region immediately
surrounding the point, or even on the distribution of land and
sea along the whole line of latitude ; we have to take into
account conditions over the whole globe, as far as they affect
the air and water circulation. He accordingly divides the
geographical factors into local, or stenomorphogenous, and
general or eurymorphogenous. Kerner has written a large
number of palæoclimatological papers, of which I have
selected three for discussion here. He subsequently put
together his results in an important book (3).

The first paper (4) deals with the winter climate of Europe
during the Tertiary period. At present this temperature is
governed on the west coast by the temperature of the Gulf
Stream Drift, and decreases eastward in accordance with the
distance from the Gulf Stream and the increasing continentality.
He accordingly represents the temperature along any latitude
by the equation :—

$$t = T - A \cdot L \cdot d - B \cdot l,$$

where t is the winter temperature of a European locality,
T the winter temperature of the Gulf Stream Drift in the
same latitude, d the distance of the point from the Gulf
Stream Drift along that latitude (the " linear continentality "),
L and l the percentages of land in large and small areas
surrounding the place. A and B are constants, which were

evaluated from the mean January temperatures reduced to sea-level taken at the intersections of every fifth degree of latitude and longitude from 35° to 55° N., and 20° W. to 70° E. The best representations of the general and local continentalities L and *l* were determined empirically, and the definitions finally adopted were : for L the percentage of land in a twenty-degree " square " of latitude and longitude surrounding the point (f_{20}), and for *l* the average of the percentages in five-degree and ten-degree squares (f_5 and f_{10}). A square in this sense is an area bounded by lines of latitude and longitude each covering the same number of degrees. The formula as finally calculated took the form

$$t = T - A(4 + 0 \cdot 2 \lambda E) f_{20} - B \cdot 5(f_{10} + f_5),$$

where λE is the longitude in degrees east of Greenwich. The values of A and B were calculated separately for each fifth degree of latitude.

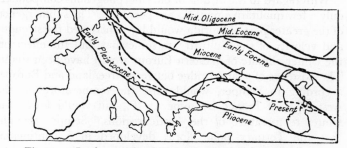

Fig. 14.—Isotherm of 32° F. in Tertiary and Quaternary.
After F. v. Kerner.

By means of this formula and reconstructions of the land and sea distribution during six stages of the Tertiary period, the " stenomorphogenous " isotherms were reconstructed. The results are summarised in a small figure showing the position of the January isotherm of 32° F. in each period, which is reproduced in Fig. 14. The results indicate a more favourable climate from the beginning of the Eocene up to and including the Miocene, a Pliocene climate differing little from the present and a less favourable climate of the early Quaternary in Western Europe. The differences from the present rarely exceed 10° F. in Central Europe, but east of the Caspian during the Middle Eocene and Oligocene the calculated January temperatures are nearly 30° F. above the present temperature. In the early Pleistocene the calculated temperatures are nowhere 5° F. below the present. The mean temperatures in January over the area covered by the figure, expressed as differences from the present mean temperature, are as follows :—

Latitude N.	55° °F.	50° °F.	45° °F.	40° °F.
Early Eocene	+2	+ 6	+10	+ 7
Eocene	+4	+ 9	+13	+10
Oligocene	+5	+12	+13	+ 9
Miocene	+1	+ 3	+ 5	+ 3
Pliocene	−2	− 3	− 4	− 3
Pleistocene	−1	+ 1	+ 1	− 1

Table 5.—Kerner's calculated temperature differences,
January.

With regard to the distant or eurymorphogenous component, only a few qualitative remarks are possible. The submergence of the greater part of Florida would have increased the strength and velocity of the Gulf Stream somewhat, but the effect on the winter climate of Western Europe would have been slight. The existence of a land-bridge between Greenland and Europe, according to Semper, would raise the temperature off the west coast of France by some 13° F., but would lower the winter temperature of the Arctic regions. Semper thought that the Tertiary polar floras developed in a continental polar climate with hot summers, but this view is not tenable. For the warming effect of the " Indian Drift " which reached Central and Southern Europe from the Indian Ocean during the older Tertiary, Heer's simple assumption of a warming effect of 7° F. (5) is probably the best approximation that can be made at present.

The winter temperatures deduced by Heer from the fossil plants in the Miocene are from 5°-10° F. higher than the temperatures due to the local geography during that period, but this difference is sufficiently accounted for by the effects of more distant changes, and particularly by the warming effect of the Indian Drift. Kerner does not consider the possible effect of a change from " glacial " to " non-glacial " conditions.

The second paper by F. Kerner (6) extends this study to the Arctic regions. The distribution of the January temperatures along each parallel of latitude in the Arctic Ocean is expressed in the form

$$t = aw - bk,$$

where w is the warming effect of a gap ten degrees of longitude broad, open to the world oceans, and k is the cooling effect of a ten-degree barrier between the Arctic and the open ocean farther south. The values of the terms w and k are deduced from the distribution of temperature in January under present geographical conditions, and from these values and the land and sea distribution reconstructed for the Middle Eocene the isotherms for that period are reconstructed. Some attempt is made to allow also for a decreased cooling of the Gulf Stream by the cold Labrador Current, but the calculations implicitly assume a large area of floating ice in the Arctic Ocean. The reconstructed temperatures are still below freezing point over most of the region north of 70° N., but they are well above the present temperatures, and it seems probable that the improvement was sufficiently great to raise the mean " non-glacial " temperature at the pole above the freezing point. In accordance with the argument of Chapter I., this would mean that there would be no floating ice-cap, and a totally different climatic régime would come into force. This question can be better investigated on the basis of some more recent work of Kerner's (7) in which he returns to the discussion of Arctic temperatures. The first part of this paper, dealing with the akryogenous or " non-glacial " marine climate of an open polar ocean, has already been referred to in Chapter I. In the second part, Kerner analyses the distribution of January temperature at present along the 75th parallel of latitude. This parallel runs mainly over the ocean, the only land which it crosses being Novaja Zemlya, the Taimyr Peninsula, the New Siberian Islands, parts of the Arctic Archipelago, and Greenland, but over the greater part of its course it has extensive land-masses—Europe, Asia, and North America—within five degrees to the southward. There are only two gaps in this surrounding land-ring, the very narrow and shallow Bering Strait, through which very little warm sea water penetrates at any season of the year and none at all in winter, and the broad Atlantic gap through which the Gulf Stream Drift makes its way. The surface of the Arctic Ocean is cooled in winter by the cold winds from the winter anticyclones over the great continents, and warmed by the Gulf Stream Drift, and the January temperature at any point along the 75th parallel therefore depends on the amount of land to the southward and the distance from the Atlantic gap through which the warm water enters. These two factors are named by Kerner the " Continental " term K, and the " Separation " term S.

To get the matter clear, imagine a one-roomed cottage with thin walls, against the outside of which snow is banked, making them very cold, while in one wall there is a fire. The temperature at any point near one of the walls is then determined by the distance from the wall and the distance which separates the point from the fire. The walls represent the cold continents and the fire the Gulf Stream Drift. Kerner finds that the January temperature of a point on the 75th parallel is given (in Fahrenheit degrees) by the expression :—

$$T = 45 - 1 \cdot 07 \, S - 1 \cdot 9 \, K.$$

The precise evaluation of the " Separation " term S and the " Continental " term K is somewhat complex and need not be gone into here.

According to this formula, the most favourable distribution of land conceivable for high polar temperatures would be a number of long, narrow islands extending from low to high latitudes, separated by wide, deep channels. This distribution may have actually occurred at some stage in the middle of the Palæozoic period, and it was approached to some extent during parts of the Mesozoic and Tertiary periods. Two examples are considered by Kerner, the Middle Eocene and the Upper Jurassic. In the Middle Eocene, according to a reconstruction by Matthew, the circum-polar ring was broken by three broad gaps, an enlarged Bering Strait, the present Atlantic gap, and the Obic Sea which separated Europe from Asia. In the Upper Jurassic, according to a reconstruction by Uhlig, the Atlantic gap was replaced by the " Shetland Strait " farther west, and there was in addition a fourth gap, the Jana Sea, between 120° and 140° E.

From his formula, Kerner calculates the mean January temperatures for the 75th parallel to have been as follows :

Present.	Mid-Eocene.	Upper Jurassic.
°F.	°F.	°F.
−20·7	+7·8	+18·5

All the figures for individual longitudes are higher than the present ones in the same longitudes, but none of them exceed 32° F., and they still represent a very severe climate. We saw in the Introduction that during both these periods, vegetation of a sub-tropical or warm-temperate aspect flourished at several points north of 70° N., and Kerner considers that either the plant evidence is not reliable or the solar heat must have been greater.

The real reason for the discrepancy is that, in spite of his discussion of akryogenous (" non-glacial ") temperatures in the first part of the paper, Kerner employs the actual distribution of temperature as the basis of his calculations, instead of the " non-glacial " distribution. Thus his calculated figures imply the presence of a large amount of floating ice and ice-cold water in the Arctic Ocean, whereas we have seen that, given a " non-glacial " temperature only five degrees above the present, there would be no ice, and the water cooled by radiation would at once sink to the bottom. Evidently we have to recalculate his figures on a " non-glacial " basis.

Kerner's Middle Eocene mean January temperature is 28·5° F. above the present mean in 75° N. Of this increase we find that 6·9° F. is accounted for by the decrease in continentality and the remaining 21·6° F. by the increased influx of warm ocean currents (Separation effect). The small continentality effect may be allowed to stand, but the Separation effect requires some modification. Suppose a warm current at temperature t introduced into a mass of water at temperature t^{I}, the resulting mean temperature of the water-mass being T. Then we can suppose that the warming effect $(T-t^{I})$ is proportional to $(t-t^{I})$, and we can write $T = t^{I} + c(t-t^{I})$ where c is a constant fraction, depending on the volumes of the mass of cold water and of the warm current. Hence the present Gulf Stream Drift would have a smaller warming effect on a non-glacial Arctic Ocean than on the present glacial Arctic Ocean. On the other hand, if there were no Arctic ice there would be no cold East Greenland and Labrador Currents. From a study of the sea surface isotherms of the North Atlantic, we find that in January the temperature along the centre of the Gulf Stream is 71° F. in latitude 30° N. and 64° F. in 38° N., a fall of 0·9° F. per degree. From 38° to 43° N. on the other hand temperature falls by about 22° F. in only 5 degrees. Of this fall, only about 5° F. can be due to the normal fall with latitude, and the remaining 17° F. is due to admixture with the cold water of the Labrador Current. That is, the present January sea surface isotherm of about 32° F. in 75° N., 10° E., would be replaced by one of 49° F. Now we have the following data for a recalculation of the warming effect of the Gulf Stream Drift in a non-glacial polar basin with the present configuration :—

	" Glacial " Conditions. °F.	" Non-glacial " Conditions. °F.
Temperature of Gulf Stream Drift, t	32	49
Temperature of Arctic Ocean, T . .	−18	25
Difference $(t-T)$	50	24

Table 6.—" Glacial " and " Non-glacial " temperatures.

The total difference $(t-t^1)$ is proportional to $(t-T)$, so that we have to multiply the coefficient 1·07 of Kerner's " Separation " effect S by 24/50 or approximately 0·5 in order to correct this factor for a non-glacial ocean with the present land and sea distribution. If, now, we introduce a second current equivalent to the Gulf Stream Drift, the additional warming effect will be proportional, not to $(t-t^1)$ but to $(t-T)$, and the resulting temperature T^1 will be equal to $t^1+2c(t-t^1)-c^2(t-t^1)$, i.e., the increase is something less than twice that due to a single Gulf Stream Drift. The constant c is small, so that the additional term is not important ; moreover, Kerner's method of calculation makes some allowance for it, but in order to be on the safe side I have reduced the factor 0·5 to 0·4. The increase in the non-glacial temperature of the Middle Eocene, due to the change in the value of S, should therefore be only 8·6° F., instead of 21·6° F., making, with the increase of 6·9° F. due to the decreased continentality, the total increase in the non-glacial January temperature during this period 15·5° F. This has to be added to the mean January "non-glacial" temperature at present, 25° F., raising the Middle Eocene temperature of the Arctic Ocean in 75° N. to 40·5° F. This is well above the freezing point of sea water, so that there is no ice, and 40·5° F. is also the real January temperature.

For the Upper Jurassic, Kerner obtains a January temperature in latitude 75° N. of 39·2° F. above the present. Of this amount, his calculations show that 6·3° F. is due to the change in the Continental effect and 32·9° F. to the change in the Separation effect. Applying the correction for a non-glacial ocean by multiplying by 0·4, the latter quantity becomes 13·2° F., making with the change in the continental effect a total increase of the "non-glacial" temperature of 19·5° F. Added to the present non-glacial temperature of 25° F., this makes the Upper Jurassic January temperature 44·5° F. in latitude 75° N., a figure which is very near the present mean January temperature of Southwest England, so that these temperatures are quite consistent with the remains of the vegetation discovered by geologists. While I make no pretence as to the absolute accuracy of these computed figures, I do claim that they are so far above the freezing point of sea water that the strictest revision is unlikely to reduce them below it, and that the case for an ice-free Arctic Ocean during some periods of geological time is thereby established.

It has been suggested to me that the figure of 17° F. adopted for the cooling of the present Gulf Stream Drift by ice and ice-cooled Arctic water is too great, and this will serve as an example of the effect produced on the calculation by a modification of the basis. Let us reduce this figure from 17° F. to 9° F., which is almost certainly too small. Repeating the calculations on this new basis, we find that the mean January temperature in 75° N. comes out as 37·5° F. during the Middle Eocene and 40° F. during the Upper Jurassic. These figures are well above the freezing point, so that the Arctic Ocean would still be non-glacial.

Another objection which may be raised to the way in which these high polar temperatures have been obtained is that they depend on the existence of powerful ocean currents, which in turn depend on the planetary wind system, and that a warming up of the Arctic Ocean without a corresponding change in the temperature of the equatorial zone would result in a decreased strength of the wind-driven ocean currents. This possible objection has already been dealt with in the chapter on Pressure and Winds, page 53, where it was shown that there is a critical point in the planetary circulation. With a temperature difference above this critical value, the winds are largely directed outwards from the pole, ocean currents have difficulty in penetrating into the Arctic basin (barely 2 per cent. of the water in the Gulf Stream off Florida reaches the Arctic Ocean), and their temperature is also greatly lowered by the cold winds. With a temperature difference below the critical value, the winds are directed inwards towards the pole, and the volume and temperature of the ocean currents are maintained with much smaller loss. Hence the oceanic circulation induced by high polar temperatures would assist in maintaining those temperatures ; similarly, the oceanic circulation with low polar temperatures would help to keep the temperature low.

The interesting question arises, has the temperature of the Arctic Ocean risen above the critical point at any stage of post-glacial time ? I think there is no doubt that it has (8). During the " Climatic Optimum " there was a rich flora in Spitsbergen, while the fossil marine mollusca indicate a coastal sea temperature much higher than the present in all the Arctic lands which are at present dominated by sea ice, including Iceland and Greenland. The " Climatic Optimum " was experienced also over most of Europe, where it has been studied more closely than in the Arctic lands. In Scandinavia the warm period appears to have begun rather suddenly during a period of increased vigour in the circulation of the Atlantic Ocean (" Atlantic " or " Maritime Phase "). During this period, owing to submergence, the Baltic lay more open to the Atlantic than at present, and a maritime climate extended as far as the coast of Finland. Depressions passed readily across Denmark and along the German coast, so that in Northern Europe the Atlantic period had a greater rainfall than the present, with milder winters and cooler summers. About 3000 B.C. the connexion between the Baltic and the Atlantic again became restricted, and there was a slight extension of the area of the British Isles. These changes were associated with a notable difference of climate. Depressions seem to have favoured a northerly track into the Arctic Ocean, and the British Isles, Western and Central Europe became much more continental, with very warm summers ; the winters do not seem to have been any colder than at present. These conditions were very marked in Switzerland, where settlements occurred at very high levels and there was much traffic over passes which are now occupied by glaciers. In Spitsbergen the " ice floor " melted completely. The mean annual temperature at Green Harbour is at present 19° F., so that there must have been a rise of mean temperature by at least 13° F. in this part of the Arctic. The favourable climate lasted until about 3000 B.C., deteriorated slowly until about 500 B.C. and then came to an abrupt end. The change of climate for the worse was very rapid, and, according to H. Gams and R. Nordhagen (9), in the Alps it " had the appearance of a catastrophe."

My reading of this history is that the increased circulation of the Atlantic period swept away the ice from the Arctic Ocean (though apparently not from the channels among the islands north of America and Greenland). After the passing of the Atlantic period, the Arctic Ocean became somewhat cooler, but, being still free of ice, was very stormy, and this storminess itself maintained the winter temperature above the critical point and prevented the ice-cap from beginning to form. Then came an unusually quiet cold winter, the ice-cap obtained a footing, and perhaps in the course of a single season covered the greater part of the Arctic Ocean. The result was a sudden great change in the climate of Europe ; the conditions of to-day came in " with the appearance of a catastrophe." The ice-cap, once formed, kept the winter temperature below the critical point by its own power of persistence.

It is possible that the Arctic Ocean again became free of ice during historic times, from about the fifth to the tenth or eleventh centuries of the Christian era. O. Pettersson (10) makes out a good case for the absence of sea ice in the East Greenland Current during the latter part of this period. His map of the old Norse sailing routes shows a track direct from Iceland to the east coast of Greenland in latitude 66° N., then down the coast to Cape Farewell, and up the west coast.

According to the documentary evidence which he adduces, this route was followed until nearly A.D. 1200, and for most of the period *there is no mention of ice* in any of the numerous descriptions. From Greenland the Norsemen sailed to Wineland (on the coast of North America), and again there is no mention of ice. Recently this question has been re-investigated by L. Koch (11). From an exhaustive study of historical records of the ice off East Greenland and Iceland he concludes definitely that from A.D. 800 to 1200 there was scarcely any summer ice near Iceland. This is very striking ; Pettersson's own inference is that the ice did not then come so far south as it does now, and it seems probable that the Arctic Ocean was, if not ice-free, at least in the intermediate or " semi-glacial " condition described in Chapter I., in which a small cap formed in winter but disappeared completely in summer. This question is discussed further in Part III.

In 1917 I made my first incursion into the subject of continentality and temperature (12). This paper dealt with the region between 40° and 60° N. and between the Atlantic coast of Europe and 90° E., that is, practically the same region as the first of Kerner's papers (4), which I had not then seen. Fifty-six stations were selected in this area, and for these were found the height above sea-level, the mean temperatures of January and July, the " continentality " and the radiation received. The " continentality " was measured as the percentage of land in a five-degree circle, in a zone between five- and ten-degree circles, and in a zone between ten- and twenty-degree circles surrounding each station, these measures being termed respectively C_{0-5}, C_{5-10}, and C_{10-20}. It was also found necessary to introduce a Gulf Stream component into the January temperatures north of 50° N., the temperature decreasing by 0·6° C. for every hundred kilometres, or 1·7° F. for every hundred miles, east of a great circle through Valentia in South-west Ireland, and touching the north-west coast of Norway.

The most important result of the investigation was to show that the January temperature of Europe is much more closely related to the land and sea distribution than to the amount of heat received from the sun. The solar heat is the same at all points on the same line of latitude (apart from the effect of cloudiness, which was not discussed but which itself depends on the land and sea distribution) and decreases rapidly from south to north, so that if the temperatures were governed by this cause only, the isotherms should run east and west. It is found that actually they run nearly north and south, the rate of temperature decrease eastwards from the coast towards Russia being much more rapid than the rate of decrease northwards from Southern to Northern Europe.

In July the effect of land and sea distribution is less marked

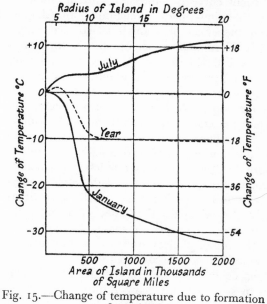

Fig. 15.—Change of temperature due to formation of an island.

and is slightly exceeded by the effect of solar radiation, so that the isotherms run more east-west than north-south.

Another interesting result concerns the effect of land-masses of different areas. Suppose a circular island were to form in mid-ocean in about latitude 60° N., and to increase gradually in size until it reached a radius of twenty degrees of arc or an area of about two million square miles. The resulting changes of temperature at a point on the edge are shown in Fig. 15. With increasing area the January temperature (lower curve) would fall and the July temperature (upper curve) would rise. From the figures obtained it appears that the fall of temperature in January would be very slow at first, being only 0·5° C. or 0·9° F. by the time the island had reached a radius of five degrees of arc (area about 375,000 square miles). As the island increased in size the fall of temperature would then become very rapid, until by the time the radius was ten degrees (area 1,500,000 square miles) it would amount to 22° C. (40° F.). After this the cooling would again increase more slowly with increasing area.

This curious curve is connected with the influence of the island on the atmospheric circulation. In Chapter II. we found that the effect of a small ice-covered island on the atmospheric circulation is slight—the storms sweep across it with very little hindrance. A larger island modifies the distribution of pressure, and an island with a radius of about 450 to 500 miles (about seven degrees) begins to develop a winter anticyclone. With a radius of ten degrees this winter anticyclone becomes semi-permanent and dominates the pressure distribution, and any further increase of radius merely results in a slow increase of intensity. While the radius is not more than five degrees (350 miles), the general meteorological conditions are unaltered and the cooling effect of land is due to its lower specific heat, which causes it to lose its summer heat more rapidly than does water, but with a radius of seven degrees (480 miles) or more, the winter anticyclone prevents the influx of heat from the neighbouring sea, and cooling by radiation proceeds rapidly.

Summer conditions are very different. As the island increases in size, the July temperature rises rapidly at first, until the radius has reached about five degrees (350 miles). Up to this point the sole effect is probably the absorption of the sun's heat by the soil and its transference to the lower layers of air by conduction. When the radius reaches seven degrees (480 miles) the increase of temperature becomes slower, and may even cease altogether ; this is probably due to a sea-breeze effect which lowers the afternoon temperature considerably. After the radius has exceeded about twelve degrees (830 miles) the July temperature begins to rise again more rapidly, but with a radius of twenty degrees the warming effect at the centre is only about 11° C. (20° F.).

The broken line in Fig. 15 shows the effect of the island on the mean annual temperature (mean of January and

July). The annual temperature is raised slightly by the introduction of an island with a radius of less than seven degrees (480 miles), the maximum effect, a rise of rather more than 1° C. (2° F.) in the mean temperature, occurring in the centre of an island of radius five degrees (350 miles). With greater areas the mean annual temperature is lowered. The distribution of land most favourable for high average temperatures is therefore a number of small islands each about 700 miles across, a condition which was frequently

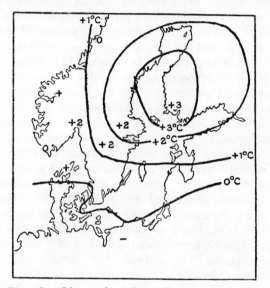

Fig. 16.—Observed, and calculated temperature changes, *Litorina* period, January.

approached in Europe during the early part of the Tertiary period.

This investigation concluded with an attempt to reconstruct from the changes in the land and sea distribution during the *Litorina* post-glacial submergence of Scandinavia (which occurred during the Atlantic period), the changes in the mean temperatures of January and July. For this purpose the present mean temperatures of those months at a large number of Scandinavian and Baltic stations were compared with the percentage of land in a five-degree circle, and it was found that a decrease of one per cent. in the continentality raised the January temperature by 0·20° C. (0·36° F.) and lowered that of July by 0·06° C. (0·11° F.). On this basis the change of temperature over the Northern Baltic was calculated as +3° C. (+5° F.) in January and −1° C. (−2° F.) in July. The calculated temperatures in different districts are compared with those deduced by various authors from the fauna and flora in Table 7 ; the results for January are shown graphically in Fig. 16, in which the calculated changes are shown by lines of equal temperature change, and the variations from present conditions required by geologists are shown by the figures. A variation of uncertain amount is indicated by the sign + or − without a figure. The agreement is good on the whole. The actual type of change in the direction of a more insular climate, warmer on the whole, is in perfect agreement. Many of the botanists comment on the prolongation of the autumn into the present winter, which is especially characteristic of the change to a more insular climate. The amounts of the change are also in good agreement except in the Christiania region and in North Denmark, where the geologists require a greater change than would be inferred from the change of continentality. This is probably accounted for by the greater freedom of ingress which the more open seas allowed to the warm waters of the Gulf Drift, but this point is discussed in detail later.

In 1918 I was able to extend the study of " Continentality and Temperature " to embrace the greater part of the world (13). The method adopted was extremely simple, perhaps rather too simple. From various sources I obtained the average mean temperature in January and July at each point of intersection of the ten-degree co-ordinates of latitude and longitude over both land and sea. On a globe[1] a ten-degree circle was drawn round each of these points, divided into east and west semi-circles, and the amount of land in each semicircle was measured and expressed as a percentage of the area of the semicircle. The area of ice in the whole circle was also measured and expressed as a percentage of the area of the whole circle ; this area included both land ice and sea ice. In winter the area of sea ice is many times the area

---

[1] The octagonal globe employed in the Meteorological Office for work connected with the *Réseau Mondial* was utilised for this purpose.

District.	Author.	Inferences from Fauna and Flora.	Calculated Differences.	
			January.	July.
Norway Helgeland (65° N.).	Rekstadt and Vogt.	No trace of warm period.	°C. +1·0	°C. −0·3
Trondhjem.	J. Rekstadt.	Climate " not greatly different from present."	+1·0	−0·3
Bergen.	C. F. Kolderup.	Climate " somewhat milder than present."	+0·2	0·0
Jaederen (S.W. Norway).	K. Bjorlykke.	No marked warm period.	+0·1	0·0
Sorland (S. Norway).	D. Danielsen. J. Holmboe.	Climate similar to present. " Somewhat warmer than present."	+0·2	0·0
Christiania Region.	C. Brögger. J. Holmboe.	Rise of 2° in mean annual temperature. Rise of 1·9° to 2·2° in mean annual temperature. Climate more maritime.	+0·6	−0·1
Denmark North.	V. Nordmann.	A damp, warm period (warmer winters, summers unchanged).	+0·3	0·0
Sweden General.	G. Andersson. R. Sernander. L. von Post.	About 2° warmer. More temperate. Insular climate. Warm, moist.	+1 to +3	−0·2 to −1
North Germany.	Several authors.	The difference, if any, was in the direction of a more continental climate.	0·0 to −0·8	0·0 to +0·2
East Baltic.	Kupfer. H. Lindberg.	Damp, warm (climate of West European coasts). Finland had a more insular climate.	+2·0 +3·0	−0·6 −1·0

Table 7.—Comparison of actual and calculated changes of climate.

of land ice, and the results are taken as applying to the former. The effects of each of these variables—land to the west, land to the east, and ice, independently of the other two—were then worked out by the method of correlation. The results can be set out in general terms as follows :—

1. In winter, the effect of land to the west is always to lower temperature. This holds in every latitude except 10° S. and 20° S.

2. In winter, the effect of land to the east is almost negligible. The only important exception to this rule is in 70° N. latitude, which may be considered as coming within a belt of polar east winds.

3. In summer, the general effect of land whether to the east or west is to raise temperature, but the effect is nowhere anything like so marked as the opposite effect of land to the west in winter.

4. The effect of ice, in the few cases in which it is possible to measure it, is invariably to lower temperature.

5. The temperature even of a point in mid-ocean in any latitude is modified by the presence of land along other parts of that parallel of latitude. The January temperature of a point in mid-Atlantic in latitude 60° N., for instance, is higher at present than it would be if the North Pacific were occupied by land.

These general conclusions could have been arrived at without a laborious statistical analysis, but the latter was necessary to reduce them to figures, and so make possible calculations of the thermal effect of changes in the land and sea distribution. This calculation is carried out by means of the formula :—

$$T = Z + aL_W + bL_E + cI,$$

where T is the temperature of the point required, Z is the " zonal temperature " (see below), L_W is the percentage of land in the semicircle to the west, L_E the percentage of land in the semicircle to the east, and I the percentage of ice in the whole circle. a, b, and c are constants for any particular latitude.

The " zonal temperature " Z is not a fixed quantity for any particular latitude ; it depends on the amount of land in the neighbourhood of that latitude, a zone which is mainly land-covered having a lower zonal temperature than a zone which is mainly occupied by water, but the relationship is not simple.[1]

The coefficients a, b, and c were obtained by purely statistical methods. They are as follows :—

January.	a Land to west. °C.	b Land to east. °C.	c Ice. °C.	July.	a Land to west. °C.	b Land to east. °C.	c Ice. °C.
70° N.	−0·43	−0·20	−0·46	70° N.	+0·02	+0·02	−0·16
60°	−0·31	−0·01	−0·07	60°	−0·01	+0·11	..
50°	−0·29	+0·09	−0·09	50°	+0·04	+0·06	..
40°	−0·17	+0·04	..	40°	+0·05	+0·07	..
30°	−0·08	+0·03	..	30°	+0·08	−0·01	..
20°	−0·01	−0·01	..	20°	+0·07	+0·02	..
10° N.	−0·01	+0·03	..	10° N.	+0·03	−0·01	..
0°	+0·01	0·00	..	0°	+0·02	−0·01	..
10° S.	+0·04	−0·01	..	10° S.	+0·04	−0·03	..
20°	+0·07	0·00	..	20°	+0·02	−0·02	..
30°	+0·06	+0·03	..	30°	−0·01	−0·01	..
40°	+0·09	−0·03	..	40°	0·00	−0·03	..

Table 8.—Effect of one per cent. of land to the west, of land to the east, and of ice on temperature.

It should be noticed that in this table the unit area of ice, one per cent. of the whole circle, is twice that of the unit area of land, one per cent. of a semicircle. The two figures can be made comparable by adding together the two land coefficients, thus :

	January.			July.
	70° N.	60° N.	50° N.	70° N.
Effect of land . .	−0·63	−0·32	−0·20	+0·04
Effect of ice . . .	−0·46	−0·07	−0·09	−0·16

The cooling effect of ice in winter is apparently less than that of land. This is because most of the area shown as occupied by ice is sea which is covered by more or less

[1] Winter :
$$Z = -70(0 \cdot 95 - \cos \phi) \log L - \frac{\tan^4 \phi \, L}{100}$$
Summer :
$$Z = 30(1 \cdot 05 - \cos \phi) \log L + \frac{\tan^2 \phi \, L}{100}$$

scattered drift ice and icebergs, that is, the surface is partly ice and partly water. In high latitudes in winter the land is usually snow covered, and it is easy to see that this snow surface must have a greater cooling effect than scattered sea ice. For ice-sheets over land it would be best to adopt the coefficients of land in winter.

These formulæ, being based on the present land and sea distribution, could not be employed in calculating the distribution of temperature during periods with a radically different distribution of the continents, but it seems legitimate to employ them to calculate the differences from the present

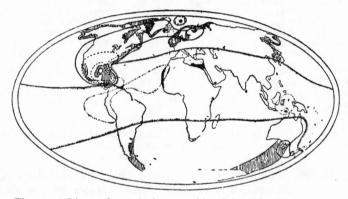

Fig. 17.—Lines of equal change of land level in Quaternary, and changes of land and sea distribution. Additional land shaded, additional sea black.

of the temperature distribution in a geological period during which the main outlines of the geography were the same as at present. In doing this it is best to measure the differences in the land and sea distribution from the present, and to calculate from these the differences in the temperature distribution, rather than to attempt to work *ab initio*. By working only with differences, we preserve intact the local peculiarities of climate and minimise the risk of error. The formulæ were applied in this way to a reconstruction of the land and sea distribution during the early part of the Quaternary period. This was the culmination of a great period of elevation in cold temperate regions, with corresponding depression in the tropics. I have represented the differences in level and land and sea distribution between that time and the present in Fig. 17. The lines of equal change of height were drawn from plotted figures accumulated from a great variety of sources. The restoration of the land and sea distribution is based mainly on the change of height used in conjunction with relief and bathymetrical maps, but in a few cases the actual ancient shore line has been traced ; it depends on the assumption that the continents held their present positions. The hypothetical restoration of Antarctica is based on well-known bio-geographical data, much of which has been admirably summarised by C. Hedley (14). Bio-geographical data have also been used as additional criteria in a few cases, such as the separation of Madagascar from Africa and New Zealand from Australia, or the union of Siberia to Alaska and of Japan to the mainland. It is also necessary to remark that there is not always evidence that the changes were strictly contemporaneous, but there is enough to show that the map is sufficiently correct to form the basis for a discussion. From this map and the formulæ it is evident that, even without an increase in the glaciation, the fall of temperature in winter outside the latitudes of 40° must have been very considerable. This fall was still further augmented by the great increase in the altitude of these regions. Now we know that, except in parts of Asia, practically the whole land surface north of 50° N. was glaciated, so that as a rough approximation we may assume that glaciers formed wherever the mean annual temperature fell below 32° F. (the problem of precipitation is dealt with in Chapter IX.). At first the increase of the land area would have raised the summer temperatures, but as the snow-cover began to persist through the year and form permanent ice-sheets, this excess disappeared and was replaced by a deficit. Taking the coefficient of land ice in summer as $-0\cdot16°$ C. for one per cent. of a ten-degree circle, and the average coefficient of land as $+0\cdot05°$ C., every increase of four per cent. in the portion of a ten-degree circle of land covered by ice lowered its temperature by $0\cdot8°$ C. ($1\cdot4°$ F.), and where the ice extended on to the sea the lowering was $0\cdot6°$ C. ($1\cdot1°$ F.) ; consequently, within the borders of the great ice-sheets the lowering of temperature in July amounted to nearly 20° C., an amount sufficient to enable the accumulation of ice to continue in summer as well as in winter. In January the change probably made little difference.

In Figs. 18 and 19 are shown the calculated differences of temperature from the present, both before the formations of the ice-sheets and at their maximum extension. The lines are lines of equal difference of temperature from present conditions in the same months. The calculated differences at a few points in the Northern Hemisphere are set out in

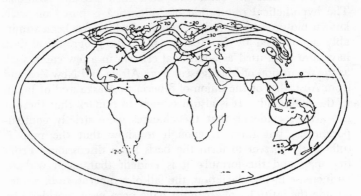

Fig. 18.—Changes of temperature due to geographical changes, January.

Table 9, with, for comparison, the differences at the same points deduced from the geological and biological evidence :—

| Locality. | Author. | Inferred Fall. °F. | Calculated Fall. | | |
			Jan. °F.	July. °F.	Mean. °F.
Scandinavia.	J. Geikie.	More than 20	36	18	27
East Anglia.	C. Reid.	20	18	13	15
Alps.	Penck and Brückner.	11	13	9	11
Japan.	Simotomai.	7	9	5	7

Table 9.—Comparison of calculated change of temperature with that inferred from geological evidence.

The agreement is quite good, and seems to show that the decrease of temperature during the Quaternary Ice-Age was completely accounted for by the changes in the distribution of land and sea and the effect of the ice itself. This does not necessarily mean, however, that an increase in the land area in high latitudes would by itself suffice to initiate a glaciation ; the winter temperatures would be lower than at present, but the summer temperatures would be higher, and on low ground the winter snowfall would not survive the hot summer—that is why Siberia is not glaciated. Probably, glaciers can only originate on an area of high ground, but the presence of a large continental region with a low winter temperature is essential if a local glaciation of the mountain-valley type is to develop into a regional glaciation of the ice-sheet type. This is clearly illustrated in the glacial history of

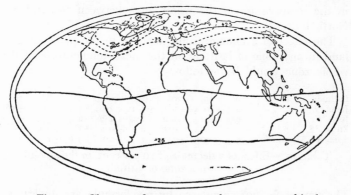

Fig. 19.—Changes of temperature due to geographical changes, July.

Europe ; the Quaternary glaciation began in the mountains of Norway, but the extension of the ice half-way across Europe was possible only because, owing to elevation, there was a large area of high continentality to the eastward. The beginning of glaciation in Norway was due to increased elevation bringing a larger area above the snow-line, and perhaps also to the shutting out of the Gulf Stream Drift. Once the glaciers had reached a certain size they became independent of the elevation of the ground in their centres, and their extension was governed, first, by the amount of snowfall available for their nourishment, and secondly, the snowfall being sufficient, by the balance between the cooling-power of the ice and the natural or " non-glacial " temperature of the regions into which they intruded.

It will be interesting to compare the results of a ten per cent. decrease in the amount of land in each zone of latitude, calculated from my formula, with those obtained by means of Spitaler's formula given on page 132. According to my formula, the change of temperature falls into two parts, a local part due to the local changes in the land and sea distribution, and a general part due to the change in the general zonal temperature. For example, if owing to local elevation the Faroe group became a single large island of the size of Iceland, the result would be two-fold—a large local decrease in the winter temperature of the site of the new island and of the surrounding area, and a small general decrease in the temperature of the whole zone between 50° and 70° North latitude.

The effect of a change from 50 to 40 in the percentage of land in any zone of latitude on the mean temperature of the whole zone would be as follows :—

Latitude	0°	10°	20°	30°	40°	50°	60°	70°
January, °F.	—0·5	—0·4	+0·1	+1·1	+2·6	+4·7	+7·3	(+16)
July, °F.	—0·3	—0·4	—1·2	—1·3	—2·2	—2·4	—3·3	—3·4
Mean, °F.	—0·4	—0·4	—0·5	—0·1	+0·2	+1·1	+2·0	(+6)
Spitaler's Mean, °F.	—3·5	—3·3	—2·7	—1·7	—0·6	+0·6	+1·7	+2·7

Table 10.—Effect of a decrease of 10 per cent. in the amount of land in a zone of latitude.

The figures for 70° calculated by my formula are uncertain, but in 60° my result is in good agreement with Spitaler's, which was obtained by quite different methods. The chief difference in the results occurs in low latitudes, where the effect of land area is much less important according to my results than according to Spitaler's.

The essential point for the theory of climatic variations is that, according to either computation, the warming effect of a decrease in the land area becomes greater as we get nearer to the poles. In summer, a decrease in the area of (unglaciated) land results in a slight decrease of temperature, and the net increase during the year is entirely due to the very large increase in winter. If the percentage land covering north of 60° N. decreased by ten, the result would be a rise in the " non-glacial " January temperature by at least 7° F., which would be sufficient to bring it above the freezing point and so introduce a " non-glacial " climate. This is the general case, of which the two examples—Middle Eocene and Upper Jurassic—discussed by F. Kerner are special instances. The conclusion to which we are brought, therefore, is that moderate changes in the land and sea distribution, such as have occurred frequently enough in geological times, are amply sufficient to bridge the gap between non-glacial and glacial climates, or between warm and cold geological periods, and that extraneous aids, such as variations of solar radiation or changes in the astronomical climate, while possible causes, are not necessary conditions.

REFERENCES

(1) FORBES, J. D. " Inquiries about terrestrial temperature." Edinburgh, *Trans. R. Soc.*, 22, 1861, p. 75.

(2) SPITALER, R. " Die Wärmevertheilung auf der Erdoberfläche." Wien, *Denkschr. K. Akad.*, 51, 1886, Abt. 2, p. 1.

(3) KERNER-MARILAUN, F. " Paläoklimatologie." Berlin (Gebr. Borntraeger), 1930.

(4) KERNER, F. " Synthese der morphogenen Winterklimate Europas zur Tertiärzeit." Wien, 1913.

(5) HEER, O. " Untersuchungen über das Klima und die Vegetationsverhältnisse des Tertiarlandes." Winterthur, 1860.

(6) KERNER, F. " Klimatogenetische Betrachtungen zu W. W. Matthews, ' Hypothetical outlines of the continents in Tertiary times '." Wien, *Verh. k. k. geol. Reichsanstalt*, 1910, p. 259.

(7) KERNER, F. " Das akryogene Seeklima und seine Bedeutung für geologischen Probleme der Arktis." Wien, *Sitzungsber. Ak. Wiss.*, 131, 1922, p. 153.

(8) BROOKS, C. E. P. " The problem of warm polar climates." London, *Q. J. R. Meteor. Soc.*, 51, 1925, p. 83.

(9) GAMS, H., and R. NORDHAGEN. " Postglaziale Klimaänderungen und Erdkrustenbewegungen in Mitteleuropa." München, *Geogr. Gesellsch. Landesk. Forschungen*, H. 25, 1923.

(10) PETTERSSON, O. " Climatic variations in historic and prehistoric time." Svenska Hydrogr.-Biol. Komm. *Skriften*, 5, 1914.

(11) KOCH, L. " The East Greenland ice." Copenhagen, *Komm. Videnskabelige Undersøgelser i Grønland*, København, 1945.

(12) BROOKS, C. E. P. " Continentality and temperature." London, *Q. J. R. Meteor. Soc.*, 43, 1917, p. 169.

(13) BROOKS, C. E. P. " Continentality and temperature." (Second paper). " The effect of latitude on the influence of continentality on temperature." London, *Q. J. R. Meteor. Soc.*, 44, 1918, p. 263.

(14) HEDLEY, C. " The palæographical relations of Antarctica." London, *Proc. Linnæan Soc.*, 124, 1911-12, p. 80.

CHAPTER IX

PRECIPITATION—RAIN, SNOW, AND HAIL

THE various forms in which water falls from the sky, of which the most frequent are rain, snow, and hail, are conveniently grouped together under the term *precipitation*. The precipitation is occasionally slightly augmented by other forms, dew and hoar-frost, and on the arid western coast of South America there are mountain plants which live on the moisture they derive from mist, but, practically speaking, the three forms first mentioned are the only ones which need be considered. Precipitation is formed by condensation of the water vapour in the air, which has been derived by evaporation from the surface of the sea, lakes, vegetation, soil, etc. Air at a certain temperature can only hold a certain amount of water vapour, and this amount decreases very rapidly with falling temperature. Hence, if saturated air is cooled, some of the water vapour which it contains is condensed to form cloud, and, if the condensation is continued far enough, rain or snow. Hail is formed when a column of air rises very rapidly to great heights, as in thunderstorms. The cooling of moist air is the only way in which an appreciable amount of precipitation can be produced. If the pressure on a mass of air or any other gas is lessened, the gas will increase in volume, and in doing so will become colder, unless heat is supplied from without. At sea-level the air is subjected to the pressure of the whole of the atmospheric column above it, equivalent in weight to about fifteen pounds per square inch. As we go to higher levels and leave part of the atmosphere below us, the mass of superincumbent air becomes less, and the pressure falls. Hence any sample of air which rises to higher levels in the atmosphere will expand and cool, while a sample which descends to lower levels will be compressed and warmed. That is the reason why, generally speaking, the air is colder the higher the level.

It will be understood from this that very nearly all

precipitation falls from air which is rising and therefore expanding and becoming cooled. From the circumstances under which the air rises, precipitation is classified into three types—Orographic, Cyclonic, and Convectional or Instability precipitation.

Orographic precipitation falls where a current of air encounters high ground and is forced to rise. The hilly parts of Western Britain, standing in the path of the moist south-west winds from the Atlantic Ocean, receive a great deal of orographic rain, and are, in fact, the wettest parts of these islands. The west coast of Norway is similarly situated ; other regions are the eastern end of the Black Sea, the mountains of Lebanon, the coast of Honduras in Central America, the Western Ghats of India, and especially the Khasi Hills of Assam, one of the wettest spots on the globe. Orographic rain may be very heavy in warm countries where the air contains a great deal of moisture, but in this country it is persistent rather than heavy. Much depends on the topography ; a long range of hills of uniform height is more effective as a rain-maker than an isolated mountain, since, unless it is already in an unstable condition, air will always go round an object rather than rise above it. Orographic precipitation in cold regions frequently falls as snow in winter.

When we study the distribution of precipitation in a mountainous region, such as the Alps, we find that the amount is moderate in the lowlands and valleys, and becomes greater as we ascend the slopes of the mountains. At a certain height, however, termed the " level of maximum precipitation," this increase with height ceases, and above this level the precipitation becomes less as we ascend. This level depends on the temperature and relative humidity of the air over the lowlands, and the vertical decrease of temperature ; in the Alps it occurs at a height of about 7,000 feet, where the precipitation is between two and three times that over the lowlands. We can also distinguish the level of greatest rainfall in the Alps between 4,000 and 5,500 feet, and the level of greatest snowfall in the Alps at about 8,000 feet. The latter must not be confused with the snow-line, which is about 2,000 feet higher in the Alps ; the level of maximum snowfall depends on the winter conditions, while the snow-line is determined very largely by the conditions in summer. The relative position of the snow-line and the level of maximum snowfall are of great importance for the development of glaciers, as we shall see in Chapter XVI.

Cyclonic precipitation falls during the passage of barometric depressions, cyclones, and other forms of atmospheric disturbance which depend on general rather than on local conditions. The hurricanes of tropical and sub-tropical regions bring torrential rains which often cause widespread

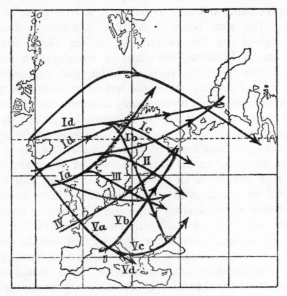

Fig. 20.—Tracks of depressions.

floods and add to the havoc of the winds, but it is only in temperate and sub-polar regions that cyclonic rain forms an important part of the total average rainfall. In the eastern half of Britain the greater part of the winter rainfall is cyclonic, and the same is true of the winter precipitation of all parts of the temperate belt except the high ground near the sea. The distribution of cyclonic precipitation is largely governed by the tracks followed by barometric depressions. Although the paths of individual depressions in temperate regions often appear to be erratic, it has been found possible to classify them into a number of tracks, which are more usually followed than the intervening regions (Fig. 20). These tracks have a preference for moist areas, especially inland seas such as the English Channel, the Baltic, and the Mediterranean, or for well-watered plains such as Hungary and Poland. Track I, the favourite track at present in all seasons but spring, runs from Iceland or the Faroe Islands north-eastward, some distance off the coast of Norway, to the Arctic Ocean, or across the north of Norway to the White Sea.

This question of the tracks of depressions is important for palæometeorology, for a considerable degree of permanence has been attributed to them. During the Quaternary Ice-Age, when the northern tracks were closed by the ice-sheets and the glacial anticyclones which occupied them, Track V, which runs south of the main glaciated area, was the favourite track. Many depressions passed from end to end of the Mediterranean, and the rainfall associated with them caused the north of Africa to be much moister than at present ; this was a " pluvial period " in that region. Track Vb, passing northwards across Central Europe, was probably also extensively followed during the glacial period, and caused heavy snowfall over the south-eastern margin of the ice-sheet. At present it is followed chiefly in spring, and is associated with high pressure and low temperature to the north-west ; it brings cold spells in Central Europe.

Marsden Manson (1) supposes that the cyclone tracks across North America have been fixed in their present position throughout the whole of geological time, and that the distribution of precipitation has always resembled that existing at present. He supports this theory by the coincidence that the pre-Cambrian glaciations of that continent occurred in the present storm belt, but the geological evidence does not warrant the generalisation, since North America had an arid climate during a large part of geological time. During the Eocene period, when the plant-bearing deposits of the Arctic Circle were formed, the winds on the west coast of Greenland in 70° N. appear to have been mainly south-westerly, indicating that the area of lowest pressure and, presumably, of heaviest rainfall, lay still farther north, and other evidence will be adduced later. It will be an important part of future work to lay down the storm tracks of past ages as closely as possible, since this will provide a large amount of information as to the barometric distribution and the position of the belts of rainfall.

The precipitation associated with barometric depressions falls as rain or snow according to the temperature at the level of condensation. Not infrequently in winter the northern half of a depression brings snow, while the southern half brings rain. During the Quaternary Ice-Age this distinction between the southern and northern halves was probably very pronounced, and depressions skirting the ice-sheets must have caused a large annual snowfall on the ice. As explained in " The Evolution of Climate," this tendency to a maximum snowfall near the margin of the ice-sheet appears to have played a large part in causing the successive development of centres of glaciation more and more to the south-west— Scandinavia, Scotland, Ireland.

Convectional or Instability precipitation is typified by the hail or heavy rain of thunderstorms. It is due to the warming up of the lower air, by contact with ground warmed by the sun. When the warming has proceeded far enough, the lowest layer of air becomes potentially lighter than the air above it (i.e., its temperature is so much higher than that of the layer above that even after the expansion and cooling consequent on lifting it to the level of the latter it would still be warmer). Under these conditions the surface air begins to rise through the air above it, at first in thin threads, but if the process goes far enough, in thicker columns. The first result of this process is the formation of the small cumulus clouds so commonly seen in England on a summer afternoon ; very often in this country the process goes no farther and the clouds die away in the evening without producing rain, but under favourable conditions of vertical distribution of temperature (and in the presence of sufficient moisture), sudden thundery showers or true thunderstorms may result. The mechanism of a thunderstorm is complex, and need not be discussed ; here it is sufficient to remark that the typical thunderstorm is essentially the product of hot, relatively calm weather and moist air. The thunderstorms associated with " cold fronts " during the passage of depressions have a different origin, and the rainfall associated with them comes under the heading of cyclonic rain.

On tropical coasts, the rising of the warmed air over the

land is facilitated by the onset of the relatively cool sea breeze, and the progress inland of the latter is marked by a line of thunderstorms. In many parts of the tropics the greater part of the rainfall is caused in this way.

Thunderstorm clouds often extend to a great height, and the temperature of the upper part of the cloud may be below freezing point. Under these conditions the precipitation often takes the form of hail, which is typically associated with thunderstorms. It is an important point in the theory of the climate of the " warm " periods that the vegetation of Europe during the Tertiary period often shows evidence of damage by hail, and of torrential rains of the instability type, even when the stage reached by the plants shows that the deposit was formed quite early in spring. During these warm periods the relief was generally low, giving little oro- graphical rain, and depressions were probably weak and sporadic, but conditions were especially favourable for the occurrence of thundery showers, and most of the precipitation was of this type.

The origin of the snow required for the nourishment of ice-sheets is a difficult problem. The snowfall of Greenland appears to be due almost entirely to the winds from the ocean, which blow from the sea on to the ice during the passage of intense barometric depressions, but for the far larger ice- sheet of Antarctica this explanation is only tenable for the margins. W. H. Hobbs (2) considered that the process was as follows : the air which flows outward on the surface of the ice-sheet must be replaced by inflowing moist air at higher levels. These upper air currents are indicated by clouds which can be seen passing inland across the coast. In the centre of the anticyclone this air descends, and is of course warmed by compression, but owing to the intense radiation the surface of the ice is intensely cold, colder in fact than the air at considerably higher levels (this Antarctic inversion of temperature is a well-known phenomenon). The descending air is cooled by contact with this intensely cold ice surface, and deposits its moisture in the form of small granular masses of ice. The formation of an ice-mist of small crystals of ice in this way has been observed in Siberia, but apparently the precipitation was not sufficient to be termed snow.

Sir George Simpson (3) showed that this ingenious

mechanism cannot be the whole story, since the amount of snowfall which it would yield would be insignificant. He therefore carries the argument a step farther. The moist air which flows in at high levels is brought down to the surface of the ice by the anticyclonic circulation, and is cooled by radiation and by contact with the ice. In this way it becomes about as cold as when it crossed the margin of the continent. If it becomes colder, some snow will be precipitated in the way supposed by Hobbs, but, in any event, it will be almost or quite saturated. Owing to local circumstances there are irregularities in the temperature distribution ; the coldest air tends to spread out on the surface and flow down slopes, undercutting and lifting up the air which is less cold. The latter is forced to rise again ; it is cooled still further by expansion, and consequently snow is formed. This account agrees with the fact that by far the larger part of the snowfall of the Antarctic continent occurs in blizzards.

We are now in a position to discuss in general outline the distribution of precipitation over the globe. The local details are so complex that it is not possible to present the distribution adequately on a small-scale map ; this distribution is intimately connected with the local geography, and for our purpose it is more important to have generalisations applicable to any land and sea distribution. In the first place, we may distinguish four zones of precipitation in each hemisphere :—

1. The equatorial belt of heavy rainfall.
2. The sub-tropical dry belt.
3. The temperate rain belt.
4. The polar cap of generally light snowfall.

These four rainfall belts correspond with the four pressure belts described in Chapter II., the equatorial belt of low pressure, the sub-tropical anticyclonic belt, the temperate storm belt, and the polar caps of relatively high pressure. The belts of pressure are best developed over the oceans, and it is probable that the same is true of the rainfall. The most detailed estimate of the zonal distribution of rainfall was made by C. E. P. Brooks and T. M. Hunt (4), and the results are shown in Fig. 21.

The rainfall in inches for the different 10° zones of latitude are as follows :—

Latitude °N.	.	.	90-80	80-70	70-60	60-50	50-40	40-30	30-20	20-10	10-0
Land (in.)	.	.	—	5·8	12·1	19·3	20·2	23·2	26·6	32·1	56·3
Oceans (in.)	.	.	(5)	8·0	27·2	56·4	51·1	44·2	38·6	48·0	63·3

Latitude °S.	.	.	0-10	10-20	20-30	30-40	40-50	50-60	60-70	70-80	80-90
Land (in.)	.	.	60·1	42·7	26·0	22·2	31·3	38·4	6·9	2·4	(2·0)
Ocean (in.)	.	.	54·3	42·5	36·7	43·4	47·9	37·8	15·0	3·7	—

Table 11.—Distribution of precipitation according to latitude.

The rainfall zones are clearly shown in the figures for the sea and in the land over the Southern Hemisphere. In the Northern Hemisphere the great continental masses of Eurasia

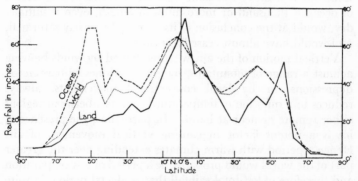

Fig. 21.—Variation of rainfall with latitude.

and North America are largely beyond the influence of winds blowing directly from the oceans, and except near their coast-lines they are relatively dry.

The total amount of precipitation over the whole earth, including both land and water, in the course of a year averages 40·9 inches. This must be equal to the total amount of evaporation, since the amount of water vapour and of water droplets or ice-crystals in the form of cloud held in the air at any time is equivalent to only a few millimetres of rain— less than one per cent. of the average annual fall. Evaporation and precipitation are intimately related, since both depend largely on vertical movements in the atmosphere. Rising air, as has been pointed out, is responsible for very nearly all the condensation of water vapour to form rain, snow, or

hail, and every ascent of air must be counterbalanced somewhere by descending air. This is warmed by compression
as it descends ; its capacity for holding water vapour is
increased, and it arrives at the surface with a low relative
humidity. Evaporation takes place chiefly into air which
has recently descended to the surface from higher levels in
this way ; it also occurs where air is blowing from a colder
to a warmer surface and is being warmed by contact with the
latter. The total amount of precipitation, therefore, is not
necessarily proportional to the area of the oceans even in low
latitudes, an assumption which is sometimes made. A
narrow tropical sea with mountainous shores, across which a
steady wind is blowing, would be subject to intense evaporation, but a wind of the same velocity blowing over a stretch
of ocean a thousand or more miles in length, even if initially
dry, would at the conclusion of its journey be nearly saturated,
and would have almost ceased to evaporate.

Vertical motion of the air, whether caused by winds blowing
against a range of mountains, by great cyclones or barometric
depressions, or by local convectional movements, always
reduces to a question of temperature contrasts between regions
either remote or near at hand. In particular, the presence of
ice is a potent factor in causing vertical movement of air.
Hence a period with warm climates extending over the greater
part of the world would probably be a period of less evaporation
and therefore of less total rainfall than a glacial period, in spite
of the increased amount of water vapour which the warmer
air can take up ; conversely, during a glacial period, even
if the temperature is lower, the total rainfall may be increased.
This was exemplified during the Quaternary Ice-Age, when
the rainfall over the non-glaciated regions was heavier than
the present rainfall, and probably much heavier than that of
any part of the Mesozoic or Tertiary periods.

We have to take account of the evaporation in another
way. The biological effectiveness of rainfall depends not
only on the total fall, but much more on the amount which
becomes available for plant life. In regions subjected to
great evaporation the effectiveness of the rainfall is greatly
diminished, and a hot country with the rainfall of South-
eastern England would be regarded as dry. The rainfall of
Jerusalem, for instance, is heavier than that of London, but

owing to the greater evaporation it is less effective. R. Lang (5) expresses the effectiveness of the rainfall as the " Rain-factor," which he obtains by dividing the annual precipitation in millimetres by the mean temperature in centigrade degrees, and he finds that the character of the soil is closely related to this factor in the way shown by the accompanying diagram (Fig. 22). According to this diagram, with a mean annual temperature of 30° C. (86° F.), desert formations may occur with a rainfall as high as 1,200 mm. (48 inches) a year, but if the mean temperature is only 10° C. (50° F.), the rainfall cannot be more than 400 mm. (16 inches). With a rain-factor

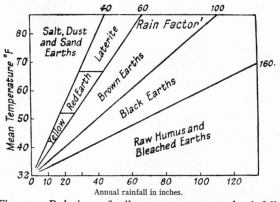

Fig. 22.—Relations of soil to temperature and rainfall.
After R. Lang.

between 40 and 60, the deposits are still coloured entirely by iron oxides, but the chemical composition and the colour vary according to the mean annual air temperature. When the latter lies between 32° F. and 54° F., yellow earths are formed ; between 54° F. and 68° F., red earths ; and above 68° F., deep red loams or laterite. The latter, therefore, requires a rainfall of at least 800 mm. (32 inches) a year, and may be taken as evidence of a fairly moist warm climate.

When the rain-factor is between 60 and 100, the colour of the deposit is due to red or yellow iron oxides and partly to black humus, the resulting mixture giving the deposits a brown colour (brown earths). With the rain-factor between 100 and 160, the colour is determined entirely by humus, and

" black earths " result ; above 160 the earth is bleached of all colouring matter by the rich vegetation and the heavy rain, and the result is a white subsoil surmounted by a deposit of pure humus.

When the mean temperature is below 0° C. (32° F.), all chemical action ceases, and the purely mechanical deposit takes the colour of the rock from which it was formed. Deposits formed from igneous rocks, or from a mixture of rocks of different colours, are generally grey, and the deposits formed near the edge of an ice-sheet are often of this colour.

A more recent table by E. M. Crowther (6) is based on a " leaching factor " $R - 3 \cdot 3$ T, where R is the rainfall in cm. and T the mean annual temperature in ° C. The results, based on work in U.S.A., may be summarised as follows :—

Leaching factor above 70	Temperature increasing Podsol ; Brown Forest Soils ; Ferruginous laterites.		
Transitional (Leaching factor below 70, rainfall above 70 cm.)	Prairie soils.		
Leaching factor and rainfall both below 70	Rainfall increasing		
	Grey desert ; soils	Chestnut soils	Brown semi-desert soils ;	Tchernosem

Table 12.—Relations of soil to temperature and rainfall.
After E. M. Crowther.

The biological effectiveness of the precipitation also depends on the proportion of it which sinks into the soil. This is governed by a number of factors, especially the character of the fall and the nature of the soil and vegetation. A persistent " soaking " rain of moderate intensity is much more effective than a torrential downpour which runs quickly off the land and floods all the streams, although the actual amounts may be the same. A snow-cover which accumulates during the winter and melts gradually in spring may be very effective. Thick vegetation covering a soft soil checks the rate of run-off and allows a larger proportion of the rainfall to be absorbed than does hard bare earth. This aspect of the rainfall has been brought out by the discussions of the desiccation of South Africa (7). In the past fifty years the country has been suffering increasingly from drought, but the conclusion from expert evidence is that this is not due to an actual decrease in the amount of rainfall, but to a change in the nature of the soil and vegetation. When South Africa was first settled, the country was covered by a rich vegetation, the rainfall was steady and persistent, and a large proportion of it was absorbed. The effect of over-pasturage has been to destroy much of the protective vegetation, and the soil has been washed away or trampled hard. The temperature contrasts have been increased owing to the heating effect of the sun on the patches of bare ground, and the rain now falls largely in heavy " instability " showers, including destructive thunderstorms. The run-off is proportionally greater, owing to the more torrential nature of the fall and the loss of the vegetation, so that with nearly the same rainfall the amount of water available for use has decreased. The possibility of changes of this nature brought about by human activities has to be remembered in all discussions of the vexed question of " desiccation " in historic times ; in fact a passage in Plato's " Critias " suggests that the decadence of Greece may have been due to such a change.

The distribution of the precipitation among the seasons is almost as important as the total amount. We have to distinguish between regions with their precipitation almost equally distributed throughout the year, regions with their rainy season in winter and dry summers, and regions with their rainy season in summer. The character of the seasonal distribution governs the type of vegetation on the one hand, while on the other hand it is intimately related to the general meteorological régime. This often enables us to derive important information as to the general meteorology of a period from a study of the plant remains ; for instance, the presence of annual growth rings in tree stems may mean a seasonal alternation of temperature, but associated with evidence of a high degree of warmth, it implies the alternation of dry and rainy seasons and a monsoon type of climate.

The limits of the various rainfall types have been set out in great detail by W. Köppen (8). Along the immediate neighbourhood of the equator is a belt of heavy rain in all seasons. This is the region of the dense forests of the Amazon and Congo River basins, and it is known as the " tropical rain-forest region." It also includes most of the East Indies and the Malay Peninsula, and extends into Cambodia and Assam. On low ground the tropical rain-forest belt does not usually extend more than 10° on either side of the equator, but under favourable conditions it may extend to 20° or even to nearly 30° latitude. These outlying areas include the eastern coast of Madagascar, the eastern slopes of the Andes, and the " everglades " of Southern Florida, another region of tropical swamp.

The soil is usually a rich dark humus. The following graphic description is given by E. Warming (9) : " Forest is piled upon forest. The trees forming the highest storey have tall thick trunks, which are unbranched up to a height of 120 to 150 feet or more. Beneath them are trees of moderate stature with branches not reaching those of the higher tier. Beneath these in turn succeed slender thin-stemmed low palms, tree-ferns, and shrubs . . . scattered about are huge herbs which reach 12 or 15 feet in height. If there still remain space available on the ground that is reached by the light, it is occupied by dark green ferns, *Selaginellæ*, mosses, and similar scrophytes. But often the light is too feeble to permit of more than a very small number of plants developing on the ground, which then may be almost bare of vegetation, with its black humus covered only by fallen decaying wet leaves, twigs, and remnants of fruits, between which only bizarre saprophytes find places . . . but there are hordes of epiphytes clothing trunks and branches . . . as well as ferns, mosses, and so forth. Trees of the forests situate in the cloud-belt of Java and the Moluccas are enveloped in a soaking mossy felt, which may be thicker than the trunks themselves and imparts to them a peculiar dark appearance. . . . Finally, there is a wealth of lianas, whose flowers and fruit one can rarely see, and whose long, often curiously shaped stems, span the distance between soil and tree-crowns, or hang down from the latter or partly trail along the ground. . . . The twilight prevailing is much less dark than in European beech-forest. All the species . . . seem to abhor a vacuum and to combine in an endeavour to utilise all the space available."

Many of the species show protective devices against the

very heavy showers, especially a smooth cuticle which cannot be wetted, drip tips and channelled nerves, but paradoxically some plants of the highest storey show xerophytic characters. A highly important feature for our purpose is that the trees show no annual growth rings.

Where the contours of the ground are favourable, as in Eastern Sumatra, great tropical swamps are formed, which appear to reproduce the conditions prevailing during the formation of the coal measures, and H. Potonié (10) believes that the coal measures were in fact formed under similar conditions in a very warm rainy climate. The forests which formed the coal measures seem to have been similar in many respects to the present tropical rain-forests. As described by David White (11), the Carboniferous forests showed " rankness of terrestrial vegetation ; great size of trees, plants, and leaves . . . great size of fronds, and absence of annual rings," while " fairly well-developed palisade tissue points towards sunlight." The lianas of to-day were represented by " many long slender clambering or climbing ferns and fern-like types."

A. Wegener, in discussing the movements of the poles according to his theory of " Continental Drift " (see Chapter XIII.), makes great use of the beds of coal for determining the position of the equator in the successive geological periods. It should be remarked, however, that although peat is forming at present in the tropics in one or two isolated regions, it is rare, while in the temperate rain belts peat bogs occur wherever there is sufficient moisture.

Nearer the poles, in temperate latitudes, are other areas in which rain or snow falls in sufficient quantities in all seasons. These areas include Northern and Eastern North America, where they pass into the tropical rain-forest area, all Northern and Central Europe, and a large part of Asia ; in the Southern Hemisphere they are limited to relatively small areas in Chile, South-eastern Australia, and New Zealand. In Europe the area occupied by these temperate rains was the site of great peat formation during the post-glacial period, and certain coal beds of earlier geological periods are attributed by Wegener to the temperate rain belt.

On either side of the equatorial " rain-forest " belt, and extensively developed on the eastern sides of the continents, is the monsoon or summer rain region, best known from its occurrence in India and China. This type of distribution is due to the extensive alternate heating and cooling of the interiors of large continents and the consequent alternation of monsoon winds ; it occurs on either side of the tropical rain-forest belt in South America and West Africa, and is very extensively developed in Eastern Africa and in Southern and Eastern Asia ; it is also found in the north and east of Australia. The type of vegetation associated with it is the savannah or meadow land, passing into open forest with increasing rainfall. Where the rainfall is especially heavy and the temperature steadily high, it may give dense tropical rain-forests.

On their poleward sides, the summer rain regions in the western and central parts of the continents usually pass into deserts, which are characterised by a slight and irregular rainfall, many months sometimes passing without even a shower. In parts of the South American desert it has probably not rained for centuries. Such plants as there are show special devices to prevent the loss of water, but in many deserts the ground is entirely bare of plants. Among animals, one of the most characteristic is the lung-fish (*Ceratodus*), which is adapted to breathe either air or water, and can remain dried up for long periods. This form, which is still living in Australia, has persisted since the Permian (allied forms are known since the Devonian), and affords some evidence of the continuity of desert conditions throughout a large part of geological time. The best geological evidence of arid climates is lithological—desert sandstones of rounded and polished grains usually red in colour, " dreikanter," and other wind-eroded rocks, deposits of gypsum and of salt ; these abound in many horizons and seem to suggest that the desert belts were greatly expanded in the past. This was probably true, but we have to remember that before land plants reached their present degree of specialisation, large areas which would now be habitable by plants must have been bare rock, subject to aerial denudation, so that " desert " sandstones do not necessarily imply a rainfall as small as that of modern deserts.

On the poleward sides of the deserts the rainfall increases again, but falls mainly in winter, while the summers are hot and dry. The best known example of this climate is the Mediterranean region, whence the type is often known as the

Mediterranean. It is found also in California and in limited regions in Chile, near Cape Town, and in Southern Australia ; it is thus practically limited to the belts between latitudes 30° and 45° and, except in Mesopotamia, it never extends far from the sea. Eastward it usually passes into a semi-arid or arid climate. It gives a peculiar type of vegetation adapted to resist the drought of summer and to maintain its moisture from considerable depths, for example, the vine.

The Mediterranean type of climate appears to have persisted in South-east Europe during a large part of the Tertiary period (12), though with variations in the total amount of rainfall. The region was an archipelago of small islands ; in the Early Eocene it had a rather moist climate with some rain in all seasons, but in the Middle Eocene the entire absence of a land flora probably indicates a semi-arid climate. In the Oligocene there were well-marked dry and rainy seasons, and in the Early Pliocene the climate was again warm and rather moist.

On its polar side, the Mediterranean climate passes into the temperate rain belts, which, except in the interior of the great continents of Eurasia and North America, have a sufficiency of rain at all seasons, and are occupied mainly by forests of conifers or deciduous trees, which in the more equatorial parts of the belts include some sub-tropical species. Nearer the poles the winter is severe, with a persistent snow-cover, and the summer is short ; this is the " boreal " climate. Near the coast there is abundant rain at all seasons ; in the interior the winter is usually dry, but the annual variation of temperature in all cases enforces a period of inactivity in plant growth. This zone is the peculiar home of peat bogs, which require a rainfall of at least 40 inches a year and a mean temperature above 32° F. The great outbursts of peat formation in the " Atlantic " and " sub-Atlantic " periods of post-glacial time gives us a measure of the raininess of these periods.

The distribution of the belts of rainfall is closely related to the distribution of the belts of pressure and wind described in Chapter II., and the temperature belts described in Chapter VIII., and we may set out the general succession of climatic belts in the way shown in Table 13. The first column gives the average latitude in which the different belts are found

Latitude.	Astronomical.	Temperature (Supan).	Winds.	Rainfall.
N.90	North Polar Cap.	North Cold Cap.	(?)	Mainly snow; relatively dry.
			North Polar East Winds.	
60	North Temperate Zone.	North Temperate Belt.	Northern Belt of Westerlies.	North Temperate Rain Belt.
30			Sub-tropical Calm Belt.	Nor Tropical Dry Belt.
	Tropical Zone.	Hot Belt.	North-east Trades.	
0			Calms (Doldrums).	Equatorial Rain Belt.
			South-east Trades.	
30	South Temperate Zone.	South Temperate Belt.	Sub-tropical Calm Belt.	South Tropical Dry Belt.
60			Southern Belt of Westerlies.	South Temperate Rain Belt.
S.90	South Polar Cap.	South Cold Cap.	South Polar East Winds.	Snow, dry.

MONSOON RAINS

Table 13.—Climatic zones.

in the two hemispheres, while the second column gives the " astronomical " zones of the old school text-books, which are limited by the tropics and the polar circles. The third column shows the belts of temperature according to Supan, who limited his " hot belt " by the mean annual isotherm of 20° C. (68° F.), which approximately fixes the polar limit of palms, and his cold caps by the isotherms of 10° C. (50° F.) for the warmest month, which forms the limit of growth of cereals and forest trees. It will be noticed that the cold cap extends into much lower latitudes in the Southern than in the Northern Hemisphere ; this is due partly to the influence of the great Antarctic ice-cap and the ice-laden Southern Ocean, and partly to the smaller annual range of temperature over the oceanic regions of the Southern Hemisphere than over the continental regions of the Northern Hemisphere.

The rainfall belts have a close relation to the wind systems. The equatorial rain belt is limited to the Doldrums and part of the region of the South-east Trades. The dry belts include most of the trade wind regions, the whole of the sub-tropical calm belts, and extend a short way into the domain of the westerly winds, where they give place to the temperate rain belts, which in turn are limited on their poleward sides by the polar east winds. We have seen reason to believe that when the temperature-difference between equatorial and polar regions was below a certain critical value, the polar caps of east winds would be suppressed and the westerly winds would extend almost to the poles. This would result in a spreading out of all the other zones into higher latitudes, and would carry the poleward margins of the sub-tropical dry belts into the regions at present occupied by the temperate rain belts, while the latter moved northward to occupy the polar " cold caps," now characterised by tundras, glaciers, and in Greenland and the Antarctic by " the climate of eternal frost."

REFERENCES

(1) Manson, Marsden. " The physical and geological traces of the cyclone belt across North America." Washington, *Monthly Weather Review*, 52, 1924, p. 102.
(2) Hobbs, W. H. " Characteristics of existing glaciers." New York, 1911.
(3) British Antarctic Expedition, 1910-1913. *Meteorology*, vol. i. Discussion, by G. C. Simpson. Calcutta, 1919.
(4) Brooks, C. E. P., and T. M. Hunt. " The zonal distribution of rainfall over the earth." London, *Mem. R. meteor. Soc.*, 3, no. 28, 1930.

(5) LANG, R. "Verwitterung und Bodenbildung als Einführung in die Bodenkunde." Stuttgart, 1920.

(6) CROWTHER, E. M. "The relationship of climatic and geological factors to the composition of soil clay and the distribution of soil types." London, *Proc. R. Soc.*, B, 107, 1930, p. 1.

(7) UNION OF SOUTH AFRICA. "Report from the Select Committee on Droughts, Rainfall, and Soil Erosion." Cape Town, 1914. "Final report of the Drought Investigation Commission, October 1923." Cape Town, 1923.

(8) KÖPPEN, W. "Die Klimate der Erde." Berlin und Leipzig, 1923.

(9) WARMING, E. "Œcology of plants." Oxford, 1909.

(10) POTONIÉ, H. "Die Entstehung der Steinkohle und der Kaustobiolithe." 5 Auf. Berlin, 1910.

(11) WHITE, D. "Upper Palæozoic climate as indicated by fossil plants." *Sci. Mon.*, New York, 20, 1925, p. 465.

(12) KERNER, F. "Bauxite und Braunkohlen als Wertmesser der Tertiärklimate in Dalmatien." Wien, 1921.

CHAPTER X

MOUNTAIN-BUILDING AND CLIMATE

IT has long been evident that the succession of climates in the various geological periods has not been haphazard, but has followed a certain ordered sequence. The last climatic episode on the grand scale has been Quaternary glaciation. Looking back beyond that, we see a long succession of genial climates in the Tertiary and Mesozoic, until we come to the great Upper Carboniferous glaciation in the late Palæozoic. Beyond that again is another long period of mainly warm climates—Lower Carboniferous, Devonian, Silurian, Ordovician, and Upper Cambrian, bringing us to another great glaciation at the close of the Proterozoic, immediately preceding and perhaps extending into the Cambrian period. Beyond that again, and almost lost in the mists of antiquity, deposits of a still earlier glaciation have been recognised in South Africa, Australia, North America, and perhaps in Scotland.

The systematic nature of these occurrences is made more obvious when we consider their probable absolute ages in years. The Quaternary glaciation is recent, it began perhaps one million years ago. From the evidence of the radio-active rocks (see Appendix I.), it is calculated that the upper part of the Carboniferous system is about 260 million years old. The base of the Cambrian is placed at 500 million years. The exact age of the first of the four great glaciations is not known, but a fair estimate would be 700 to 800 million years. Thus the great glaciations seem to have occurred at almost regular intervals of a quarter of a thousand million years. The last two of the long genial intervals, and for all we know, the first also, have been interrupted by minor deteriorations of climate, as in the Silurian and the Cretaceous to Lower Eocene, which produced local valley and sometimes piedmont glaciers but not regional ice-sheets, suggesting a shorter cycle superposed on the longer one.

The same ordered sequence has been observed in the

evolution of the earth's surface features, where it has been termed " the rhythm of geological time." As was remarked in the Introduction, there have been long periods in which the earth's crust was at rest, while the denuding agencies gradually lowered the surface almost to a uniform plain and the waves of the sea bit deeply into the continents. Alternating with these have been relatively short periods of intense disturbance during which the earth's surface was thrown into great folds and ridges, when the mountain ranges which form the articulated skeletons of the continents were brought into being. The greatest periods of mountain-formation occur in close relation with the greatest periods of glaciation ; thus the Alpine period of folding in the Tertiary preceded the Quaternary Ice-Age ; the Hercynian folding in the Upper

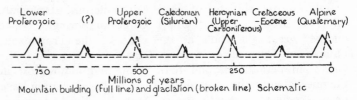

Fig. 23.—Mountain-building and glaciation (schematic).

Carboniferous preceded the Upper Carboniferous glaciation. It is known that there was a period of great disturbance and mountain-building preceding the Cambrian period, and another, lower in the Proterozoic, probably preceded the first of the four great glaciations, the deposits of which, in Australia at least, rest on great outflows of lava. The minor cold period of the Silurian was also associated with a period of folding and mountain-formation, the " Caledonian," which, however, did not reach the intensity of the Hercynian and Alpine foldings. Thus we can represent the variations of mountain-building activity and of climate during geological time as a series of waves (Fig. 23) in which the long troughs represent the periods of stability and genial climate, the sharp crests the periods of mountain-building and climatic stress.

The last two ice-ages at least were not synchronous with the maximum of mountain-formation, but followed them after some millions of years, as has been indicated in the diagram.

This lag has been attributed to two causes. After a long quiescent warm period the whole mass of the oceans is warm, and has to be cooled down before general glaciation can begin. This process may occupy thousands of years, and smooth out climatic fluctuations within an ice age, but could not cause a lag of millions of years. The second cause is that mountain ranges are first elevated as smooth domes, which are worn into irregular contours by the ordinary processes of erosion. The lightening of load caused by the removal of this eroded material causes further isostatic elevation and a greater effective height. A. Wagner (1) has put forward a third explanation. The steady warming of the earth's crust by radio-activity is much greater than the normal escape of earth-heat at the surface so that the crust becomes continually hotter and more plastic. This allows folding, mountain-building and volcanic outbreaks, in which the accumulated earth-heat is liberated. At present earth-heat raises the mean temperature of the earth by only 0·3° F., but Wagner thinks that during the mountain-building of the first half of the Tertiary this figure may have been nearer 10° F., more in high than in low latitudes. This radio-active heat warms the earth's crust, and melts the base of any nascent glaciers. The latter flow rapidly to lower levels where they are dissipated.

After the end of the main epoch of mountain building the crust becomes solid and quiescent and cools again. This allows ice to freeze to the ground and so pile up great ice-sheets. When the ice reaches a certain thickness the vertical temperature gradient in the upper part of the crust increases (at cost of lower layers) and melts the bottom of the ice-sheet, causing it to flow out to the warmer periphery where it melts. This is the phase of maximum extension of the ice-sheets. After the ice-sheets have melted back the earth-flow decreases again and the cycle recommences. In this way Wagner accounts not only for the lag of glaciation behind mountain-building but also for the succession of glacial and inter-glacial periods. The theory is mentioned here for the sake of completeness, but it seems improbable that the surface temperatures can have fluctuated so greatly solely because of earth-heat, especially during the course of the Quaternary Ice-Age. H. Jeffreys (2) states definitely that the surface temperature of the earth must have been almost wholly maintained by solar radiation practically ever since it became solid at the surface, and certainly throughout geological time. Conduction from the interior is by comparison quite unimportant.

The dates of the pre-Cambrian glaciations are still uncertain, but have been provisionally put at 500, 1,000 and probably 1,500 million years ago, with more doubtful occurrences round 600 or 800 million years ago. The duration of a period of folding and mountain-building is about 50 to 80 million years, while that of an ice-age is much shorter, hence the peaks representing mountain building have been shown broader than those representing glaciation.

As G. Manley (3) pointed out, the problems of why the Quaternary glaciation began just when it did, and of why it was interrupted by interglacial periods, is the same, on a larger scale, as the problem of the fluctuations of the glaciers during the historical period. So far as is known, these have been broadly similar over the whole world. Especially during the past 100 years glaciers have been in rapid retreat in both hemispheres alike. The causes of these minor advances and retreats of the glaciers are still unknown.

The close parallelism between mountain-building and climate has naturally attracted the attention of geologists. For example, in 1899 J. Le Conte (4) pointed out that the Quaternary glaciation was preceded by a period of " almost universal continental elevation and enlargement," and that the Quaternary ice-sheets developed on the surfaces of the plateau so formed. The greatest exponent of the importance of elevation, however, has been W. Ramsay (5), who, under the title " Orogenesis and Climate," attempted a complete solution of the climatic problem on these lines.

In the second paper (5), Ramsay summarises his theory of the climatic importance of elevation : " It is well known that a climate is modified by the relief and the elevation of a region, so that with increasing altitude it gradually becomes cooler and even glacial. Further, mountains and highlands exercise a great effect as condensers of precipitation, as boundaries of climate, etc. But meteorologists and geologists reflect less often that the relief of the continents influences, not only the local or regional climate, but the whole economy of the calories which the sun supplies."

We have seen in Chapter VI. that the atmosphere, owing to the selective absorption which it exercises on radiation of different wave-lengths, acts like the glass of a greenhouse in raising the temperature of the earth's surface. Ramsay points out that this action takes place under most favourable conditions over extensive low plains where the air mantle is thickest and most dense. Over the mountains and highlands, where the air is thinner and less dense, the loss of heat by terrestrial radiation escaping to space is greater, "the lofty parts of the continents can be regarded as holes in the glass. They not only chill the place just beneath them, but more or less the whole hotbed."

Loss of heat also takes place owing to the vertical movements which high ground introduces into the atmosphere. "The air currents, passing over mountains, high coasts, and other elevations in their way, are forced to rise, and at the higher position their loss of heat is greater than if they had flowed at a greater level." It may also be remarked, though this consideration is not mentioned by Ramsay, that high ground favours the formation of cloud which, as shown in Chapter VII., lowers the temperature owing to its reflection of solar radiation. But the most important effect of high ground, according to Ramsay, is that it serves as a gathering ground for glaciers. High land is necessary for glaciation, and high land in rather high latitudes is especially favourable for the occurrence of a general cold period. "If there only exist high enough islands and continents, ice-caps will appear, extend their glaciers down to the sea, and send out their armadas of icebergs. To melt them, enormous quantities of the heat reserve of the sea will be consumed. Cold water forms extensive superficial layers, and gradually fills the depth of the ocean right to the equator." As the ice-caps grow, more and more water is bound up in them ; this water is taken from the sea and gradually lowers the sea-level, thus accentuating the elevation of the land. Antevs (6) calculates that during the Quaternary glaciation the average elevation of the land above the sea was increased by more than 300 feet owing to the locking up of water in the form of land ice.

Thus Ramsay has built up a very complete qualitative theory of geological climates on the basis of changes of elevation alone. His theory contains a very considerable amount of truth ; he has realised the importance of vertical motion in the atmosphere for the loss of heat by radiation, and he has also introduced the effect of changes in the general temperature of the ocean. On the other hand, I think he has overestimated the importance of the " holes in the glass," since at the present day only relatively insignificant areas of land are high enough to make an appeciable difference to the proportion of their radiation which is absorbed by the atmosphere. The loss of heat by radiation from air-masses forced to rise in order to pass over high ground is no doubt appreciable, but the amount of air raised in this way must be much less than that raised in cyclones, thunderstorms, convection currents, and other phenomena of atmospheric instability. In fact, as pointed out in Chapter IX., air, unless it is already unstable, much prefers going round a mountain to going over it. One would be inclined to say that the " polar front " in the North Atlantic is responsible for more vertical air movement than the whole of the mountain regions of the globe put together, but there are no figures available for a quantitative estimate. On the other hand, we have seen in Chapter VI. that the loss of radiation by reflection from ice-sheets, at the maximum of the Quaternary Ice-Age, would suffice to cool the whole earth by at least 4° F.

The weakness of Ramsay's theory, as of so many other theories of climatic change, is the lack of a quantitative basis. Let us examine the various effects of elevation which he postulates and attempt to evaluate them numerically. The climatic effects of high ground are, as we have seen, complex. The most obvious effect is the decrease of mean temperature with height, at the rate of about 0·3° F. per 100 feet. This is due to the fact that air which rises from a low level to a higher level expands and cools by expansion. But when the air descends again it is warmed by compression, so that the direct effect of high ground in causing the air to cool by expansion is purely local, and cannot influence the temperature of the sea surface or of the lowland plains. The effects which we have to consider are those which cause a net loss of heat from the earth's surface as a whole. These include :—

   1. Radiation to space from the high ground which is not absorbed by the air.

2. Reflection of solar radiation from the surface of clouds (not postulated by Ramsay).

3. Cooling power of the surfaces of ice and snow.

4. Loss of heat owing to increased evaporation.

Reasons have already been given why the loss of heat under the heading 1 is probably negligible, and this factor will not be considered further.

2. Except in winter in high latitudes, clouds lower the mean temperature by reflecting the rays of the sun back to space. Where high ground forces the air to rise above the level of condensation, clouds will be formed, and the local temperature will be decreased. The level of condensation varies according to the humidity of the air, but 5,000 feet would be a fair estimate of the average level over the land. The air above the sea and the lowlands is not by any means free from cloud, but it is noticeable that there is a marked increase in the cloudiness in the neighbourhood of mountain ranges, and especially on the windward sides, which is not entirely compensated by the somewhat clearer skies on the leeward sides. Let us say that the net effect of land above 5,000 feet is to increase the local cloudiness by three-tenths of the sky. Now we find (7) that about 12½ per cent. of the land, or 3·6 per cent. of the whole surface of the earth, is at present above 5,000 feet in height. The effect of this high ground is therefore to increase the cloudiness, expressed as an average over the whole earth, by one per cent. or 0·1 tenth of the sky. We saw in Chapter VII. that an increase of one-tenth in the mean cloudiness results in a decrease of the mean temperature by 6° F. The effect of the introduction of the present topography into a nearly base-levelled world would therefore be to decrease the mean temperature by about 0·6° F. owing to the increase of cloudiness over the mountain ranges.

During the Quaternary, the average elevation of the land was considerably greater than it is now ; we do not know exactly how much, so we must guess. Let us put the elevation at 3,500 feet instead of the present average of 2,500 feet. Further, let us suppose that the present elevations were all increased in the same proportion, so that instead of there being 12½ per cent. of the land above 5,000 feet there was the same amount above 7,000 feet. From the data given by de Martonne

(7) we then find that the area above 5,000 feet would have been 21 per cent. instead of 12½ per cent.   The resulting increase of cloudiness would have averaged 1·8 per cent. instead of 1 per cent. over the whole earth, giving an average cooling of 1·0° F. compared with a warm period or 0·4° F. compared with the present.

3. The cooling power of the surface of ice and snow is difficult to evaluate.   The mean temperature of the earth's surface reduced to sea-level is at present 59° F., varying from below freezing point in high latitudes to above 80° F. in parts of the tropics.   The " snow-line " accordingly varies from sea-level to above 16,000 feet.   But this low snow-line in high latitudes is really due to the presence of the ice-sheets and floating ice ; the glaciers creeping down the sides of the hills have brought the snow-line down with them.   The cooling power of these ice-surfaces belongs partly to the account of elevation, but partly to other causes, such as increased continentality, shutting out of ocean currents, volcanic action.   What we are seeking for here is the direct effect of elevation alone, apart from the other geographical factors which are connected with elevation.   Let us put it that, starting with a warm period in which there was no ice, a burst of mountain-building rapidly raises the average level of the land to 2,500 feet ; we require to find the area brought above the snow-line.   In order to do this it is necessary to make some assumptions as to the distribution of temperature and land.   We will suppose that during the warm period preceding the elevation the mean temperature over the land areas is 32° F. at the poles and 83° F. at the equator ; further, that from the poles to latitude 45° N. and S. it rises at a uniform rate of 8° F. in each 10° of latitude, while from 45° to the equator the mean temperature is

$$33° \text{ F.} + 50 \cos \phi, \text{ where } \phi \text{ is the latitude.}$$

Now, suppose that the whole area is elevated in such a way that the ratio of land to sea and the percentage areas of land above different heights are the same in all latitudes, having everywhere the present average value for the whole world, the average elevation being 2,500 feet.   Then a rough calculation shows that rather more than three per cent. of the land surface, or about one per cent. of the total surface of the earth, would be brought above the snow-line, more than half this area being between latitude 70° and the poles. The cooling power of ice compared with that of unglaciated land varies according to the latitude ; from the data given in Chapter VIII. it appears that if a land surface in high latitudes which was formerly bare of ice have one per cent. of its area ice-covered, the mean annual temperature will be lowered by about 0·3° F. The ice-covered area is not limited to the area initially raised above the snow-line, since valley and piedmont glaciers can descend to low levels. The development of great ice-sheets probably requires a general cooling of the seas in addition to simple elevation, but we may take the areas of mountain glaciers below the snow-line as equal to the areas above the snow-line. This gives us for the total cooling, averaged over the whole earth, due to the development of snowfields and glaciers as a result of elevation, an amount of 0·6° F.

When the calculation is repeated, supposing the average elevation of the continents to be raised to 3,500 feet, the heights of all parts being increased proportionally, the cooling due to the snow and ice surfaces, when averaged over the whole earth, is rather more than doubled, becoming 1·3° F. If, further, we suppose that the land is concentrated in high latitudes, forming two continents extending from the poles to about latitude 50°, the cooling becomes still greater, exceeding 6° F. when averaged over the whole earth. Since during the Quaternary Ice-Age the elevation was greatest in high latitudes we may assume an intermediate figure, say 3° F.

4. The last term to be discussed is the loss of heat due to increased evaporation. During the great ice-ages the total precipitation was probably more than at present ; during the warm periods it was very much less. The actual figures can only be guessed at, but judging from the widespread aridity of periods like the Triassic, we may not be far out if we suppose that the average precipitation during the warm periods was about half that during the ice-ages. Since the amount of water held in the air as vapour or clouds at any time is equivalent to only a fraction of an inch of rain, the precipitation gives us a measure of the evaporation, so that the evaporation also during the ice-ages may have been twice that during the warm periods. Most

of the evaporation takes place from the surface of the sea. Evaporation cools the evaporating surface, and an increase in the elevation of the land, by increasing the evaporation, lowers the temperature of the sea.

The loss of heat due to evaporation at present averages about 100 gram calories per square centimetre per day over the oceans, or perhaps 90 gram calories over the land and sea together. The loss of this amount of heat would lower the mean temperature by about 15° F., so that the mean temperature during the ice-ages may have been lowered by as much as 8° F. compared with that of the warm periods from this cause alone. But we cannot attribute all this amount directly to elevation. As explained in the preceding chapter, the greater part of the world's precipitation is probably due to the "polar fronts." During the warm periods the polar fronts were not developed, and the greater part of the rain fell in instability showers of a thundery nature. Such showers, though they are sometimes very heavy, are extremely local, and do not deliver such a large total of rain as the extensive rain areas associated with cyclonic depressions. Much of the difference between the total precipitation of the warm periods and that of the ice-ages was probably due to this cause, and not directly to the differences of elevation. Let us put down half the difference to the account of the polar fronts and half to the elevation. Then we have an elevation of 3,500 feet, resulting in a cooling of 4° F. due to increased evaporation.

We can now sum up our three guesses. Comparing an ice-age with a warm period, we have :—

Cooling due to increased cloudiness . . . . 1° F.
Cooling due to area above the snow-line . . 3° F.
Cooling due to increased evaporation . . . 4° F.

The total cooling due to the elevation alone is about 8° F., and from the difficulties of the calculation may be anything from 5° F. to 10° F., but is not likely to be outside these limits. Evidently, elevation is by itself and directly an important factor in climatic changes, but it cannot account entirely for the changes of some 30° F. in the mean temperature of the Arctic lands at sea-level, or of the surface of the Arctic Ocean.

The truth is that mountain-building is so intimately

associated with changes in the land and sea distribution that it cannot really be considered alone. When the continents are thrown into folds and ridges, so too are the ocean floors, at least near the continents. If the continents are more elevated, the oceans are deeper. The waters retire into these submarine deeps and expose large areas of the continental shelves as new land ; this at once increases the continentality and limits the ocean currents. At the same time volcanoes break out in the mountain ranges, and by ejecting dust into the atmosphere add to the general cooling. The polar oceans, cooled below their freezing point, develop floating ice-caps, while ice-sheets spread from the mountains over the surfaces of the continents. The temperature difference between low and high latitudes passes the critical point for the atmospheric circulation, and the polar front appears. The oceanic circulation is radically changed. The whole process is in fact essentially cumulative, and the final lowering of temperature is out of all proportion to the small initial causes. That is why the whole problem of climatic changes is so baffling.

The time has therefore come to gather up the scattered threads which we have been patiently disentangling in the preceding chapters, and to attempt to weave them into a tapestry which will give us a connected picture of the mechanism of climatic changes. As ultimate causes, we have a choice between variations of solar radiation, sunspots, astronomical changes, ocean currents, continentality, mountain-building, volcanic dust, and carbon dioxide, though the possibilities of the last-named are limited. As auxiliary causes, we have variations of water vapour, of cloudiness, of the wind circulation, and, most important of all, of floating ice—surely enough, between them, to give us sufficient material for our purpose.

Let us start with the conditions of the present day and try to forecast the meteorological changes during the next quarter of a thousand million years, which, I believe, is the longest-range forecast ever attempted. We are still living in an ice-age, though not at its maximum ; whether, before the forces of mountain-building which began their activity in the late Cretaceous have worn themselves out, there will be a return of the ice-sheets in Europe and North America, we cannot say, so we will skip a million years or so and begin our forecast with the period when the subterranean fires are banked and the earth begins again to settle down. The Arctic Ocean is still filled with ice, and the summits of the high mountains are above the snow-line. We have our regular succession of storms born on the edge of the polar front and travelling eastward into Europe—altogether, the weather is pretty much as we know it to-day (unless by that time the problem of the artificial control of weather has been solved—who can foretell the triumphs of science ?).

When the mountain-building forces cease to repair the ravages of denudation, the average level of the land begins to fall steadily. The process is most rapid in the high mountains, where frost is still active, but everywhere the rivers are carrying sediment into the oceans and the waves are grinding into the land. The first noticeable change is the gradual disappearance of the mountain glaciers as their gathering grounds are worn down beneath the snow-line ; the ice-sheets of Greenland and Antarctica still persist, but are beginning to wane. Bering Strait has been widened and deepened a little ; a small warm current penetrates through it into the Arctic Ocean for longer and longer periods in each summer, and finally succeeds in keeping clear of ice a " bridge-head " on the northern side, and in maintaining its flow throughout the year. The Atlantic gap is a little wider— Iceland is smaller and the southern end of Greenland has retreated a short distance northward, while at the same time the gap between Newfoundland and Labrador has widened, allowing much of the ice and cold water of the Labrador Current to flow directly down the coast of America, instead of mingling with the waters of the Gulf Stream off the Grand Banks. Thus the water of the Gulf Stream Drift is a little warmer when it reaches the Arctic Ocean, and Spitsbergen is now ice-free throughout the year. Attacked on two sides, the Arctic ice-cap melts farther and farther back each summer, and the Palæocrystic ice begins to disintegrate. Finally, there comes a succession of years in which the sun's radiation is unusually powerful—the solar constant remains steadily at 2·0 calories per square centimetre per minute—and one summer the Arctic ice-cap breaks up completely and disappears. That summer the polar east winds are greatly weakened and the " polar front " hardly develops ; the Azores anticyclone extends persistently across the British Isles and Western Europe almost to the Urals ; hardly any cyclonic depressions occur south of the Arctic Circle, and the summer is almost rainless. The cold East Greenland Current carries a little ice in spring, but this fails to reach Cape Farewell, and the west coast of Greenland is bathed by a warm branch of the Gulf Stream instead of by cold Arctic water. The Greenland ice-sheet makes a record retreat, but the level of the sea is rising still more rapidly, and the ice-edge still reaches the sea in many places.

The Arctic Ocean remains open far into the following winter, and when it finally freezes, the ice is thin and easily scattered ; it begins to break up quite early in spring, so that by the middle of summer it has completely disappeared. The " semi-glacial " condition has been reached in the Northern Hemisphere. In the following years the ocean becomes steadily warmer ; the Arctic ice-cap, and with it the polar east winds, the " polar front," and the Atlantic cyclones develop later and later each winter and break up and disappear earlier and earlier each spring, until finally there comes a winter in which the ice-cap does not form at all, and even in the middle of winter the Atlantic Ocean is almost free of storms. The air in high latitudes is not cold enough for its greater density to counterbalance the relatively warm and light polar stratosphere, and west winds prevail everywhere outside the tropics, with, at the surface, a component towards the poles. The late summers in the British Isles and Western Europe are now intensely hot and dry ; day after day the skies are almost clear of cloud, while even the slight hindrance to the sun's rays offered by volcanic dust is absent, and a powerful sun warms the surface of the land and of the seas. The temperature becomes very high, exceeding 100° F. on several days at midsummer ; owing to the high temperature the air contains a large amount of water vapour which absorbs the terrestrial radiation. Part of this radiation is given back to the earth at night, and in spite of the clear skies the nights are not cool. (In the dry summer of 1921, which made some approach to these conditions, the mean daily minimum temperature at Kew Observatory in July was nearly 5° F. higher than the mean daily minimum during the much cloudier month of July 1924.) There are a number of thunderstorms in spring, but by the end of June the higher layers of the troposphere have become warmed up to such an extent by this absorption of terrestrial radiation that the air is stable, and in the late summer and early autumn there are not even thunderstorm rains to break the drought. The general warming spreads even to the equator, though the rise of temperature is less there than in other parts of the world, and the temperature difference between low and high latitudes is greatly diminished.

The Greenland ice-sheet is now retreating very rapidly, and soon breaks up into a number of isolated masses of dead ice in the valleys. The warming up of the oceans breaks up the ring of pack ice surrounding the Antarctic ice-mass, and that, too, has withdrawn within the limits of the continent and is now in full retreat. All this ice-water added to the sea raises its level still further, and helps the influence of the warm oceans to penetrate into the land. As the mountains are worn down they offer less hindrance to the winds, and there is less forced ascent of air and less orographic cloud and rain. Cyclones are now fewer (though destructive hurricanes still occur in the tropics and sometimes travel into temperate latitudes), and altogether the general rainfall is decreasing. Since vertical motion is the most effective agent in causing the air to condense its moisture, and the driest air is that which has parted with its water vapour by being raised, and then has again descended to the surface of the earth, this general decrease of vertical motion gives rise to a general increase in the relative humidity, so that evaporation decreases, and the surface of the oceans does not lose so much heat in this way. The wind velocity also is generally smaller, and this again decreases the evaporation. When all the ice has finally disappeared, the surface of the oceans is everywhere warm, and owing to the prevailing poleward component of the surface winds, this mass of warm water is almost everywhere moving from lower to higher latitudes, carrying genial conditions into the neighbourhood of the poles. At first the depths are still filled with cold water—a relic of the ice-age— but this cold mass receives no fresh supplies, and is gradually warmed up by earth-heat and by conduction from the surface. The earth has now entered on the " non-glacial " stage, and conditions will remain sensibly unaltered for millions of years, until a fresh manifestation of the internal forces of the earth raises new mountain ranges to begin the cycle afresh. The climatic conditions at the height of a " non-glacial " or warm period are of such great interest that a fuller description of them is reserved for the next chapter.

For a fuller description of the cycle of compression, mountain-building and erosion, the reader is referred to a book by J. H. F. Umbgrove (8).

REFERENCES

(1) WAGNER, A. " Klimaänderungen und Klimaschwankungen." Braun-schweig, *Die Wissensch.*, Bd. 92, 1940.
(2) JEFFREYS, H. " The earth, its origin, history and physical constitution." 2nd ed., Cambridge, 1929.
(3) MANLEY, G. " Glaciers and climatic change, some recent contributions." London, *Q. J. R. Meteor. Soc.*, 72, 1946, p. 251.
(4) LE CONTE, J. " The Ozarkian and its significance." *J. Geol.*, Chicago, 7, 1890, p. 525.
(5) RAMSAY, W. " Orogenesis und Klima." *Ofversigt af Finska Vetenskaps Soc. Forh.*, 52, 1910.
——. " The probable solution of the climate problem in geology." *Geol. Mag.*, London, 61, 1924, p. 152, and Washington, *Smithsonian Rep.*, 1924, p. 237.
(6) ANTEVS, E. " The last glaciation." New York, *Amer. Geogr. Soc., Research Ser.*, no. 17, 1928.
(7) MARTONNE, E. DE. " Traité de géographie physique." 2nd ed., Paris, 1913.
(8) UMBGROVE, J. H. F. " The pulse of the earth." 2nd ed., The Hague, 1947.

CHAPTER XI

The Weather of the Warm Periods

WE have seen that during long periods of geological time the earth seems to have been free from ice, so that even if we accept the theory of Continental Drift (set out in Chapter XIII.) it is still necessary to believe that the climatic zones were much less marked than at present. These periods include a large part of the Palæozoic, almost all the Mesozoic, and about half of the Tertiary. Remembering that they were also periods of low relief and very slow denudation, we must suppose that the time intervals represented by the rocks of these periods were very much greater than the time intervals represented by the rocks of the more active periods, and that this type of climate has, in fact, prevailed during the greater part of geological time. Its most striking feature is the appearance of vegetation of sub-tropical or warm temperate aspect in very high latitudes. These periods have been termed " pliothermal," though I prefer the simpler word " warm," which means the same.

In the preceding chapters the various meteorological features of the warm periods have been deduced, and it is only necessary here to collect these results and so present a more or less connected picture of their climate and weather.

Let us imagine that we are voyagers in one of these favoured periods of antiquity, sailing northward from the equator on a voyage of discovery towards the pole. We have set out from a low marshy western shore, through which a broad but shallow river moves sluggishly towards the sea, winding in endless curves over a vast plain which stretches as far as the eye can reach. Under the influence of a light easterly breeze we sail slowly towards the north-west. The sky is half covered by woolly cumulus clouds, which now and again thicken and darken to a passing shower, with perhaps a burst or two of thunder and a slight squall. The air is warm and moist, but not unpleasantly so, though we are conscious of a feeling of lassitude which makes us disinclined for effort, either physical or mental. As we pass northward, the barometer rises, the wind backs to north-east, the sky becomes ever clearer, and the air more bracing. Detached cloudlets are now the rule. Showers still fall at long intervals, but mostly in the early hours of the night. On one or two islands that we pass we find evidence in the fallen trees that occasional hurricanes occur, but they are very rare, and we are not greatly troubled by the chances of meeting one of them. The sea is very warm and blue, and is teeming with life ; the nights are almost as warm as the days. Still farther north, and we enter a region of light variable winds and calms. At times we are becalmed for several days, while overhead the sun blazes through a cloudless sky. Then there is a breath of wind from the north-east or from the south-east, and a few clouds gather, with perhaps a shower of rain. We are now in the western half of the ocean, and all the time we are drifting slowly northward on a great ocean current. The barometer falls very gradually, the southerly winds become more frequent and stronger, they draw round to south-west, and, finally, with all sails set, we bear away to the north-east. The weather, however, does not differ appreciably from that experienced farther south ; it is still very warm and for the most part sunny, with a few scattered clouds. The only difference is that now and again, perhaps once a month, the barometer drops a few tenths of an inch, the southerly wind freshens, and a uniform but not heavy cloud canopy covers the sky, while for a few hours it rains almost steadily. Then the wind veers to west or north-west, the sky clears, and after a few showers the steady fine weather sets in again. It is noticeable, however, that even when the wind blows straight out of the north it is never cold, although the season is only late winter.

For some time now we have been sailing northward, with a low palm-fringed coast in view to the eastward. Beyond the belt of palms, which marks the limit of the occasional thunder rains brought by the daily sea breezes, we catch glimpses of a series of low rounded sandy hills, which seem to be almost devoid of vegetation. Here and there a line of reeds marks the course of a stream bed. Mostly, they are dry and withered, and the channel opens out on to a glittering level of white, where a layer of salt or gypsum encrusts the dried-up floor of a temporary pool, but occasionally the reeds are fresh and green, marking either a more permanent river, or the channel taken by the waters of a recent storm. Only once, however, do we see it actually raining over the land. On that occasion, the low rounded hills which everywhere form the background of the landscape, and which usually stand out clearly against a sky of intense blue, seem to support a mighty column of cloud, in which we can see the play of lightning flashes and the deluging rain. We know that some of the dry channels will soon be rushing torrents of water, and that the apparently lifeless hollows of the plain will wake to swarming life. One of the fishes—the *Ceratodus* or lung-fish—is specially adapted to the chances of this life, since it can breathe air or water at will, but how the other animals survive the periods of desiccation is a mystery.

As we pass northward, the desert character of the land becomes less marked, and in about the latitude of London we decide to land and carry out some explorations on foot. It is now early in March, but the weather has almost the character of a fine English June. The vegetation is very fresh and green. As we go inland across Europe the air becomes cooler and crisp though not cold. Now and again, after a few hotter days, there follows a heavy thunderstorm, and this is the only rain that falls at this season. Later in the year all this expanse of country will be burnt brown in the steady heat of a rainless summer ; then as autumn draws on, the sun will lose some of its power, and the year will draw to its close in a spell of perfect weather, such as even now we occasionally meet with in September. Just about midwinter perhaps one or two mild storms from the ocean to the westward will pass across the land.

Here we must interrupt our voyage for a moment to explain that the narrative would differ slightly according to whether we were in the Mesozoic or in the Early Tertiary world. A vivid picture of Jurassic Oxfordshire has been drawn by W. J. Arkell (1). In a clear shallow sea grew a complex of true coral reefs intersected by narrow channels. The reefs, which are nowhere very thick, are interspersed with beds of limestone resulting from their erosion, and it is inferred that the corals grew on a rising sea floor. The corals are obviously found where they grew, and it is very improbable that the sea temperature was less than 60° F., more than 5° F. warmer than the present temperature of the Atlantic west of Ireland, which itself has the highest sea temperature for its latitude anywhere in the world. In the Early Mesozoic even frost seems to have been unknown, and in Europe the rainfall was not heavy enough to balance the evaporation. The prevailing continental deposit was a red desert sandstone, and in our own country we have in Leicestershire a true " fossil desert," where wind-worn pillars of granite stand up into the Triassic sands which ultimately buried all but their highest summits. A very good picture of the conditions in the type of country which prevailed over a large part of the Mesozoic world is conjured up by the finds of dinosaur eggs in Mongolia. The term " desert " is probably not strictly applicable, for there must have been enough vegetation to support these great reptiles. Probably, however, the vegetation was limited to the larger water-courses which received the drainage from a considerable area, and the intervening country was a sandy waste, in which the dinosaurs buried their eggs to be hatched by the heat of the sun.

Still earlier, in the Permian, a large part of Europe had been occupied by a great salt inland sea with desert shores, a European Caspian. This lake was saturated with salt, and a striking witness to the stability of the climate is the fact that year by year for ten thousand years the excess of salt was deposited in regular layers. In summer the deposition was mainly in the form of gypsum, but this mineral is much more soluble in cold than in warm water, and as the sea cooled in winter the deposit of gypsum ceased and rock salt took its place. From the evidence of similar evaporating solutions in the Sahara, Kubierschky, quoted by Köppen and Wegener (Chapter XIII.), estimates that the temperature of the lake varied from about 60° F. in winter to as much as 95° F. in summer. In addition to the main mass of salts there are some isolated layers, apparently formed when adjacent hollows were flooded at high lake stages, the water afterwards evaporating and the deposited salts being covered by desert sands.

In the Early Tertiary, after the brief cold spell which ushered in the Eocene had passed, there was another long period of warm climate, though not nearly so hot and dry as the Early Mesozoic. In fact, although summer was probably dry, the spring appears to have been decidedly wet in Europe north of the Mediterranean, and beds of brown coal were formed in many localities. The rainfall was still mainly of the instability type, however, falling in violent thunderstorms accompanied by heavy rain or hail, which stripped leaves and twigs from the trees and washed them into the lakes. Many of the trees were evergreens, and the deciduous trees were in leaf by the end of March, as shown by the relation of the leafing to the flowering period. Early in the Miocene, however, the leaves of some of the beech trees show signs of the action of frost. H. von Ihering (2) estimates the mean temperature of central Europe as about 68° F. in the earliest Eocene, 74° F. in the main part of the Eocene and Oligocene, 72° F. in the Miocene and 60-55° F. in the Pliocene.

In western U.S.A., according to various papers summarised by R. W. Chaney (unpublished), the climate in the lower part of the Upper Eocene was warm and moist, similar to that now found on the borders of the tropics. The mean annual temperature was about 68° F. and the annual rainfall 70 inches, rather uniformly distributed through the year. The high rainfall was probably due to the neighbourhood of the growing Sierra Nevada. These conditions persisted through the Lower Oligocene, but in the Upper Oligocene a progressive slow cooling and desiccation set in.

We must now resume our interrupted voyage towards the pole, but this time we will suppose definitely that we are in the Upper Eocene period. As we pass towards the Arctic Circle, we are still in a great northward-flowing warm current, and the vegetation along the shores continues to be very rich, but its character gradually becomes more temperate. The skies become cloudier, and steady cyclonic rain replaces more and more the thunder rains of Central Europe. By the time we have passed the latitude of 70° N., we find that misty rainy weather forms the rule and fine sunny weather the exception, while the dense forests along the eastern shore are frequently hidden by fog. The wind is now mainly from the west, and the western shores of the ocean, and still more the country some distance inland, begin to present a bleak appearance. In winter, the low hills are probably snow covered, but there are no glaciers. So we come to the pole, in a great open basin filled with warm sea water from the south, which circulates slowly round and round until it cools and sinks to the bottom.   There is no great mass of Palæocrystic ice such as we find to-day, no icebergs even, and although the great rivers sometimes bring down a few fragments of drift ice in the winter, these soon melt. If we had made this journey in the Jurassic period we should have seen no ice at all ; instead, we should have found coral reefs almost to the Arctic Circle and isolated corals even farther north.

This is perhaps a somewhat fanciful picture of conditions during the great warm periods, but it is based almost entirely on geological evidence, or on logical meteorological deductions from the slight differences of temperature which we know to have prevailed between different latitudes during those periods.

REFERENCES

(1) ARKELL, W. J.   " On the nature, origin and climatic significance of the coral reefs in the vicinity of Oxford."   London, Q. J. Geol. Soc., 41, 1935, p. 77.

(2) IHERING, H. v.   " Das Klima der Tertiärzeit."   Zs. Geophys., Leipzig, 3, 1927, p. 365.

PART II

GEOLOGICAL CLIMATES AND THEIR CAUSES

CHAPTER XII

THE GEOGRAPHY OF THE PAST

WHATEVER view we take of the major cause of the great climatic fluctuations of geological time, there can be no doubt that the geographical conditions have always played an important part in at least the local distribution of climate. By geographical conditions, we imply not merely the bare distribution of land and sea, but also other variables, such as the height of the land, the presence of important mountain chains, the vegetative covering, the movements of the sea in ocean currents, and the existence of volcanoes. Consequently, before we can pass to a discussion of the meteorology of the different geological ages, we must consider to what extent these geographical factors have varied in the past.

Palæogeography, or the reconstruction of former geographical conditions, is a very difficult science. If we find a marine deposit, we know that that particular region must have been at sea at the time ; similarly, the presence of a deposit obviously laid down on land or in fresh water indicates the presence of land, but unequivocal evidence of this kind is the exception, especially for the earlier periods, the deposits of which have largely been destroyed during the course of geological history, or buried so deeply as to be inaccessible. We are almost completely ignorant of the sequence of deposits on the floors of the oceans. Hence most of the reconstruction must be based on inference from a variety of facts. A break in a series of marine deposits, otherwise undisturbed, indicates that for part of the time, at least, the region was above the sea. The gradual change of a bed of marine deposits from fine clays to coarse sands as we follow them from one region to another, points to the existence of land not far beyond the latter. The discovery of the same marine fauna in two different localities indicates a sea connexion between them, while the presence of different marine faunas of the same age suggests a land barrier. Similarly, the presence of similar land faunas or floras in distant regions points to a land connexion, while the presence of different land faunas or floras close together points to a barrier which may be a sea channel, but may equally be a range of mountains or a desert. It is by the gradual collection of such diverse facts as these that the science of palæogeography has grown up.

The earlier geographical reconstructions presupposed that the various deposits were laid down practically where they are found to-day. A few minor exceptions were recognised ; for instance, the intense crumpling of the rocks in the Alps and Himalayas shows that the lands on either side of these chains were formerly at a greater distance from each other, and their approach has forced the intervening rocks to lie over one another in great heaps, but these shiftings were matters of only a few hundred miles. Some geophysicists, notably A. Wegener, have not accepted this limitation, but consider that continents have drifted like floating islands over the face of the earth, and that the positions of the poles have changed greatly during geological time. Discussion of this theory is postponed to the next chapter ; here it will be assumed that the various deposits were formed where they are now found, and that the positions of the poles relatively to the continents have remained practically unchanged throughout geological time.

We saw in Chapter VIII. that the mean temperature of any latitude at the present day depends to a considerable extent on the percentage of the area along a belt 20° wide centred on that latitude, which is occupied by land. Near the equator the effect of a large land-mass is to raise the temperature slightly, but in high latitudes land lowers the mean temperature, and especially the winter temperature, much more effectively. For reasons to be given later, it is necessary to limit the discussion to the mean temperature of the regions between the poles and latitude 40°. Hence the area of land north of 40° N. and south of 40° S. is an important climatic factor. Owing to the large area of the oceans south of 40° S., we are almost entirely ignorant of the land and sea distribution in that part of the world during geological times, so that this variable becomes in effect the area of land north of 40° N. This area has been measured from the composite charts given by Th. Arldt (1). Numerous reconstructions of land and sea distribution in different periods have been published by various authors, which naturally differ widely, and Arldt has combined these, indicating the areas shown as land by all authors and also those shown as land by some authors and as sea by others. On examining the measurements, it was found that in the different geological periods the areas shown as land by all authors were on the average about the same as the present area of land. Since we have no reason to suppose that during the Miocene period, for instance, the land area in the northern temperate and polar regions was much greater than at present, the conservative measure given by the areas shown as land by all authors seems the best measure to adopt.

The effect of land areas on temperature increases very rapidly as we approach the poles. This was allowed for by weighting the land areas in different latitudes according to a scale derived from the investigation of " Continentality and Temperature " (Chapter VIII.). Giving a unit area of land in latitude 60° a weight of 1, the scale adopted was

$$80°, 3 ; 70°, 2 ; 60°, 1 ; 50°, 0·6 ; 40°, 0·2.$$

The figure which would be obtained if the whole hemisphere north of 40° N. were occupied by land was given the value 100, so that the figures under (1) of Table 14 are percentages of a full land covering.

The second variable is the average height of the land. There is a consensus of opinion among geologists that this has varied extensively during geological history. Thus T. C. Chamberlin (2) writes :—

" It is generally agreed that the present altitude of the continents is greater than their mean elevation during geologic history. Geologists recognise at least two stages in which the continents were exceptionally high and broad ; that which attended the transition from the Palæozoic to the Mesozoic Era, and that which attended the transition from the Tertiary to the present epoch. The existing stage thus falls in one of the most notable stages when continental elevation and breadth were greatest, though perhaps not at its climax. The latest estimate of the present mean elevation of the land gives 2,500 feet. The mean elevation of the great

Period.	(1) Continentality. Per cent.	(2) Elevation (unit 100 feet.)	(3) Ocean Currents. Per cent.	(4) Volcanic Action (0-10).	(5) Mean Temperature, 40°-90° N.	
					"Observed."	"Calculated."
Upper Proterozoic . .	52	40	(45)	9	29° F.	40° F.
Eocambrian	43	26	53	4	33	53
Late Cambrian . . .	21	5	61	2	43	69
Ordovician	15	8	58	2	50	67
Silurian	14	31	65	2	51	60
Lower Devonian . .	36	33	49	5	42	48
Middle Devonian . .	28	14	52	5	50	59
Upper Devonian . .	30	8	51	5	53	61
Lower Carboniferous .	49	17	39	3	53	50
Middle Carboniferous .	49	35	41	3	42	42
Upper Carboniferous .	40	50	42	6	32	37
Lower Permian . . .	52	38	31	6	32	36
Middle Permian . . .	51	17	26	10	42	42
Upper Permian . . .	52	8	46	6	45	56
Lower Trias	45	7	46	5	48	57
Middle Trias	49	5	27	5	51	50
Upper Trias	40	5	34	5	53	54
Rhaetic	47	7	28	3	53	50
Lias	69	8	11	3	48	39
Middle Jurassic . . .	60	10	39	1	53	52
Upper Jurassic . . .	36	26	43	1	51	50
Lower Cretaceous . .	51	21	46	3	44	51
Middle Cretaceous . .	45	29	38	5	41	44
Upper Cretaceous . .	51	29	29	6	41	39
Lower Eocene . . .	47	26	46	8	45	48
Upper Eocene . . .	46	31	42	4	47	45
Oligocene	53	28	46	3	45	47
Miocene	53	23	43	3	46	48
Pliocene	60	36	24	6	37	33
Pleistocene	62	35	6	3	28	26
Recent	50	24	17	2	33	37

Table 14.—Values of geographical elements and of mean temperature.

peneplains is a matter of judgment rather than of knowledge, but no one would probably put the elevation at much more than a third of this. Probably a third is too high."

Similarly, H. Jeffreys (3) in discussing the discrepancies between the age of the earth calculated from the rate of accumulation of sedimentary rocks or of salt in the ocean and the much greater age calculated from the data of radio-activity, points out that the former " amount to a proof that the present rate of denudation is several times greater than the average of the past."

It has been pointed out in Chapter X. that the earth's surface has passed through a series of cycles, each cycle consisting of a relatively short stage of intense mountain-building, in which the rocks were thrown into great folds and ridges and the average elevation of the land above the sea became very great, followed by a long stage of quiet conditions, in which the forces of denudation lowered the level of the land, rapidly at first, and then more and more slowly. We are at present living shortly after one of the periods of mountain-building and elevation, and the average level of the land is consequently high, though not so high as it was during the Quaternary period. At the close of one of the long quiet periods the average level must have been very much less than it is now, and was probably only a few hundred feet. At such times, conglomerates and coarse marine sandstones are almost entirely absent, and the bulk of the sedimentary rocks is composed of limestones and very fine clays or shales.

Since we cannot hope to know the average height of the land in the different geological periods directly, we have to take as a measure the amount of disturbance of the rocks caused by mountain-building during that period. The column under (2) in Table 14 and the curve of height in Fig. 24 are based on a diagram given by E. Dacqué (4, p. 449). Dacqué places the chief periods of mountain-building in the Algonkian or Late Proterozoic, the Late Carboniferous, and the close of the Tertiary, with minor periods at the close of the Silurian, in the Cretaceous, and in the Early Tertiary. The longest orogenetically quiet period appears to have fallen in the interval from Middle Permian to Middle Jurassic. There are, however, two points in which Dacqué's curve seems to require modification. Arldt points out (1, p. 711) that the " alpine " character of a mountain chain—a large number of separate peaks of approximately the same height—occurs only as a result of glaciation ; before glaciation, the system, though it may be of considerable height, possesses only rounded contours of the foot-hill type. The conversion of the rounded contours into the broken alpine type, and the

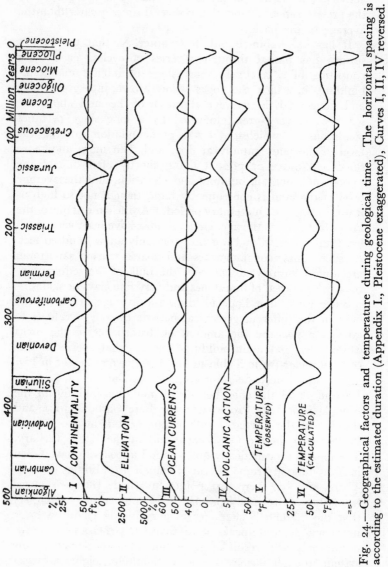

Fig. 24.—Geographical factors and temperature during geological time. The horizontal spacing is according to the estimated duration (Appendix I., Pleistocene exaggerated), Curves I, II, IV reversed.

removal of the resulting detritus, represents a great decrease in the load on the underlying plastic layer of the earth, and consequently leads to a further elevation. Hence it seems probable that the greatest height of the mountains occurred after the main periods of mountain-building. Of course this does not mean an increase in the average height of the earth's surface, but for the meteorological processes involving forced ascent of air, a broken " alpine " surface is as effective as a rounded surface of much greater area. The second point is concerned with the depression of the sea-level during glacial periods owing to the accumulation of water in the great ice-sheets, which may add several hundred feet to the effective height of the land above sea level. For these reasons, and also because Dacqué's curve represents mountain-building activity, while what we require is the effective average height of the land, which naturally lags somewhat behind the process of uplift, I have modified Dacqué's curve slightly, especially in the Quaternary period. The height of this modified curve at the end of the Quaternary was 19 (the unit being one-thirtieth of an inch on Dacqué's diagram), and this figure represents the present height of 2,500 feet. The lowest value on the scale, 1, was considered to represent a mean elevation of 500 feet, and the average height of the land during the Quaternary, 44 on the scale, was taken as 3,500 feet. Through these three points (1, 500; 19, 2,500; 44, 3,500) a smooth curve was drawn, and this curve was employed to convert the scale elevations into estimated mean heights in hundreds of feet. Of course these figures have no pretension to any great degree of accuracy ; the conversion was undertaken merely because Dacqué's curve of mountain-building, if taken directly as a curve of height, seemed to exaggerate the heights of the disturbed periods relatively to the present far too much.

The third geographical variable which we have to consider is the oceanic circulation. At present the Arctic basin is nearly surrounded by a land-ring, which is effectively broken only by the Atlantic gap, the Bering Strait being too narrow and shallow to admit an appreciable current. Even the whole of the Atlantic gap is not occupied by the warm current, its western side being occupied by the cold ice-bearing East Greenland Current, but this condition is probably only present when the Arctic Ocean is ice covered. During the warm periods, the surface cold currents were probably very limited in area ; the gaps in the circum-polar land-rings were probably occupied almost entirely by warm currents, directed towards the poles, strongest near the eastern sides of the gaps and diminishing in strength towards the west. Near the western shores of the oceans there might be local cold currents of slight intensity issuing from rivers or narrow channels between islands, which would be able to maintain their identity for a time owing to their slight salinity. In order to obtain comparable measures of the amount of heat carried to the north polar lands during the different geological periods, I measured the width at 60° N. of those gaps in the circum-polar land-mass which were directly connected with tropical or sub-tropical seas. For each gap double weight was assigned to the first 20° of breadth, single weight to the second 20°, and half weight to breadths beyond 40°. The results are shown in Table 14 and in the curve labelled " ocean currents " in Fig. 24. The figures are expressed in percentages of the oceanic effect which would be produced if the only land consisted of five long narrow islands directed from south to north.

After the publication of the first edition a series of experimental reconstructions of ocean currents were described by P. Lasareff (see p. 78). These bear out the general lines of the estimated effect of ocean currents given in Table 14.

The fourth factor which has been considered as " geographical " is the amount of volcanic action. Numerous climatic rôles have been assigned to volcanoes by different investigators, some of which are favourable and others unfavourable to high temperatures. The cooling effect of volcanic dust postulated by W. J. Humphreys (Chapter VI.) seems to be the best founded of these rôles. As Humphreys points out, the amount of volcanic dust discharged into the upper air depends on the explosive eruptions and not on the total amount of volcanic action, but it does not seem possible to obtain a curve of explosive volcanic activity only. Consequently, we have to be content with a measure of the total volcanic activity in each period, based on the thickness of volcanic rocks, especially lavas. For this purpose the admirable summary of volcanic activity given by Arldt (1) was converted into figures on a comparative scale of 0-10.

Finally, values were assigned for the mean temperature, based on a curve given by Dacqué to show the zonal differentiation of climate. While it is not difficult to accept Dacqué's curve as having some value for the Mesozoic and Tertiary periods, we get into difficulties as soon as we go back to the Palæozoic. The Upper Carboniferous especially, with its apparently tropical forests in temperate latitudes accompanied by an enormous glaciation near the present equator, is a meteorological paradox. For the purposes of comparison, figures were assigned, but they are very doubtful. In considering Dacqué's curve, it was assumed that the mean temperature of the equatorial regions (apart from the Permo-Carboniferous period) had remained constant, and that the curve, therefore, gave the variations of temperature in the temperate and polar latitudes in the Northern Hemisphere ; to be precise, in the area between 40° N. latitude and the North Pole. This cuts out the areas which were most heavily glaciated during the Upper Carboniferous, and the temperatures to be assigned to that period were therefore not so low as those of the Upper Proterozoic and Quaternary periods, although the total amount of land ice present during the Upper Carboniferous was probably greater than the amount at any other stage of the earth's history. On the other hand, the Upper Carboniferous presents evidence of a considerable amount of glaciation even in North temperate latitudes, a point which is discussed further in Chapter XV. Moreover, the faunal changes at this time, and especially the great extinction of corals, indicate a great lowering of the temperature of the sea. It was therefore assumed that there was a considerable fall of temperature during the Upper Carboniferous even in North temperate and polar latitudes, though not so much as in the Upper Proterozoic or the Quaternary.

In order to obtain numerical measures it was necessary to find some means of calibrating Dacqué's curve. The mean temperature of the area between 40° N. and the North Pole is at present 33° F. For the Middle Jurassic, the January temperature of the north polar basin calculated in Chapter VIII. was 44·5° F. and the annual range was 13·5° F., giving a mean annual temperature of 51° F.

The variation with latitude over the oceans was small ; on the other hand, the winter temperatures over the interior of the continents must have been several degrees lower than those near the oceans. As a rough approximation, a value 20° F. above the present, or 53° F., was accepted as the mean temperature of the whole region in Middle Jurassic times. This value seems reasonable from a consideration of the biological evidence.

The mean temperature in the Pleistocene period was taken as 5° F. below the present mean. The maximum decrease in the mean annual temperature calculated from the lowering of the snow-line was more than 20° F. in Scandinavia, 20° F. in East Anglia, 11° F. in the Alps, and 7° F. in Japan ; on the other hand, the Pacific Ocean was little affected, and it is not improbable that over the interior of Asia the winter temperatures were higher than now. When the interglacial periods are taken into account, a mean decrease of 5° F. over the whole area north of 40° N. seems to be a reasonable estimate. It happened that the differences of 5° F. between the Pleistocene and the Present, and 20° F. between the Present and the Middle Jurassic, were actually proportional to the differences measured on Dacqué's curve, and while this is probably nothing more than a coincidence, it greatly facilitated the conversion of the scale of this curve into a temperature scale.

The next step was to determine how far the various elements —continentality, elevation, ocean currents, and volcanic activity—were responsible for the mean temperature. For this purpose the figures in the column of Table 14 headed " Mean Temperature, 40°-90° N., Observed," were " correlated " with the figures in the columns headed " Continentality," " Elevation," " Ocean Currents," and " Volcanic Action." The figures were divided into two groups, the doubtful figures for the Upper Proterozoic and Palæozoic being separated from the much more reliable figures for the Mesozoic, Tertiary and Recent. The correlation coefficients are given in Table 15.

| | Temperature with | | | |
	Continentality.	Elevation.	Ocean Currents.	Volcanoes.
Palæozoic	— ·52	— ·66	+ ·37	— ·50
Mesozoic to Recent	— ·37	— ·72	+ ·51	— ·11

Table 15.—Correlation coefficients between temperature and geographical conditions.

A correlation coefficient of +1 indicates that the fluctuations of the two variables considered are exactly proportional ; a coefficient of −1 indicates that the relationship is exactly inverse.

These coefficients agree with our expectations in showing that extensive land areas, a high level, and extensive volcanic eruptions are all associated with low temperatures, while open connexions between the Arctic Ocean and equatorial seas are associated with generally high temperatures. The chief difference between the two periods Palæozoic and Mesozoic to Recent, lies in the importance of volcanic eruptions, which appear to have been much more effective in the former than in the latter. This is largely due to the very great volcanic activity which prevailed during the Permo-Carboniferous glacial period, which dominates the first half of the climatic curve.

Correlation coefficients show how closely two variables are connected, but they do not give immediately the quantitative effect which one variable has on the other. This is given by the " regression coefficient," which is the average amount of change in one variable associated with a change of one unit in the other variable. For instance, it was found that the regression coefficient of mean temperature (Fahrenheit degrees) in terms of continentality (per cent.) during the Palæozoic period was −0·31. This means that an increase in the continentality by one per cent. is associated with a decrease in the mean temperature north of 40° N. by 0·31° F. The regression coefficients calculated from the correlation coefficients in Table 15 are shown in Table 16.

Factor	Continentality.	Elevation.	Ocean Currents.	Vulcanicity.
Unit	1 per cent.	100 feet	1 per cent.	1 (scale 0-10)
Change of Temperature—				
Palæozoic	−0·31	−0·38	+0·28	−1·68
Mesozoic to Recent	−0·32	−0·47	+0·28	−0·42
" Theoretical " (see below)	−0·35	−0·3	(+0·3)	−0·5

Table 16.—Effect of a change of one unit on the mean temperature in °F.

With the exception of the figures for vulcanicity, there is remarkably good agreement between the two periods.

Let us now consider the factors separately. An increase of one per cent. in the land area north of 40° N. is found to decrease the mean temperature by 0·31° F. In discussing the effect of continentality on temperature at the present day (Chapter VIII.), I obtained expressions for the effect of land in different latitudes on the mean temperature in January and July. The effect of an increase of the land area in any region is made up of two parts, a general decrease in the mean temperature over the whole belt of latitude, and an additional local decrease in the neighbourhood of the new land area. The average effect of an increase of one per cent. in the land area, calculated from the present distribution of temperature in relation to the distribution of land and sea, is a decrease of 0·22° F. in the "zonal" temperature (p. 150), and the local effect, if spread out over the whole zone, would be equivalent to an additional decrease of 0·13° F., making a total lowering of temperature by 0·35° F. This is the "theoretical" figure of Table 16 ; it is in good agreement with the figures 0·31° F. and 0·32° F. obtained from the regression equations.

The effect of elevation on temperature at present is well known. It is very close to an average decrease of 0·5° C. per 100 metres or 0·3° F. per 100 feet of elevation, which is given as the "theoretical" figure in Table 16.

The effect of ocean currents is complicated by the great cooling power of floating ice described in Chapter I. For our representation of the warm periods, we may take the arithmetical mean of the temperatures over the Arctic Ocean calculated in Chapter VIII. for the Upper Jurassic and Middle Eocene periods, viz., 43° F. Thus we have the following data :—

Warm Period.		Present.		
Ocean Currents.	Arctic Temperature.	Ocean Currents.	Arctic Temperature. Glacial.	Non-Glacial.
44 per cent.	43° F.	17 per cent.	−18° F.	24° F.

If we take the glacial temperature at present, we have a difference of 61° F. corresponding with a difference of 27 per cent. in the ocean currents ; if we take the non-glacial, we have a difference of 19° F. Now, in geological time, non-glacial conditions in the Arctic Ocean have been the rule and glacial conditions the exception. If we take the ratio of occurrence of the two conditions as five to one, and combine the two figures 19° F. and 61° F. in that proportion, we obtain a weighted mean of 26° F. for a difference of 27 in the ocean currents. This figure refers to winter over the Arctic Ocean ; the difference between summer and winter is probably not great, but the effect diminishes rapidly southward, and is also less over the land than over the oceans. If the average effect over the whole area north of 40° N. is one-third of that over the Arctic Ocean itself, or 9° F., we obtain an increase of temperature of about 0·3° F. for an increase of one per cent. in the effect of the ocean currents. The amount given in Table 16 is 0·28° F.

The effect of volcanic action is difficult to discuss because of the arbitrary nature of our scale of 0-10. W. J. Humphreys (Chapter VI.) considers that during the past 160 years the mean temperature of the earth has been lowered 1° F. by volcanic dust. If the value of 2 for the present vulcanicity is correctly assigned (a very large " if "), this is equivalent to a decrease of temperature by 0·5° F. for an increase of one in the scale of vulcanicity. The corresponding value found for the Mesozoic to Recent periods is 0·42° F., which is a good agreement. On the other hand, the figure for the Palæozoic, 1·68° F., is very much greater, and suggests that the amount of volcanic dust present during the Upper Proterozoic and Upper Carboniferous glaciations was much greater than is indicated by the values of 9 and 10 on the linear scale. The figure for the Upper Carboniferous should probably be nearer 40 than 10.

From the correlation coefficients given in Table 15, we see that the variations of climate during geological time have been associated to some extent with the variations of all four of these factors—continentality, elevation, ocean currents, and volcanic action. But the curves in Fig. 24 show also that these factors of climate have a close relationship among themselves. When the continents were generally lofty they were also extensive, and the passages between them along which ocean currents could penetrate into high latitudes were few and narrow, while volcanic action was greatest in periods of mountain-building. Hence part of the effect of great continentality in lowering temperature may be due to the great elevation, weak ocean currents, and great vulcanicity which accompany it. In order to determine the effect which would follow a change in one factor only, while the others remained constant, it is necessary to calculate "partial" correlation coefficients. This was done, and the results are shown in Table 17.

	Continentality.	Temperature with Elevation.	Ocean Currents.	Volcanoes.
Palæozoic	— ·25	— ·22	— ·22	— ·53
Mesozoic to Recent	— ·08	— ·82	+ ·63	— ·15

Table 17.—Partial correlation coefficients with temperature.

In the calculation of partial coefficients, small errors in the original data are apt to make a great difference in the final result. The figures of continentality and ocean currents for the Palæozoic are uncertain, and the partial coefficients for this period have very little value. In particular, the negative coefficient between ocean currents and temperature is obviously wrong, since one cannot conceive a state of affairs in which a wide connexion between the polar and equatorial oceans brings about a low temperature in temperate and polar regions. Probably the only significant figure is the high correlation between volcanic activity and temperature. For the Mesozoic and later periods our data are more exact and the partial coefficients show that the most important geographical conditions which determine temperature are the average elevation of the land and the volume of the ocean currents. From these partial coefficients we obtain the following formula for calculating the temperature in any part of the Mesozoic or Tertiary :—

Temperature (°F.)=48—0·13 × Continentality (per cent.)—0·45 × Elevation (hundreds of feet)+0·43 × Ocean Currents (per cent.)—0·26 × Vulcanicity (0-10).

It will be noticed that these coefficients differ somewhat from those given in Table 16. The effect of continentality appears to be greatly reduced ; this is because the cooling power of land is due partly to the mountain systems usually found somewhere in a large land-mass, partly to the barriers which large land-masses place in the way of ocean currents, and only partly to actual cooling by radiation from the surface of the land. All three of these effects are included in the coefficient in Table 16, but in the equation given above, the first two have been eliminated, leaving only the purely local radiation effect. It may be only a coincidence, however, that the value of this local effect in the equation given above ($-0 \cdot 13$) is exactly the same as the local part of the total effect of continentality at present, as described on page 212.

The theoretical temperatures given by this equation are shown in column (6) of Table 14. The calculation was extended to the Palæozoic, although the equation is based only on the data since the beginning of the Triassic, because we cannot suppose that the physical laws of climate have changed, and the equation deduced from the later periods agrees with what we know of those laws. The discrepancies shown by the Palæozoic are no doubt due partly to our incomplete knowledge of the geographical conditions of this era, but probably partly also to errors in the temperatures deduced from the records of the rocks. The most noticeable feature is that until the Middle Triassic the "calculated" curve is generally above the "observed" curve. The three great glacial periods of the Upper Proterozoic, Upper Carboniferous, and Quaternary stand out clearly; following each one of them the "calculated" curve rises more rapidly than the "observed" curve, as if the earth took a long time to warm up again after the crisis of the glaciation had passed. We have good reason to believe that this is true of the Recent period, the temperature being kept lower than it should be by the relics of the Quaternary ice-sheets in Antarctica and Greenland, and by the low temperature of the great body of sea water. The delay in warming up after the other ice-ages may be due to similar causes, in which case the statement sometimes made, that until the Quaternary the oceans had never been generally cooled, is incorrect. There is, in fact, a large amount of biological evidence that the oceans became cold during the Upper Carboniferous. It is possible, however, that the delay in these cases is more apparent than real. It is often difficult to determine the exact horizon of a glacial deposit, and if a few such deposits are placed too high in the series, they will make the stage following the glacial period appear colder than it actually was. This explanation may apply to the apparently low temperature of the Early Cambrian, but scarcely to the discrepancy of the Upper

Permian and Lower Trias, and on the whole, I believe this delay in warming up after an ice-age to be a real phenomenon. For the Pliocene, the "calculated" temperature is much lower that the "observed." It is a very striking fact, which has often been commented on, that the Quaternary glaciation did not coincide with the period of greatest elevation, but lagged considerably behind it. The cause of this lag was discussed on page 179.

There is a steep drop in the "calculated" curve in the Lias (Lower Jurassic) which is barely shown on the "observed" curve. The Liassic period, so far as we know, had no glaciers ; probably the distribution of mountain ranges in relation to moist winds was not suitable. In the absence of ice, the other factors of low temperature fail to produce their full effect. The drop in the "calculated" curve during the Cretaceous, on the other hand, is almost equally marked on the "observed" curve ; in this instance we have evidence in the erratics of the English chalk that either shore ice or glacier ice occurred somewhere in the Northern Hemisphere and that floating ice was present on the chalk seas.

The relations between the "observed" and "calculated" temperatures, given in Table 14, since the beginning of the Carboniferous have some points of interest. It will be noticed that when the "calculated" temperature is below 39° F. the "observed" temperature tends to be below the "calculated" temperature, the mean values for the five cold periods (Upper Carboniferous, Lower Permian, Pliocene, Quaternary, Recent) being : "observed," 32·4° F. ; "calculated," 33·8° F. When the "calculated" temperature lies between 39° F. and 50° F., the "observed" temperatures tend to be higher than the "calculated," the mean values for fourteen moderate periods being : "observed," 47·4° F. ; "calculated," 45·4° F. When the "calculated" temperature is above 50° F., the "observed" temperatures are again lower than the "calculated," the mean values for the warmest periods being : "observed," 50·0° F. ; "calculated," 52·3° F. A decrease of the "calculated" temperature from 52·3 to 45·4° F., or 6·9° F., is associated with a decrease of the "observed" temperature from 50·0 to 47·4° F., or only 2·6° F., a fall of less than 0·4° F. in the "observed" temperature for a fall of one degree in the "calculated" temperature. On the other hand, a decrease of the "calculated" temperature from 45·4 to 33·8° F., or 11·6° F., is associated with a decrease of the "observed" temperature from 47·4 to 32·4° F., or 15° F., a fall of 1·3° F. in the "observed" temperature for a fall of one degree in the "calculated" temperature. This result may be due to one of three causes :—It may be accidental, due to the chance run of the figures, or it may be due to an error in the scale of the "observed" temperatures, owing to which the temperatures of the moderately warm periods are overestimated compared with those of both the warm and the cold periods. On the other hand, it may represent a real phenomenon, a unit change in the geographical factors making less difference to the mean temperature of a warm period than to that of a cold period. Chapter I., and especially Fig. 2, show very strong reasons why such a difference should actually occur. So long as the climate remains "non-glacial," the change of temperature due to a change in the geographical factors is limited to the direct effect of those factors. Land in high latitudes has a lower mean temperature than sea, so that an increase in the land area lowers the mean temperature somewhat, but this effect at present is partly due to the winter snow-cover. The more free the oceanic communication between high and low latitudes, the more heat is carried by ocean currents, but the effect is also proportional to the difference between the temperature of the warm currents and that of the main mass of cooler water in the Arctic Ocean, and therefore two ocean currents during a warm period do not raise the mean temperature twice as much as one of them during a cold period. In Chapter X. we saw that the effect of an increase of elevation by 100 feet becomes greater the higher the average level of the land ; the change from an average elevation of 2,500 feet to one of 3,500 feet is responsible for nearly as much cooling as the change from 500 to 2,500 feet. When the temperature of the polar regions falls below a certain level, the climate becomes "glacial," and the cooling power of ice is added to the direct effect of the geographical factors. The relations between the "observed" and "calculated" temperatures north of 40° N. may be due to this introduction of the cooling power of ice when the mean temperature falls below about 45° F. over the whole area, which may imply a

January mean of 27° F. at the pole. If this explanation is correct, the ice present in the oceans in high latitudes, and the Greenland ice-sheet, lower the mean temperature north of 40° N. by nearly 10° F., which is in sufficiently good agreement with the results of the theoretical investigation in Chapter I. The changes of mean temperature which take place during the transition from a warm period to an ice-age are

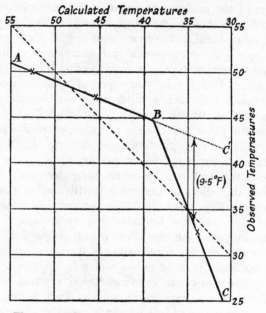

Fig. 25.—The transition from a warm period
to an ice-age.

shown diagrammatically in Fig. 25. The mean temperature north of 40° N. is initially 52° F., as shown at A. As the continents emerge and the ocean currents become weaker, the temperature falls slowly, as shown by the full line, until the point B is reached, when it is 45° F. At this point the Arctic Ocean becomes glacial. The temperature now falls much more rapidly along the full line BC. The dotted line BC′ indicates the " non-glacial " temperature due to the action of the geographical factors alone, without the intervention of the ice, and the distance between the dotted line and the full line shows the additional cooling due to the ice itself. The arrow indicates the amount of cooling by ice at the present time. The broken line shows the corresponding temperatures calculated from the regression equation given above, which assumes that the relations are linear throughout. The three crosses mark the three points determined from the comparison of the " observed " and the " calculated " temperatures. The diagram seems to agree well with all the results previously obtained ; for instance, it indicates that the " non-glacial " temperature is now only about two degrees below the critical point, and that a permanent increase in the general temperature by more than this amount would result in the breaking up of the Arctic ice. The good quantitative agreement between the effects of the different geographical factors calculated from purely geological data and those deduced from existing conditions, the coincidences in points of detail between the observed and calculated curves of temperature in Fig. 24, and the fact that the discrepancies between the two curves are what we should expect from the combination of " glacial " and " non-glacial " periods in the same equation, combine to form a very strong body of evidence that throughout the greater part if not all of geological time the major variations of climate have been entirely controlled by changes in the geographical factors. Since a careful calculation of the effects of known causes suffices to explain the facts, it is unnecessary to introduce hypothetical causes such as variations of solar radiation or continental drift to explain the *long-period* oscillations of climate. The more rapid oscillations within the major climatic chapters, such as the succession of glacial and interglacial periods within the Quaternary, may not be explicable by changes in the geographical factors. Zeuner's reconstruction of the Quaternary sequence (p. 107) fits in well enough with the facts to lend some support to the theory that such secondary oscillations are due to astronomical causes, but variations of solar radiation also remain a possibility.

The topsy-turvy Permo-Carboniferous period, in which the greatest glaciation occurred not far from the equator, demands a special investigation, which is given to it in Chapter XV.

REFERENCES

(1) ARLDT, TH. "Handbuch der Palæogeographie." 2 vols. Leipzig, 1919.
(2) CHAMBERLIN, T. C., and Others. "The age of the earth." Philadelphia, *Proc. Amer. Phil. Soc.*, 61, 1922, and Washington, *Ann. Rep. Smithson. Inst.*, 1922, p. 246.
(3) JEFFREYS, H. "The earth, its origin, history, and physical constitution." Cambridge, 1924.
(4) DACQUÉ, E. "Grundlagen und Methoden der Palæogeographie." Jena, 1915.

CHAPTER XIII

THE THEORY OF CONTINENTAL DRIFT

IN the calculations discussed in the last chapter, the assumption was made that deposits were laid down not far from where we now find them, or in other words, that the positions of the continental massifs relative to each other and to the poles have not changed during geological time. That assumption has been challenged from time to time, but was not seriously countered until A. Wegener (1) published his well-known theory of continental drift, and supported it with a wealth of detail and acute reasoning. For a time Wegener's theory was in considerable favour, but it has been found to introduce so many difficulties that opinion now seems to be that if continental drift ever occurred on the scale postulated by Wegener it was long before the beginning of the geological record. The final acceptance or rejection of Wegener's theory is a matter for geologists, but inasmuch as palæoclimatological evidence plays a considerable part in the working out of the theory, which in turn, if accepted, completely alters the aspect of the problem of climatic changes, this book would not be complete without a discussion of what the theory implies. The theory of continental drift falls into two parts, and the truth of one part does not necessarily imply the truth of the other part. The first contention is that the positions of the continents have changed relative to each other during the course of geological time ; that at first there was a single large continent (" Pangæa "). This original continent split up into various parts which gradually drifted asunder, the latest division being the separation of America from Europe. The second contention is that there have also been radical changes in the positions of these land-masses relative to the poles. Of course, if the relative positions of two continents change in any way except by means of a direct east-west movement, there must necessarily be some change of latitude, but the movements postulated by Wegener go far beyond this. Regions like Brazil and Central Africa, now near the equator, are supposed to have been formerly near the South Pole, while other regions now far to the north were once close to the equator.

This power of free movement of the continents depends on the difference of constitution between the continental massifs and the mass of the earth's crust. The former are composed mainly of silicates of alumina, termed *Sial* for short, and are lighter than the rest of the crust, which is mainly composed of silicates of magnesia (*Sima*). The sima is continuous and thick, and forms the floor of the deep oceans ; the floor of the Atlantic, however, is believed to be covered by a thin layer of sial, and this difference between the Atlantic and Pacific has given rise to much speculation. The sial now consists of a number of separate masses (continents and continental shelves) with some smaller detached portions (oceanic islands). Under the action of any long continued force, however small, the sima acts as a very viscous fluid, while the masses of sial are rigid, and may be compared to slabs of wood floating in a sea of treacle. This distinction between the continental masses and the rest of the crust is based on a number of converging lines of evidence, and is now generally accepted.

The argument that the individual continents have been formed by the splitting up of an original Pangæa, starts with the notable similarity in the shape of the opposite coasts of the Atlantic Ocean. Not only is there a great similarity of shape, but the structural features on either side of the ocean show a considerable degree of resemblance, and throughout geological time there has been also a strong likeness between the animals and plants. Hence it is supposed that America split off from Europe during the Tertiary period, the rift beginning in the south and gradually extending northwards. This movement is considered to be still in progress, and to have been sufficient to become evident in the successive determinations of the longitude of Sabine Island in North-east Greenland, which is thus shown to have moved westward by nearly a mile in eighty-four years. These differences in the longitude, however, as Sir Charles Close has pointed out (2), are all within the limits of the probable error, and are not sufficient to constitute a proof of the westerly drift of Greenland.

It is true that the distribution of a number of animals and plants which are found on either side of the Atlantic could be explained much more readily on the hypothesis that America and Europe-Africa were in contact not long ago, but this involves the supposition of a formerly wider Pacific Ocean, against which must be set the distribution of a number of animals and plants common to both shores of this ocean.

The second part of the theory is that independent of the drift of the continents the earth's axis of rotation is undergoing a progressive change, which since early Palæozoic times has brought the North Pole from high southern latitudes via the Pacific Ocean to its present position. The arguments rest entirely on deductions from the distribution of climatic zones in past times, and especially on the location of the Permo-Carboniferous glaciation. We will return to this point in the next chapter.

The chief weakness of Wegener's theory is the inadequacy of the forces which he postulates to move the continents. These are twofold—a force directed towards the equator and a force directed towards the west. The force directed towards the equator depends on the facts that the earth is not a true sphere, and that a continent consists of a floating mass of sial, the centre of gravity of which is higher than the centre of gravity of the displaced sima, or centre of buoyancy of the sial. Anywhere except on the equator and at the poles, a plumb-line on the surface of the earth points, not directly towards the centre of the earth, but to a part of the equatorial plane at a somewhat lesser depth than the centre. If the earth were a true homogeneous and non-rotating sphere, a plumb-line in latitude ϕ, if produced downward, would make an angle ϕ with the plane through the equator, but on the earth as actually constituted the angle would be greater than ϕ, say, ϕ^1. In a very deep mine the plumb-line, if produced, would pass nearer to the equator than if produced from the surface, that is, it would make with the plane through the equator an angle between ϕ and ϕ^1. In a large mass of sial floating in a layer of sima, therefore, the downward force due to the attraction of the main mass of the earth on the sial, which can be considered as concentrated at the centre of gravity of the sial, is not exactly opposite in direction to the upward force due to the displaced sima, which can be considered as concentrated at the centre of buoyancy. The resultant of these two forces is a small component towards the equator, which reaches its maximum in latitude 45° and vanishes at the poles and the equator.

There is no doubt that such a force actually exists, and if the sima is a true fluid, however viscous, it would produce a slow movement of the continents towards the equator. The movement has been calculated from the not very complete data available, and has been found to amount to 20 cm. (8 inches) a year in latitude 45°, where it is greatest. But all this rests on the assumption that the sima really is a fluid, and that this fluidity persists even in the uppermost coldest layers. The equatorward force is so small that the resistance of a quite thin non-fluid layer at the surface would suffice to overcome it. Jeffreys (3, App. C.) is of opinion that while the earth may be considered to be a plastic body of zero strength at depths greater than 450 miles, there is some evidence that the cooler surface rocks have in fact a finite though small strength to depths of a few hundred miles. For instance, the floor of the ocean is strong enough to maintain the Tuscarora deep. This surface strength would probably be great enough to overcome the force due to the difference between the centres of gravity and of buoyancy. According to Jeffreys, then, it is doubtful whether the force available is strong enough to move the continents at all, and it becomes highly improbable that such a small force can raise enormous mountain-chains. To raise a slice of the earth's crust thousands of feet against gravity requires an enormous force, much greater than the small forces due to the difference between the values of gravity in different parts of the continents. The forces postulated by Wegener are of the order of one hundred-thousandth (10^{-5}) of a dyne per square centimetre, whereas to elevate the Rocky Mountains a force of about one hundred million (10^9) dynes per square centimetre would be required. According to Jeffreys' calculations, therefore, the forces available on Wegener's theory are about one hundred billion times too small for the effect which is attributed to them. Jeffreys sums up " Secular drift of continents relative to the rest of the crust . . . is out of the question. A small drift of the crust as a whole over the interior, which would be shown as a displacement of the poles relative to the earth's surface, is not impossible, but the maximum amount seems too small to be of much interest."

The forces which tend to produce motion in an east-west direction are much less clearly defined by Wegener than the force directed towards the equator ; they appear to depend mainly on the effects of tidal friction both on the floor of the sea and within the earth's crust, and to be of the same order of magnitude as the forces directed towards the equator. From the point of view of palæoclimatology, the question of the east-west movement is of less importance than the question of the north-south movement. East-west movements of some of the continents relative to others may affect the annual range of temperature or the distribution of rainfall to some extent, but cannot radically change the mean temperature of the whole belt in any latitude, but rearrangements of the positions of the continents relatively to the poles can obviously be made to produce almost any changes of mean temperature which may be required to fit the evidence.

This, in effect, is what Wegener does. Taking as a basis the power of free movement of the continents over the face of the earth, Wegener and Köppen (4) proceed to study the distribution of climates during the various geological periods, and to map out the positions of the continents relative to each other and to the poles which will best fit in with the distribution of climates, while preserving some continuity from one period to the next. It is assumed that the earth has been under solar control throughout, and that the distribution of climatic zones relative to the poles has always been similar to that found at present. On either side of the equator there has always been on the land a belt of rich vegetation represented by thick coal formations, and in the oceans a high-water temperature represented by reef-building organisms such as corals and *Rudistes*. On either side of this tropical belt there has always been over the land a zone of deserts. Nearer the poles another belt of vegetation in temperate latitudes has formed other coal beds less well developed, with annual growth rings in the tree stems. Finally, the sites of the poles have generally, if not always, been occupied by ice, either inland ice-sheets or floating ice-caps according as the pole lay on land or in the ocean. These zones at present do not run strictly parallel with the lines of latitude, and no doubt there were similar irregularities in the past, but Köppen and Wegener consider that these were never sufficient to mask the zonal distribution. It is proposed, first, to run briefly through the distribution of zones in the different periods according to this theory, reserving criticism to a later stage.

The Late Carboniferous is the earliest period for which a good cartographical basis is available according to the continental drift theory, but some attempt is made to reconstruct the earlier periods. Thus, in the Algonkian (Late Proterozoic), there was an intense glaciation of North America, which is therefore considered to have been the site of one of the poles, while the corresponding dry belt is represented by the Torridon Sandstone in Scotland and the Dala Sandstone in Central Norway. In the Early Cambrian [now believed to be mainly Late Proterozoic] there are more or less doubtful glacial deposits in Norway, Yangtse (China), South Australia, India (Salt Range), and South Africa. In Australia the glacial deposits are followed by thick limestones with reef-building *Archæocyathinæ*, indicating a rapid warming. The authors find that they cannot indicate the position of the poles and equator during the Cambrian and Ordovician periods.

In the Silurian, the evidence is a little more definite, but orientation is still difficult. The occurrence of glacial deposits is doubtful, but there may have been ice in South Africa. The equatorial belt, represented by corals in the marine deposits and poor coal seams over the land, passed through North America, the British Isles, Central Europe, Northern Siberia, and possibly Australia. The northern desert zone lay near Leningrad, in Baffin Bay, and especially in North America. In the Devonian, there was still ice in South Africa, but Europe lay farther north than in the Silurian, the equatorial belt passing through France and Spain, Central Asia and China. Desert formations such as the Old Red Sandstone were extensively developed in the northern continent, and the fauna includes the famous " lung-fish " (*Ceratodus*), and a " lung-snail." This continent must therefore have " had a hot desert climate, whose dry periods were only occasionally interrupted by thundery rains."

During the Carboniferous period the conditions are known in much greater detail, partly because of the interest aroused by the great glaciation which occurred during this period, and partly because the majority of the workable coal beds are of Carboniferous age. The greatest development of ice deposits occurred in South Africa, India, Australia, the Argentine and Eastern Brazil, and the Falkland Islands. These are so thick and extensive that they must be due to great inland ice-sheets, which Köppen and Wegener consider can only have formed in the neighbourhood of the poles. If we suppose the South Pole to have been in South Africa at that time, while the continents still had their present positions relative to each other, the most remote of them would still lie too near the equator to be readily glaciated. Hence the authors suppose that at this period Pangæa, the original continent, had not yet been split up into its component parts, and the glaciated continents were all grouped in contact with each other round the South Pole. The area covered by the ice-sheets was so extensive, however, that even this rearrangement does not suffice, and it is considered that the various glacial deposits are not all of the same age, but that glaciation followed the location of the moving pole. The Brazilian deposits are the oldest, the Australian and Indian the youngest. The South Pole travelled from Antarctica *via* South America to South Africa, and thence in a great arc across Australia back to Antarctica. The position of the equator is determined by the great coal beds in North America, Europe, and China, and it is found that these lie on a great circle the centre of which falls in the glaciated region. These coal beds, therefore, represent the tropical rain-forest, a conclusion which will be discussed later. Other coal measures in Alaska, South America, South India, Australia, and Antarctica are attributed to the temperate rain belts. Between these temperate coal beds and the main mass of coal are a number of desert deposits.

During the whole of the Mesozoic period there was little if any ice action, and the development of coal was also restricted. On the other hand, there was a great expansion of the dry belts, especially in North America and Africa. In the sea, great coral reefs were formed. The authors tacitly admit the generally accepted opinion that during the Mesozoic the development of climatic zones was less marked than at any other period since at least the middle of the Palæozoic. It was during the Jurassic that the continents first began to drift apart, India and Antarctica splitting off from Africa, and Australia separating from Farther India. South America did not become an independent continent until the Cretaceous. Throughout the Mesozoic, the South Pole lay in Antarctica and the North Pole in the North Pacific. It may be remarked here that according to the reconstruction of Cretaceous geography the British Isles lay in about 20° N., and therefore had a tropical climate. The British chalk, however, contains a number of erratic pebbles, which are so alien to the general character of that deposit that it is difficult to attribute them to any other agency than floating ice, and this fact seems to be fatal to Wegener's reconstruction of that period.

The interval of rest during the Mesozoic did not extend into the Tertiary, which was a period of great mountain-formation and of great and rapid shiftings of the earth's axis, which brought the North Pole over the land and caused a great ice-age in the Northern Hemisphere. There are no certain traces of ice in the Eocene ; there was a considerable development of brown coal formation, especially in North America and Europe, but whereas the European beds are attributed to the equatorial rain belt, the North American beds are placed in the north temperate zone. (This should be compared with Berry's description of the Eocene floras quoted in the Introduction.) The Oligocene was generally similar to the Eocene, but in the Miocene we have the beginning of the ice-age in Alaska, North-east Siberia, and the New Siberian Islands. This is a very important point which is referred to again later. In the Pliocene, the Alaskan glaciation spread over the greater part of North America, including Greenland. At this time the North Atlantic existed only as a very narrow rift, and Greenland lay to the north or north-east of the British Isles. The east winds shown by the late F. W. Harmer to have blown across the North Sea are attributed to the glacial anticyclone associated with the American and Greenland ice-sheets. In the Miocene, the pole lay just north of Alaska. In the Pliocene it moved rapidly across the Canadian Arctic Archipelago, and at the beginning of the Pleistocene it lay near Disco Island off West Greenland. During the greater part of the Quaternary the North Pole lay in Central Greenland. In the Baltic Readvance it was near Spitsbergen, and then gradually assumed its present position. Ice-sheets were formed over the parts of the continents which lay nearest to the poles. The theory, therefore, requires a revision of the generally accepted correlation of the Quaternary deposits ; the Alaskan glaciation, as we have seen, is placed in the Miocene, along with a somewhat hypothetical glaciation of British Columbia. In the United States the two earliest glaciations, Kansan and Jerseyan, are attributed to the Pliocene, the Illinioan is paralleled with the Gunz, the Iowan with the Mindel, the Earlier Wisconsin with the Riss, and the Later Wisconsin with the Wurm. Correlation between American and European glacial deposits is admittedly difficult, but this arrangement of the American sequence cuts right across the present opinions of most American geologists, which are set out on page 242.

While the general course of the ice-age depended on the successive positions of the moving pole, the alternation of glacial and inter-glacial periods cannot be explained in this way, and the authors have recourse to the variations in the obliquity of the ecliptic and in the eccentricity of the earth's orbit. This part of the theory, and the objections to it, were dealt with in Chapter V.

We may sum up the results of the investigation of climatic changes by Köppen and Wegener by giving the positions which they assign to the North Pole relative to the present position of Africa. Since Europe has always had almost its present position in relation to Africa, these figures give also the various positions of the North Pole relative to Europe.

Period.	Position of North Pole.		Latitude of England.	Latitude of Antarctic.
Carboniferous	30° N.	145° W.	0°	75° S.
Permian	35° N.	115° W.	15° N.	70° S.
Trias	50° N.	125° W.	20° N.	85° S.
Jurassic	47° N.	132° W.	20° N.	90° S.
Cretaceous	47° N.	140° W.	15° N.	85° S.
Eocene	45° N.	160° W.	15° N.	90° S.
Miocene	75° N.	150° W.	40° N.	ca. 90° S.
Late Pliocene and Early Pleistocene	70° N.	60° W.	60° N.	ca. 85° S.

Table 18.—Changes of latitude according to Wegener.

In the third column I have added the corresponding latitude of England, and in the fourth column the latitude of the centre of the Antarctic continent.

The zonal arrangement of climates has persisted throughout geological time, and though there have probably been minor fluctuations in the average rainfall over the globe, the variations of climate in any one area have been governed almost entirely by the variations of latitude which it has undergone. In the next chapter we will examine the latter contention in greater detail.

REFERENCES

(1) WEGENER, A. " The origin of continents and oceans." Transl. by J. G. A. Skerl. London, 1924.
(2) CLOSE, SIR CHARLES. " The geodetic evidence for the supposed westerly drift of Greenland." London, *Geogr. J.*, 63, 1924, p. 147.
(3) JEFFREYS, H. " The earth, its origin, history, and physical constitution." 2nd ed. Cambridge, 1929.
(4) KÖPPEN, W., UND A. WEGENER. " Die Klimate der geologischen Vorzeit." Berlin, 1924.

CHAPTER XIV

AN EXAMINATION OF THE CLIMATIC EVIDENCE
FOR CONTINENTAL DRIFT

WE will begin the critical discussion of the views set out in Köppen and Wegener's book, "Die Klimate der geologischen Vorzeit," with some further analysis of the climatic conditions during the Upper Carboniferous period, beginning with the United States, the British Isles, and Central Europe. According to the "drift" theory, these coal beds represent a luxurious tropical rain-forest, and the equator is therefore drawn as nearly as possible through the middle of them. The evidences of glacial action which have been adduced from time to time in close proximity, both in space and time, to these coal beds are dismissed out of hand as not genuine. The American evidence, however, seems to be too well founded to be dealt with in this summary fashion. Thus S. Weidmann (1) describes conglomerates of Upper Carboniferous to Permian age in the Arbuckle and Wichita Mountains of Oklahoma and in Kansas, associated with all the paraphernalia of glaciation—scratched boulders, erratics, fluted and polished floors, and U-shaped valleys. Some of the boulders in marine deposits have apparently been carried by icebergs, and the author attributes the phenomena to islands in the Late Palæozoic sea bearing local valley glaciers. J. A. Taff (2) found boulders up to 20 feet across and 5 or 6 feet thick, 50 miles or more from their source, in the marine Caney shales of Eastern Oklahoma. "No other competent means of their transportation than ice—presumably heavy shore ice—has been suggested." Similarly, A. P. Coleman (3) considers that there is good evidence for glaciation in Oklahoma, Nova Scotia, and Alaska (Thousand Isles). As regards Nova Scotia, Coleman writes, "It is . . . probable that there were moderate elevations from which, under a cool climate, glaciers spread out over the plains on which coal forests had been growing not long before."

In the Squantum tillite near Boston, Mass., there are massive conglomerates 2,000 feet in thickness, which cover a considerable area. The chief interest of these beds, apart from the presence of striated boulders, lies in the associated " varve " beds (4)—banded clays which are similar in all respects to those formed during the retreat of the Scandinavian and North American ice-sheets at the close of the Quaternary glaciation, and also similar to those formed in Australia during the Upper Carboniferous glaciation. These clays owe their banding to the seasonal variations in the rate of melting of glaciers, and are therefore incompatible with an equatorial climate. In places the banding is disturbed during the deposition of the shales, probably by the grounding of floating ice-masses.

Wegener recognises that the Squantum tillites demand serious consideration, and the effort he makes to explain them away tacitly implies that they form a very serious objection to the " continental drift " theory. He considers the possibility that they are real, but were formed at a very high level in the Appalachian mountain system, then young and vigorous, and agrees with the general opinion of geologists that the chances of preservation of high-level glacial deposits over a wide area would be very slight. As we have seen, there are other indications that the glaciers reached low levels. He therefore concludes that as smoothed floors have not been found beneath the moraines, the remaining phenomena, although very suggestive of glacial action, could have originated in other ways. As all the other evidence indicates that Boston lay in the equatorial rain zone during the Carboniferous and in the region of hot deserts during the Permian, " *the glacial nature of these tillites is in irreconcilable opposition to the numerous climatic traces of another kind which surround it both in space and time.*" He therefore says that the burden of proof that the deposits are really glacial rests with the opponents of the " drift " theory. The Squantum tillites are, however, accepted by all American geologists, while the Caney shales in the Arbuckle and Wichita Mountains were examined by an impartial observer, J. B. Woodworth, who concluded (5) that while the striæ on the boulders which he observed were probably not glacial, the transport of the boulders was almost certainly effected by floating ice.

Farther west, in Colorado (6), there are Middle Carboniferous conglomerates some 6,000 feet in thickness, said to contain boulders up to 50 feet in diameter, and although no striated blocks have yet been found, the great size of the boulders strongly suggests ice action.

C. A. Süssmilch and Sir T. W. E. David (6) discuss in detail the deposits in Europe which suggest ice action. The evidence, while not so strong as that from North America, yet has some interest. Sir Andrew Ramsay was of opinion that glaciated pebbles occurred in the Permian conglomerates of England, but this interpretation is not now accepted. The Millstone grit, although it contains no striated material, points to enormous denudation. In France, M. Julien has described large masses of angular breccia in the St Etienne coal basin, with a thickness up to 800 feet. Striæ are extremely rare, but some have been found. "Vertical roots of Calamites are seen in the sandstones underlying the breccias, while their stems, as they pass upwards into the breccia, are crushed, a phenomenon very suggestive of glacial action. . . He considers that these 'morainic breccias' were deposited by glaciers having their origin in the great early formed folds of the Hercynian ranges which were already rising to the north." In Germany there is also some rather doubtful evidence of glacial action, the strongest being a shale bed containing occasional boulders up to a foot or more in diameter, suggesting the action of shore ice or river ice. The European evidence, taken by itself, is not very convincing, but in conjunction with the much stronger American evidence it throws considerable doubt on the theory that the coal measures are the remains of equatorial rain-forests.

So great an authority as A. P. Coleman (7) examined the distribution and sequence of both Permo-Carboniferous and Pleistocene glacial deposits from the point of view of the continental drift hypothesis, and concluded that the latter fails completely to account for them. The extent of the Permo-Carboniferous glaciated area was so great that even if the continents were joined up round the South Pole ice-sheets would still extend into sub-tropical latitudes. There is, however, evidence that in all the glaciated areas the ice reached the sea, and in South America and Australia there was open sea on both sides of the continent. In any case such a gigantic continent as that postulated by Wegener would be highly unfavourable to glaciation owing to the difficulty in the supply of moisture.

There is one other piece of evidence in connexion with the climatic zones of the Permo-Carboniferous which may be referred to, not so much for its intrinsic importance, which is small, as because it illustrates the methods too often adopted by Köppen and Wegener in dealing with items which do not quite fit their theory. Salt beds occur in Angola, formerly attributed to the Carboniferous. The authors class these as Permo-Triassic, because during the Carboniferous " Angola was too near the South Pole for salt beds to form." The drift hypothesis has certainly not reached a stage of proof in which it can be asserted that evidence which does not fit it is thereby proved to be false.

The next point in the discussion of the " drift " theory concerns the relative ages of the ice-sheets in the Southern Hemisphere. According to Köppen and Wegener, the Brazilian deposits are the oldest, the Australian and Indian the youngest, and L. Waagen is quoted as the authority for a statement that the glaciation of Brazil and South Africa occurred before the development of the *Glossopteris* flora, that of Australia after it. The age of the glacial deposits in different parts of the world is discussed in great detail by Süssmilch and David (6), and they arrive at very different conclusions. The succession in New South Wales is taken as the standard of reference, and from the generalised section given, we may make out the following simplified series :—

Permian	{ *Glossopteris* coal measures, etc. Crinoidal shales.
Upper Carboniferous	{ Branxton glacial horizon. Greta coal measures. Sandstone. Basalt and tuff. Horizon of *Eurydesma cordatum*. *Gangamopteris* mudstone. Brandon conglomerate. Tillites, etc. Varve beds. *Rhacopteris* horizon. Tuffs, fluvio-glacial conglomerates, etc.

There are thus two main glacial series, the older and more important falling in the lower part of the Upper Carboniferous, while the upper horizon falls probably at the top of the Upper Carboniferous. Between the two glacial series we find *Gangamopteris* and *Eurydesma*, while *Glossopteris* first occurs *above* the upper glacial horizon. In Victoria, Tasmania, and South Australia the main glacial horizon lies below the *Gangamopteris* horizon, and probably on the same horizon as the lower tillites of the New South Wales series. Tillites on the Irwin River in Western Australia are probably somewhat younger, falling in the Upper Carboniferous near the zone of *Eurydesma cordatum*. In Western Australia there are two glacial horizons, both older than the *Glossopteris* flora.

In India (Salt Range) the tillites are associated with *Eurydesma*, and may be referred to the Upper Carboniferous. In South Africa the Dwyka tillites are likewise associated with *Eurydesma*, and the thick series probably belongs mainly to the Upper Carboniferous. In the Falkland Islands the boulder beds are Upper Carboniferous and occur beneath beds containing *Gangamopteris* and *Glossopteris*, and appear to come at the base of the Permo-Carboniferous. Finally, the South American tillites occur just beneath coal measures with an exclusively *Gangamopteris* flora, and may be attributed to Upper Carboniferous. There was very little glaciation in the Lower Permian anywhere.

Thus the evidence, as set out in 1919 by two unbiassed observers, and still further emphasised by Sir T. W. Edgeworth David in a series of lectures in 1926, is to the effect that there was very little difference in age between the main glaciations of the different areas in the southern group. All of them are certainly older than the *Glossopteris* flora, and the only feature which supports the theory of a moving pole is the slight recrudescence of glaciation in the highest Carboniferous or possibly lowest Permian of New South Wales (Bolwarra and Branxton beds).

Süssmilch and David note the close relationship between the areas of folding and the glaciated areas, and also the enormous thicknesses of volcanic tuff. They further note that the period of glaciation of Eastern Australia was almost exactly synchronous with the period of great orogenic movements, also in Eastern Australia, and they suggest that the glaciations were due largely, if not entirely, to the presence of high mountains and of great quantities of volcanic dust.

The recognition of glacial deposits and of ice floating in the sea in the middle of the northern belt of coal measures, profoundly modifies the problem of the habitat of coal-building plants. For a long time it was believed that peat, which is the first stage in the formation of coal, could not form in a hot country owing to the rapidity of decomposition at high temperatures, and that the modern representatives of the coal measures are the peat-bogs of moist temperate regions. The reversal of this view is mainly due to the influence of H. Potonié, who described a swamp in Eastern Sumatra in which peat is actually being formed at the present moment, almost entirely from the fallen leaves of evergreen trees. Köppen and Wegener consider that the coal beds extending through the Eastern United States, the British Isles, Central Europe, and China represent the remains of similar peat formed by tree ferns and other highly developed vegetation in equatorial swamps of the Carboniferous period, while the coal beds of Spitsbergen in the north, and of Australia, South Africa, South America and Antarctica in the south, represent the peat of the temperate rain belts. It seems very doubtful whether we are entitled to draw this conclusion solely from the nature of the vegetation composing the coal beds. The absence of annual rings of growth in the Carboniferous vegetation of the north temperate belt may be due, as E. Antevs (8) points out, to the comparatively low organisation of the flora, which only formed annual rings under extreme conditions. If, as suggested in Chapter V., the coal beds of the Northern Hemisphere were formed during a period of high eccentricity of the earth's orbit, at the times when northern winter was in perihelion, we need not expect to meet annual rings until we reach very high latitudes, such as Spitsbergen. Throughout the Lower and Middle Carboniferous, almost to the beginning of the ice-age in Australia, the flora was extraordinarily uniform over the whole world (9), Europe, Asia, Africa, America, Australia all had the same flora. The *Glossopteris* flora is of later date than the northern coal beds ; it represents a different stage in the evolution of plant life, but not necessarily a difference of climate. This flora was able to spread across the equator into the Northern

Hemisphere, a fact which suggests a higher organisation than that of the older flora rather than a special adaptation to cold climates. The position is similar to that at the end of the Mesozoic, when a flora of modern type originated in high northern latitudes and spread over the world, and in the latter example there is no suggestion of a sweeping change of latitude.

The distribution of desert deposits—salt, gypsum, and desert sandstone—also requires a closer examination. In Fig. 26 the occurrence of these indications of desert action in

D—desert sandstones.   S—salt.   G—gypsum.   Dotted area—present deserts.

Fig. 26.—Carboniferous and Permian deserts in relation to the present desert areas.

the Carboniferous and Permian periods is shown by their initial letters S, G, and D. The distribution of present deserts is indicated by stippling. It is seen that with the exception of a few salt deposits in Europe, Western Asia, and the east of North America, the late Palæozoic desert deposits in the Northern Hemisphere fall entirely in the present desert areas. The present deserts of the Southern Hemisphere were, however, unrepresented in the Carboniferous and Permian. This presumably indicates that during those periods the present Southern Hemisphere was moister than at present, *either* because of the shifting of the continents relative to the poles, *or* because the existence of the ice-sheets caused a " pluvial period " in the surrounding lands.

These considerations show that the theory of " continental drift " is not so complete and irresistible an explanation of the peculiar distribution of climate in the Carboniferous and

Permian periods as Köppen and Wegener seem to think. The greatest difficulty is presented by the glacial deposits of North America, especially the " varve clays," and these form almost as great an obstacle to the " drift " theory as the glacial deposits of equatorial Africa form to the assumption that the continents held their present positions. The only complete answer to the " drift " theory, however, would be a demonstration that the climatic events of the Carboniferous and Permian were the logical result of the distribution of land, especially high land, and sea during that period, the poles being supposed to have kept their present positions.

In the Mesozoic and Tertiary periods the discrepancies between the past and present distribution of climatic zones

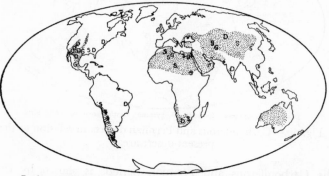

D—desert sandstones. S—salt. G—gypsum. Dotted area—present deserts.

Fig. 27.—Mesozoic deserts in relation to the present desert areas.

are far less striking. There was little ice, except towards the close of the Tertiary, and on any theory the climatic zones were much less developed than at present. It is not possible to discriminate between equatorial and temperate coals, and hence the desert deposits offer practically the only evidence of zoning. But as Fig. 27 shows, the distribution of deserts during the Mesozoic was very similar to the present distribution. The chief exceptions were in Europe and the South-eastern United States. The rainfall of Europe would be greatly decreased by the displacement of the Icelandic minimum far to the north, while the present comparatively heavy rainfall of the South-eastern United States is due almost entirely to moisture derived from the warm Gulf of Mexico.   In fact, the distribution of deserts shown in Fig. 27 is very nearly what we should expect during a period in which low continents and a free oceanic circulation reduced the temperature gradient between low and high latitudes almost to its least possible value.   The distribution of *Rudistes* in the Cretaceous period is the strongest point in favour of the displacement of the poles.   Reef-building forms reach their greatest development in the deposits of the great Tethys Sea, the Mediterranean of the Cretaceous, which, in contrast with the present Mediterranean, was open to the Indian Ocean and received a constant supply of warm water.   They are also found in the north of South America, Mexico, and the West Indies, the range extending to the north coast of the Gulf of Mexico, the limits in America so far discovered being 5° and 30° N.   The most interesting point is that outside the normal range dwarf forms are found, in latitudes 50° to 55° N. in Europe, but in latitudes 5° to 20° S. in East Africa.   If the distribution of *Rudistes* was really governed by the sea temperature—and we have no reason to suppose that it was not—the assumption of the present position of the continents requires a much greater northward displacement of the thermal equator than occurs at present, and especially a cold current along the east coast of Africa instead of along the west coast (see Fig. 28).   Probably the only cause which could produce such a great displacement of the thermal equator would be an extensive glaciation of the Antarctic continent, while the north polar regions were practically free of ice.   The glaciation of the Antarctic continent, according to C. S. Wright and R. E. Priestley (10), began at least as early as the beginning of the Tertiary, since immediately above the Cretaceous beds at Cape Hamilton, Graham Land, there occurs a "moraine-like mass, some metres in thickness," which contained angular fragments of crystalline rocks foreign to the locality.   This deposit is evidently glacial, though it may mean nothing more than a local valley glacier ; it cannot be younger than Eocene and may be uppermost Cretaceous.   The onset of cold conditions during the Cretaceous is also indicated by the Cockburn Island sandstone, which was " crowded with pygmæan forms of life such as might be expected to result from the encroachment of colder conditions upon a marine fauna long developed in, and

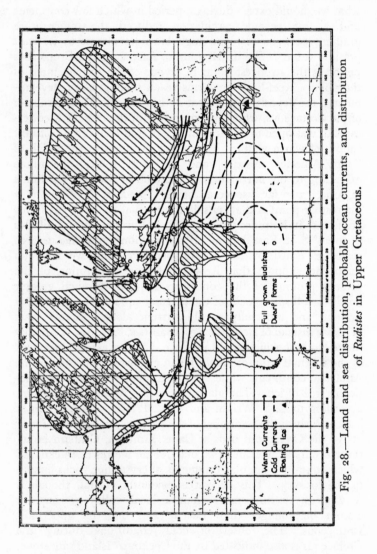

Fig. 28.—Land and sea distribution, probable ocean currents, and distribution of *Rudistes* in Upper Cretaceous.

habituated to, more genial conditions." There was shore ice or perhaps icebergs in the Cretaceous sea over South Australia, the erratics covering a wide area.

In the Northern Hemisphere there is no direct evidence of the occurrence of extensive ice-sheets during the Cretaceous. There is, however, some evidence of drift ice in the remarkable erratic blocks found in the English upper chalk. These blocks are probably due to transport by shore ice rather than by icebergs, but in any event their presence is a serious obstacle to the belief that at that time England lay in about 20° N. latitude. The problem is very similar to that of the Upper Carboniferous, since in both periods the biological evidence points to high temperatures, while the character of the deposits indicates ice action in the neighbourhood.

The most complete refutation of the drift hypothesis is given by the fossil plants of the Mesozoic and early Tertiary. E. W. Berry (11) states that " the distribution of the known fossil Arctic floras with respect to the present pole proves conclusively that there could have been no wandering pole." R. W. Chaney (12) gives maps of Eocene " isoflors " in the northern hemisphere, showing the limits of sub-tropical, temperate and cool temperate floras as irregular lines surrounding the pole. These enclose elliptical areas which approach the pole most closely round about longitude 0° and 160° W., i.e., in the openings between the circum-polar continents. The limit of the " cold-temperate " flora has an average latitude of about 78° N. centred four degrees from the pole. The limit of the " temperate " flora has an average latitude of 62° N. and is also centred 4° from the pole, that of the sub-tropical flora has an average limit of 42° N. and is centred 3° from the pole. The displacements of the first two are towards Bering Strait, that of the third towards Asia. The present mean annual isotherm of 40° F. is also displaced about 4° from the pole in the direction of western Siberia. The agreement could hardly be better, and practically amounts to proof that in the Eocene the North Pole occupied its present position and not a point in the North Pacific as shown by Wegener.

The correlation of the glacial stages during the last ice-age in America and Europe according to Köppen and Wegener differs from that generally accepted. The most important point deals with the date of the main glaciation in Alaska, which Köppen places in the Miocene. W. H. Dall, the representative of the American Geological Survey in Alaska, does not accept this correlation as possible. He gives the following sequence at Nome (13) : coal formation during Eocene and Oligocene, locally covered by marine fossiliferous Miocene, indicate a mild climate in the early Tertiary. During the Miocene, the land sank for the most part below sea-level, and there was much volcanic activity ; the climate became cool temperate in the Early and Middle Miocene, but warmed up again in Late Miocene. Since the Miocene, the land has risen continuously. In the Pliocene the climate was moderate, and it was not until the Quaternary that Arctic temperatures set in, to persist until the present day.

The direct correlation of glacial deposits in North America and Europe is difficult, and we have to rely mainly on such features as the determination of ages by the relative depths to which the various deposits have been weathered. A careful study of all lines of evidence has recently been made by Osborn and Reeds (14), who accept in the main the results of F. Leverett (15), based on a comparison of the depth of weathering of glacial deposits and on the texture and fauna of the loess. This correlation gives :—

    I.   Glaciation—Gunz, Scanian, Nebraskan, Jerseyan.
           1.   Interglacial — Gunz - Mindel, Norfolkian, Aftonian.
    II.   Glaciation—Mindel, Kansan.
           2.   Interglacial—Mindel-Riss, Yarmouth.
    III.   Glaciation—Riss, Illinoian.
           3.   Interglacial—Riss-Wurm, Sangamon.
    IV.   Glaciation—Wurm, Wisconsin.

This correlation is in fact almost inevitable. In Europe, the Mindel-Riss interglacial is distinguished from the Gunz-Mindel and the Riss-Wurm by its much greater length, and by a temperature which probably rose higher than the present temperature, by being in fact an " interglacial " rather than an " intraglacial " period. Similarly, in America, the interval between the Kansan and Illinoian was much greater in length than the remaining interglacial periods. Since the work of de Geer and Antevs has shown that the latest glaciations in North America and Sweden were contemporaneous, this estimation of the age of glacial deposits by the depth of weathering seems to require that the long Yarmouth stage be correlated with the long Mindel-Riss interglacial. The astronomical correlation of the glacial stages has been discussed in Chapter V. ; it is not an essential part of the theory of continental drift and need not be referred to again here.

It is always a useful test of a new theory to examine how far facts which come to light after the theory has been completed fit into place. An opportunity for such a test is afforded by the study of the climatic history of Antarctica given by Wright and Priestley in the volume of results of the British Antarctic expedition dealing with glaciology (10). According to Köppen and Wegener's reconstructions, Antarctica as a whole has been in high latitudes since at least the beginning of the Carboniferous period. Of course the Antarctic continent is large, and if the pole was at one side of the continent the opposite side would extend into temperate latitudes, so I have tried to pick out on Wegener's charts the particular point to which Wright and Priestley's climatic indications refer. In this way I have obtained Table 19 (see page 244).

From this table we see that during the Upper Carboniferous Antarctica had more or less the climate appropriate to its latitude according to Wegener. In the Permian and Triassic, however, while the continent was drifting into continually higher latitudes, the climate was steadily ameliorating. The flora of the Jurassic rocks of Graham Land is extremely rich, and closely resembles that of the Jurassic rocks of Europe, which according to Wegener then lay in about 15° N. latitude. The Cretaceous fauna, while still rich, contains at least one bed of dwarf forms suggesting the oncoming of colder conditions. The only determinable plant fossil, a *Sequoia*, finds its nearest relative in the Cretaceous of Europe and Greenland. It is evident that the climate of Antarctica throughout the Mesozoic was quite incompatible with its latitude according to Wegener's theory, unless we assume in addition a great amelioration of the polar climate. For the Mesozoic, therefore, the theory of continental drift presents no advantage over any other theory. The same applies to the warm climate of the Oligocene. The doubtful glaciation of the Eocene

Period.	Locality.	Climate.	Latitude according to Wegener.
Upper Carboniferous.	South Victoria Land	Temperate to hot or cold desert. No definite evidence of glacial conditions,[1] but strong evidence of seasonal climate.	65° S.
Permian.	and		75° S.
Triassic.	Adelie Land.		85° S.
Jurassic.	Graham Land.	Sub-tropical to warm temperate.	70° S.
Cretaceous.	Graham Land.	Temperate to warm temperate.	75° S.
Eocene.	Graham Land.	First Glaciation (?).	60°— 80° S.
Oligocene.	Seymour Island. Cape Hamilton. Cape Adare.	Sub-tropical to temperate becoming frigid.	
Miocene.	Campbell Island.	Temperate (?).	50° S.
Pliocene.	Cockburn Island.	Frigid.	65° S.
Pleistocene.	General.	Maximum extension of ice.	—

Table 19.—Variations of climate in Antarctica.

is not very good evidence ; from the description given it resembles the moraine of a mountain glacier descending a narrow valley, and it is certainly of far less importance than the Carboniferous glacial deposits of North America. Campbell Island is not really Antarctic at all, and the cold climate of the Pliocene and Quaternary would be expected on any theory.

We may also compare the climatic history of Australia according to C. A. Süssmilch (16) with its position according to Köppen and Wegener. Taking the mean of their figures for Perth, Cape York and Hobart, we find that the present latitude is 29° S. In the Carboniferous the latitude of Australia was 59° S. and the climate, warm at first, became very cold by Mid-Carboniferous. In the Permian the latitude was 71° and the climate cold at first, becoming

[1] Wegener fits the Antarctic continent into the Great Australian Bight. According to Sir T. W. Edgeworth David, the Australian ice-sheet radiated from a point to the south-west of Tasmania, which according to Wegener's reconstruction would be in the Antarctic. The non-glaciation of the Antarctic in the Upper Carboniferous, if confirmed by further research, will be a strong point against Wegener's reconstruction.

warmer. So far the agreement is satisfactory. In the Trias the latitude was 68° and in the Jurassic 65°, but according to Süssmilch the climate was probably warmer than to-day. In the Cretaceous Australia had moved north, to 57°, but the climate was colder. In the Miocene the latitude was 43° but the climate was at least 10° F. warmer than to-day. In the early Quaternary Australia moved to 54° S. and the climate became colder, but not very cold. There is in fact no agreement between the climatic changes in Australia and the path of the South Pole according to Köppen and Wegener.

The evidence of the pre-Carboniferous deposits is still very meagre. The constitution of the slate-greywacké formation of Robertson Bay, South Victoria Land, which is of Late Proterozoic or very early Palæozoic age, strongly suggests the action of alternate freezing and thawing, and these deposits may be the Antarctic representatives of the Late Proterozoic-Early Cambrian glaciation. Later in the Cambrian we have evidence of a moderately warm sea stretching nearly or right across Antarctica, in the form of thick limestones very rich in reef-building *Archæocyathinæ*. Compared with the forms from Australia, however, all the Antarctic forms are either embryonic or dwarfed, indicating that they lived in colder and presumably more southern seas. No reliable evidence is yet available from the Silurian or Devonian periods. It seems, therefore, that the chief, perhaps the only, justification for the theory of continental drift rests on the distribution of climatic zones during the Upper Carboniferous period.

REFERENCES

(1) WEIDMANN, S. " Was there a Pennsylvanian-Permian glaciation in the Arbuckle and Wichita mountains of Oklahoma ? " *J. Geol.*, Chicago, 31, 1923, p. 466.
(2) TAFF, J. A. " Ice-borne boulder deposits in Mid-Carboniferous shales." *Bull. Geol. Soc. Amer.*, 20, 1908, p. 701.
(3) COLEMAN, A. P. " Late Palæozoic climates." *Amer. J. Sci.*, 9, 1925, p. 195.
(4) SAYLES, R. W. " Seasonal deposition in aqueoglacial sediments." Cambridge, Mass., *Mem. Mus. Comp. Zool.*, 47, 1919, No. 1.
(5) WOODWORTH, J. B. " Boulder beds of the Caney shales at Talihina, Oklahoma." *Bull. Geol. Soc. Amer.*, 23, 1912, p. 457.
(6) SÜSSMILCH, C. A., and T. W. E. DAVID. " Sequence, glaciation, and correlation of the Carboniferous rocks of the Hunter River district, New South Wales." Sydney, *J. R. Soc. N.S. Wales*, 53, 1919, p. 246.

(7) COLEMAN, A. P. " Ice ages and the drift of continents." *J. Geol.*, Chicago, 41, 1933, p. 409.

(8) ANTEVS, E. " The climatologic significance of annual rings in fossil woods." *Amer. J. Sci.*, 9, 1925, p. 296.

(9) ZEILLER, R. " Les provinces botaniques de la fin des temps primaires." *Rev. gen. sciences*, 8, 1897, p. 5.

(10) BRITISH (TERRA NOVA) ANTARCTIC EXPEDITION, 1910-1913. " Glaciology," by C. S. WRIGHT and R. E. PRIESTLEY. London, 1922.

(11) BERRY, E. W. " The past climate of the North Polar region." Washington, *Smithson. Misc. Coll.*, 82, no. 6, 1930.

(12) CHANEY, R. W. " Tertiary forests and continental history." New York, *Bull. Geol. Soc. Amer.*, 51, 1940.

(13) DALL, W. H. " Pliocene and Pleistocene fossils from the Arctic coast of Alaska and the auriferous beaches of Nome, Norton Sound, Alaska." U.S. Geol. Survey, *Prof. Papers*, 125 C., 1920, p. 23.

(14) OSBORN, H. F., and C. A. REEDS. " Old and new standards of Pleistocene division in relation to the pre-history of man in Europe." *Bull. Geol. Soc. Amer.*, 33, 1922, p. 411.

(15) LEVERETT, F. " Comparison of North American and European glacial deposits." *Zs. Gletscherk.*, 4, 1910, p. 241.

(16) SÜSSMILCH, C. A. " The climate of Australia in past ages." *J. Proc. R. Soc. N.S. Wales*, 75, 1941, p. 47.

CHAPTER XV

THE CLIMATE OF THE UPPER CARBONIFEROUS GLACIAL PERIOD

WE have now to look more closely at the geographical and climatic conditions of the Upper Carboniferous, in order to see if the distribution and elevation of the land-masses were such that ice-sheets might conceivably have developed in low latitudes, while at the same time a comparatively mild climate obtained farther north. Fig. 29 gives a rough reconstruction of the geographical conditions on the supposition that the continents were in their present positions. This reconstruction is based on that given by Th. Arldt (1) with some alterations to include the results of later work. In the Northern Hemisphere we find three small continents : Nearctis, a primitive North American continent ; North Atlantis, including Greenland and Western Europe ; and Angaraland, occupying part of the present Siberia. Nearctis and North Atlantis were connected by a land-bridge in about latitude 50° N. South of these three continents the Tethys Sea, the forerunner of the Mediterranean, extended east and west from New Guinea to Central America, sending an arm between North Atlantis and Angaraland to the Arctic Ocean. This Tethys Sea was bounded on the south by the great continent of Gondwanaland, extending in a huge irregular crescent from South America to Australia, and from 20° N. to 40° S. In connexion with the great extent of Gondwanaland from west to east, it may be remarked that the Upper Carboniferous marine fauna of South-eastern Australia resembles that of South Africa, while the fauna of Western Australia is quite different and resembles that of the Tethys Sea. This indicates that there was a continuous land barrier separating the gulf west of Australia from the seas south-east and east of Africa.

The principal difference between the land and sea distribution of the Middle and that of the Upper Carboniferous seems to be that in the former Gondwanaland was not

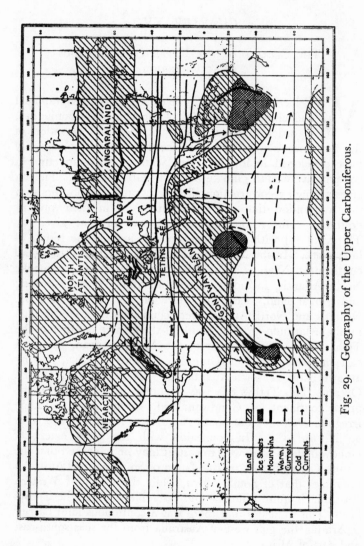

Fig. 29.—Geography of the Upper Carboniferous.

continuous from South America to Australia, but was probably broken up into three or perhaps four separate land-masses by straits leading from north to south. These are indicated by the broken lines of Fig. 29. By allowing free circulation between the waters of the Tethys Sea and those of the Southern Ocean, these breaks in the land barrier would raise the temperature of the Southern Hemisphere considerably, and help to account for the great climatic difference between the Middle and Upper Carboniferous.

We know that the Carboniferous was a period of great mountain-building. The mountain ranges followed two main directions, north-south and east-west. A range followed the south coast of North Atlantis into the Mediterranean region. The site of the Alps was occupied by the Carnic range of Mont Blanc and there were other ranges in the Caucasus and the Dobrudja. In the west, the Pyrenees and Asturia were mountainous. Farther north a range or a series of ranges ran from Bohemia first northward through the Sudetes and then westward through Germany. In the west we have the Armorican chain through Brittany, South England, Wales, South Ireland, and beyond into North Atlantis. In North America the Appalachian Mountains were forming, turning westwards in the south to the mountains of Oklahoma and South Arkansas. Other north-south ranges formed the west coast of Nearctis, from Colorado northwards. In Asia there were a number of east-west ranges somewhat similar to the present systems, bounded on the west by the Urals and the Volga Sea, on the south by the Tethys Sea. The mighty continent of Gondwanaland was bounded on the west by the Proto-Cordilleras and the mountains of the Pampas, extending to the Falkland Islands. In the south of Africa there was another chain, and a great range ran along the whole eastern coast of Australia and the present East Indies. These mountain systems are shown by the heavy lines in Fig. 29. Between the mountain ranges in the Northern Hemisphere there were, for the most part, wide moist valleys open to the sea, the home of a rich vegetation.

Opinions differ about the structure of the main area of Gondwanaland, i.e., whether it consisted of an extensive high plateau or a series of mountain ridges. It is generally agreed, however, that the Upper Carboniferous was a period of great mountain-building and the general elevation was probably high. The great thickness of the Upper Carboniferous of South Africa, for example, points to rapid denudation, suggesting a large area of high ground in the interior of that continent. The fact that the ice-sheets spread out from a line near the equator shows that initially at least the ground was highest there, and may well have been a ridge 10,000 or 15,000 feet above sea-level.

Finally, it has to be remarked that the Upper Carboniferous was a time of intense volcanic activity, and especially in Australia, great thicknesses of agglomerates point to numerous explosive eruptions from which we may infer the presence of great quantities of volcanic dust in the atmosphere, forming a veil which, as Humphreys has shown, would be very effective in shutting out the solar radiation, while it would allow the terrestrial radiation to escape with little hindrance (see Chapter VI.).

At the beginning of the Upper Carboniferous there appears to have been a general decrease in the temperature of the seas, indicated by an impoverishment of the fauna and flora resulting from the extinction of a number of animals and the withdrawal of the coral boundaries towards the equator. The land plants also suffered changes, and the introduction of holometabolism in insects (*i.e.*, the pupa stage) is attributed by A. Handlirsch (2) to this decrease of temperature. Sir T. W. Edgeworth David considers that the mean temperature of the tropical oceans decreased by about 10° F., and it is probable that the oceans were ice-covered in high latitudes.

We can now attempt to reconstruct the distribution of the meteorological elements. Over the open Pacific Ocean there is no reason to suppose that the system of pressure and winds was appreciably different from that prevailing now. Hence we postulate a great equatorial current setting westward towards the eastern coast of Gondwanaland, with a temperature of about 70° F. The configuration of this coast of Gondwanaland appears to have been very favourable for concentrating the warm current and directing it into the Tethys Sea, more favourable even than the present configuration of the coast of America for concentrating the equatorial currents of the Atlantic in the Northern Hemisphere. A very warm current, with a temperature initially in the neighbourhood of 70° F., must have flowed through this narrow inter-continental sea. The supply of warm water would have been large enough to give this warm current a considerable velocity —perhaps fifty miles a day—enabling it to conserve its heat over a long distance. Part of this warm current turned northward through the Volga Sea and brought favourable conditions to Northern Russia, where the Fusulinæ appear to indicate a temperature similar to that of the present Mediterranean (3), and to Spitsbergen and the Arctic Ocean. The remainder of the warm Tethys current travelled on between the Americas, and finally emerged again into the Pacific Ocean. Evidently there is no difficulty in accounting for the corals of the Tethys and Volga Seas or the rich vegetation of the valleys opening off them.

In the Southern Ocean also, between the horns of the great crescent of Gondwanaland, there is no reason to suppose that the surface temperatures differed greatly from those prevailing at present in the same latitudes of the South Indian Ocean. The absence of the great masses of floating ice derived from the Antarctic would have tended to raise the temperature of the whole ocean, but, on the other hand, the Southern Ocean was mainly limited in the north by land in the neighbourhood of 40° S. instead of extending to the equator, and the absence of the supply of equatorial warm water would tend to balance the absence of the supply of ice. We have also to take into account the volcanic dust veil. At present the mean temperature of the ocean surfaces in latitude 40°-50° is about 50° F., and we shall probably not be far out if we take the same figure for the temperature of the surface water south of Gondwanaland at the beginning of the Upper Carboniferous glacial period. When the land ice reached the sea over wide fronts, the temperature of the surface must have fallen much lower.

One other point about the distribution of the seas is worthy of notice, namely, the long gulf extending from the Arctic regions into the heart of Eastern America, that is, into the only region outside Gondwanaland which appears to have been indubitably and severely glaciated during the Upper Carboniferous.

P. Lasareff (see Chapter III.) made an experimental reconstruction (see page 78) of the ocean currents of the

Middle Permian (Fig. 30). This agrees in general with Fig. 29, especially in the warm current through the Tethys Sea and the branch across the North Pole. The Gulf between Nordatlantis and Nearctis is not shown (this of course may be due to the time-difference) but instead there is a cold current from the pole running down the western coast of Nearctis,

Fig. 30.—Ocean currents of Middle Permian. After Lasareff.

which would be equally effective in giving a more severe climate to the present North America than that of Europe.

We have now to discuss the system of the winds over Gondwanaland and the neighbouring seas. The " planetary " circulation of the atmosphere (Chapter II.) requires a belt of low pressure over the equatorial regions, a series of anticyclones in about latitudes 30° north and south, followed by belts of low pressure and storms in temperate or sub-polar latitudes. This system is modified by the land and sea distribution, which gives a tendency for high pressure over the land in winter and over the sea in summer. The geographical disturbance of the planetary circulation is now so great over Asia that the anticyclone does not develop there at all in summer, while in winter it attains a great intensity and is displaced some distance north of 30° N. The deflection of the entire Equatorial Current northwards into the Tethys Sea would probably suffice to maintain the temperature of middle latitudes north of Gondwanaland permanently above that of middle latitudes south of Gondwanaland, introducing an effect of permanent summer in the Northern Hemisphere and permanent winter in the Southern Hemisphere. We should expect to find a permanent low-pressure area over the Tethys Sea and the Volga Sea, while the normal sub-tropical anticyclone was developed off the southern coast of Gondwanaland.

Under these conditions, with an area of high pressure to the south of the equator separated by a very long and lofty continent from an area of low pressure north of the equator, what would be the wind system over the plateau ? We have no close parallel at present for guidance ; the nearest approach is found over Asia and the Indian Ocean during the south-west monsoon, but most of India is now at a comparatively low level, forming a plain out of which the Himalayan ridge rises steeply to a great height. The highest temperature and lowest pressure are found near Jacobabad to the south of this ridge. A large portion of the air which enters India from the high pressure area to the south is accordingly deflected to flow westward parallel with the mountain barrier ; it is only in the north-east that the air crosses at least the Khasi Hills (giving the enormous rainfall of Cherrapunji) and possibly the main ridge of the Himalayas. If the area of lowest pressure were to the north of the latter range, would the stream lines run directly across it ?

The power of air currents to cross high ridges of land probably depends to a large extent on the steepness of the slope. The west-south-west winds from the Arabian Sea are able to cross the Western Ghats, for the greater part of their length more than 4,000 feet in height (4), and descend on the other side as a dry wind. The obstacle presented by the Himalayas, exceeding 12,000 feet, may be due quite as much to the steepness of their southern slopes as to their great height. If the ground sloped more gradually from Southern India or beyond to the Tibetan plateau, it seems probable that the height would be a less serious obstacle.

Let us now suppose that the discussion of this first problem has shown that the air, starting from the Southern Ocean as a powerful south-east trade wind and changing to south-west as it crosses the equator, will climb steadily up the surface of the high ground, and, crossing the crest, will descend its northern slope. Under these conditions, what would be the general temperature and weather over the area? The air starts at a temperature of about 50° F. and a humidity approaching saturation ; if the temperature of the plateau surface is mainly above 50° F., the air will not part with its moisture readily and the cloud amount will be small, but if the temperature of the surface is below 50° F., the sky will be mainly overcast and the precipitation heavy.

Here again much seems to depend on the topography. If the southern margin of the continent was formed by a wide plain at a low level, the air would not at once be forced to rise sufficiently to develop an extensive cloud layer, the low land surface would therefore be exposed to intense insolation and would be very hot, and the air would be warmed up by contact with it to such an extent that it might rise to very high levels before it again became saturated. If, on the other hand, the ground rose fairly steeply from the sea to an elevation of 2,000 feet or so, the formation of a thick cloud layer would begin before the air had time to warm up.

In North-eastern India, where the air crosses the Khasi Hills and enters at least the foot-hills of the Himalayas, the cloudiness during the south-west monsoon is very great. The mean cloudiness (8 a.m.) at Cherrapunji, Darjiling, and Mercara during June, July, and August is nine-tenths of the sky, while the relative humidity is 95 per cent. Let us suppose that, the southern coast of Gondwanaland being sufficiently steep to raise the incoming trade wind above the saturation level, the mean cloudiness was nine-tenths.

The temperature of the air at any point is governed by

the quantity of heat which it originally contained, plus the heat which it gains mainly from the surface, minus the heat which it loses, mainly by radiation. In the conditions postulated, the cloud would probably be rather thick ; let us assume that it had a mean density of 3. Then according to the measurements of B. Haurwitz (see Chapter VII.) the amount of radiation penetrating a cloud cover of nine-tenths would be barely half that received with a cloudless sky. This is less than the average amount received at present in latitude 40°-50°. If we take into account the loss by scattering, especially great during the Upper Carboniferous because of the great amount of volcanic dust, which would be more or less proportional to the total solar radiation and therefore greater in low than in high latitudes, we see that the amount of solar radiation available for warming the earth's surface was probably less over Gondwanaland than over the ocean to the south of it.

To this it might be objected that the under surface of a cloud layer is also effective in reflecting the terrestrial radiation back to the surface of the earth, and that this would redress the balance. But it was shown in Chapter VII. that the reflecting power of clouds for long-wave terrestrial radiation is much less than their reflecting power for short-wave solar radiation.

From this it appears that if the air entering Gondwanaland from the south immediately formed a cloud layer, it would not gain any heat from the surface of the plateau. If it continued to rise along a sloping surface, it would continue to cool by expansion and form fresh cloud. Taking the initial temperature of the air as 50° F. and the vertical temperature gradient as 3° F. per 1,000 feet, the snow-line would be reached at a height of 6,000 feet. There is no evidence against the supposition that at the beginning of glaciation a large part of Gondwanaland was above this height, so that under the conditions postulated, extensive snowfields could develop. The supply of snow would be ample, since precipitation would go on throughout the year, instead of for a few months only as in India, although, of course, the monthly totals would not equal those recorded at the wettest stations during the height of the monsoon. Hence there would be a plentiful supply of ice to extend below the snow-line on the southern slopes, and to reach the sea along broad fronts.

It may be remarked here that owing to the absence of strong seasonal contrasts, the formation of ice-sheets would probably be very susceptible to slight changes of temperature and snowfall. At present glaciers form on high mountains near the equator, but owing to the steepness of the mountain slopes they descend rapidly to warmer levels and melt. Given larger gathering grounds, suitable high-level basins in which the ice could accumulate, and weakened solar radiation to lessen melting, the rapid development of glaciers into ice-sheets would appear to be inevitable. Once the snow cover had been formed and inland ice-sheets had begun to develop, conditions would at first be very favourable for their rapid growth. The ice surface would reflect a considerable part of the weakened radiation which penetrated the cloud, so that the surface would be very cold. On the southern slopes, south of the equator, there would be a tendency for the air drainage to form a north-west wind, which would come into opposition with the south-east monsoon-trade wind. At the surface both winds would probably prevail in turn—first the relatively mild and moist south-east wind with a light snowfall, then an interval of calm, followed by a blizzard from the north-west, lifting the moist air of the south-east wind bodily and bringing a burst of heavy snow. But these changes would be limited to a shallow surface layer, and over all the south-east wind would blow steadily and the skies would be heavily clouded. The ice-sheet would merely add to the effective height of the land.

But we can go much further than this. Once the ice-sheet had been formed, it would by its own cooling power and by the reflection of solar radiation from its surface back to space, effectively lower the snow-line over the glaciated area and its immediate neighbourhood, and thus enable the ice to spread over low ground previously unglaciated, or to survive a subsidence of the ground on which it rested below the original snow-line of 6,000 feet. Once a large ice-sheet has been formed, its persistence is probably almost independent of the latitude, and it can only be destroyed by the cessation of the supply of snow, by a great increase of the ablation, or by a subsidence of its bed below sea-level. Hence, during the later stages of the Upper Carboniferous glaciation, the land surface, worn down by glacial erosion and depressed by the weight of the ice, may actually have been a low plain, instead of the lofty plateau which we have supposed necessary for its inception. So long as the temperature difference was maintained between the Tethys Sea and the Southern Ocean, the snowfall would have been sufficient and the ablation small, so that the destruction of the Upper Carboniferous ice-sheets probably came about by subsidence—and, in fact, the boulder clays are generally overlain by marine beds.

The greater part of the Gondwanaland ice-sheet was apparently formed on the southern slopes of the continent, where conditions were most favourable. There can have been little, if any, snowfall on the northern slope, and for a glacier to reach the Tethys Sea an exceptional topography would be required. In fact, it happened in only one region, India, and if we may judge from the thinness of the deposits, for a short time only. We may suppose that there the main watershed lay very far north, and that the high ground formed a sort of funnel through which a large amount of rising air was forced to pass. Somewhat similar conditions give rise to the abnormal rainfall of Cherrapunji at present. This high ground would form a very rich gathering ground for snowfields. If, then, a valley on the northern slope cut back deeply into this high ground, it might well receive sufficient ice for a glacier to reach the sea.

The high temperature of the Tethys Sea and the Volga Sea would give a very favourable climate to Angaraland and the eastern parts of North Atlantis. The warm seas would keep the temperature high and the great evaporation would give rise to a heavy rainfall, making the low broad valleys among the mountains, open to the warm air from the sea, very favourable for the growth of a rich vegetation, able to give rise to the coal measures. These conditions would extend to the southern parts of Nearctis, where a small inlet seems to have existed in such a position that it carried the oceanic influences into the heart of the American coal district. The glaciers of Europe, if they existed at all, were probably not more than small mountain glaciers, such as can develop in any latitude provided the mountains are sufficiently high and the precipitation sufficiently heavy, and we know that in Late Carboniferous Europe there were both mountain ranges and heavy precipitation. The more extensive glaciation of North America can be associated with conditions in the Arctic Ocean.

We have seen that the Volga Sea was occupied by a warm current which carried mild conditions as far as Northern Russia, while even in Spitsbergen there was sufficient vegetation to form workable coal beds. Evidently, where the Volga Current entered the Arctic Ocean, there was an ice-free area similar to that now formed by the Gulf Stream Drift, but larger. The Volga Current was more powerful than the Gulf Stream Drift, as we should expect it to be from the more favourable topography. Conditions between Angaraland and Nearctis are not sufficiently known to decide whether a similar warm current entered the Arctic between these continents, or whether they joined or approached so closely—as at present—as to prevent any warm current from passing between them. The map of this region is rather hypothetical, but the climatic conditions suggest that there was no such current. The North American glaciation seems to require the presence of a floating cap of sea ice to the north of Nearctis. In Chapter I., however, we saw that at present the Arctic Ocean is only a few degrees below the critical temperature required for the formation of such an ice-cap. At present, heat is carried into the Arctic by the Gulf Stream Drift and by the powerful warm south-west winds associated with the Icelandic minimum and its north-easterly extension. In the Upper Carboniferous the Volga Current probably supplied more heat than the Gulf Stream Drift, but the barometric distribution was probably less favourable that at present for the warm south-west winds, since the alignment of the Volgan cyclone was probably west to east instead of south-west to north-east. The importance of the latter point for the Quaternary glaciation has been well brought out by the late F. W. Harmer (5). We may regard the balance as about even, but the addition of a Bering Sea Current would—other things being the same—certainly raise the Arctic temperature above the critical point and bring about an ice-free Arctic Ocean. Hence it seems probable that there was little or no flow of warm water into the Arctic Ocean through a channel between Nearctis and Angaraland.

Fig. 29 shows a long gulf extending from the Arctic far into the temperate zone between Nearctis and North Atlantis. This gulf, open to the Arctic but closed to the south, must have exerted a very powerful effect on the climate of the neighbouring land. It is doubtful if it froze in winter into a continuous surface of ice ; the influence of the warm sea to the south of the narrow land barrier may have been sufficient to keep it open, but if, as we suppose, the Arctic Ocean to the north was ice-covered, the waters of this gulf must have carried a great deal of loose floating ice in winter and spring. That this was so is shown by some of the North American glacial deposits, which contain large boulders apparently transported by icebergs or shore ice. Even in summer the water must have been very cold. There was probably a slow circulation, southward along the western side of the gulf and northward along the eastern side, similar to that now found in Baffin Bay, except that the latter being open to the south, the ice can escape into the Labrador Current. The close neighbourhood of this cold water to the north of the isthmus and the warm water of the Tethys Sea to the southward must have given rise to a tendency for great storminess, heavy rain, and dense fogs, weather similar to that now prevailing on the Newfoundland Banks. Given some mountainous country, such as we know to have been present, conditions here were very favourable for a moderate glaciation.

Finally, we have to consider the climate of the Antarctic. According to Wright and Priestley (6), the Antarctic continent was not glaciated during the Upper Carboniferous ; instead, there was a dry and wind-swept plateau subject to severe frost action, and more favourable conditions in sheltered coastal lowlands in which a fairly rich flora was able to develop. Of course these conclusions refer only to the coastal parts of the continent ; the interior may have been glaciated to some extent, but even so the fact is surprising ; the general conditions over the Antarctic appear to have been no more favourable than at present, and we should have expected to find it glaciated. Probably it was not that the temperature was too high, but that the snowfall was too low compared with the ablation. With regard to present conditions, it is not certain that if the Antarctic ice-sheet could be melted entirely away, it would be re-established under the present climatic régime ; it may be simply a survival from the Quaternary Ice-Age. It is quite possible that if the greater part of the surface of the Antarctic during the Upper Carboniferous consisted of a fairly level plateau, the climatic conditions would resemble those of Northern Siberia, the very cold but comparatively dry winter not giving enough snowfall to persist through the summer.

Thus we see that starting from a restoration of the distribution of land and sea during the Upper Carboniferous on the basis of the existing positions of the continental massifs, and deducing the system of winds and ocean currents and the local climatic conditions in accordance with meteorological experience as represented by the nearest modern analogies, we arrive at a very fair reconstruction of the peculiar climatology of this period. The critical assumption is the considerable elevation of the central parts of Gondwanaland, but this is not entirely unsupported by evidence, and is at worst not less hazardous than the extensive migrations of the continents through some fifty degrees of latitude. We find that outside Gondwanaland the only extensive glaciation occurred in Nearctis exactly where we should expect it, while on the theory of continental drift this region would lie on the equator and would be unglaciated. The main difficulty, the non-glaciation of the Antarctic, is common to both theories.

There is one peculiarity in the action of volcanic dust which deserves mention. In any latitude its cooling power is about proportional to the solar radiation received at the limit of the atmosphere, while its warming power, though much smaller, is proportional to the radiation from the earth's surface and the lower layers of the atmosphere. The cooling varies regularly with latitude, being greatest at the equator and least at the poles, while the warming is greatest in the warmest regions. Hence the effect is to increase the abnormalities of temperature brought about by the distribution of land and sea ; Gondwanaland was cooled much more effectively than the Tethys, and in Spitsbergen during the polar night the effect was pure gain.

Summing up, it appears that extensive glaciation in low latitudes required at least three, and possibly four conditions :—

1. The diversion of the whole of the equatorial ocean current into the Northern Hemisphere, which thereby became abnormally warm.

2. An extensive elevated continent along the equator, but extending much farther into southern than into northern latitudes.

3. A southern ocean shut off by land barriers from all warm currents.

4. Possibly, a general refrigeration which might be due to the presence of abnormally large quantities of volcanic dust.

So far as I can discover, these conditions occurred only once in geological time, and that occasion coincided with the only occurrence of extensive glaciation in low latitudes. Further, the climates of other parts of the world are such as would be expected from them. The Coal Measures become, not a violent negation of the possibility of glaciation, but a necessary complement to it. It seems to me that this geographical explanation is simple and natural, and does not violate probabilities as does the arbitrary shifting of continents and poles.

In the same year (1926) that I first published this geographical hypothesis of the Carboniferous glaciation, a somewhat similar theory was put forward by C. Schuchert (7) who believed, however, that the date of the glaciation was definitely Permian. He writes : " It is in the youthful topography, the enlarged continents and the peculiar connexions of the lands that seemingly are to be sought the reasons for the Permian Ice-Age. . . . This holding in of so much of the waters of the Antarctic Ocean, combined with the moist climates in the Southern Hemisphere and the general highland condition of much of the world in early Permian time, will be the explanation for the peculiar position of the continental ice-masses of the Southern Hemisphere." Later, Bailey Willis (8) attributed the glaciation of South America and South Africa to refrigeration of the South Atlantic, shut off from warm currents. The centres of glaciation were near the warm seas, which provided moisture by rising over wedges of colder air, giving cold foggy summer weather, favourable to a low snow-line. He considers that the glaciation of India was due to local high mountain ranges.

REFERENCES

(1) ARLDT, TH. "Handbuch der Palæogeographie." Leipzig, 1919.
(2) HANDLIRSCH, A. "Die Bedeutung der fossilen Insekten für die Geologie." Wien, *Mitt. Geol. Ges.*, 3, 1910, p. 503.
(3) STAFF, H. v. "Zur Entwicklung der Fusuliniden." *Zentralbl. f. Min., Geol. und Pal.*, 1908, p. 699.
(4) SIMPSON, G. C. "The south-west monsoon." London, *Q. J. R. Meteor. Soc.*, 47, 1921, p. 151.
(5) HARMER, the late F. W. "Further remarks on the meteorological conditions of the Pleistocene epoch." London, *Q. J. R. Meteor. Soc.*, 51, 1925, p. 247.
(6) BRITISH (TERRA NOVA) ANTARCTIC EXPEDITION, 1910-1913. "Glaciology," by C. S. WRIGHT and R. E. PRIESTLEY. London, 1922.
(7) SCHUCHERT, C. "The palæogeography of Permian time in relation to the geography of earlier and later periods." *Proc. 2nd Pan-Pacific Sci. Congr.*, 1926, p. 1,079.
(8) WILLIS, BAILEY. "Isthmian links." New York, *Bull. Geol. Soc. Amer.*, 43, 1932, p. 917.

CHAPTER XVI

THE CLIMATE OF THE QUATERNARY

OF the four great ice-ages, the first two, the Lower Proterozoic, the Upper Proterozoic-Lower Cambrian, and the last, the Quaternary, were developed mainly in what are now temperate latitudes, while the third, the Upper Carboniferous, found its maximum extent in regions now not far from the equator. The Upper Proterozoic-Lower Cambrian glaciation was apparently similar in many respects to the Quaternary, but as yet we know so little about it that no detailed discussion is possible. In this chapter it is proposed to refer briefly to the meteorology of the Quaternary period.

At the maximum of the Ice-Age, E. Antevs (1) estimates that the ice-sheets occupied an area of about 13 million square miles. Of these 4½ million were in North America, 1¼ million in Europe, 1⅓ million in Asia, and about 5 million in the Antarctic. The remainder was made up of the expanded ice-sheet of Greenland and relatively small areas in Australia, New Zealand and South America. The present ice-covered area of about 6 million square miles is almost entirely in the Antarctic and Greenland. It is not certain that all areas reached their maximum at about the same time, but two lines of evidence suggest that this is nearly true. The first is the lowering of sea-level by the abstraction of water, which according to Antevs's estimate would amount to 305 feet below the present if all the ice-sheets reached their maximum extent and thickness together. Various estimates have been made from the present depths of shore deposits, coral reefs, etc., which give a minimum figure of about 260 feet. The second line of evidence is the depression of the snow-line in North, Central and South America during the latest (Wurm) glaciation, which has been remarked on by many authors, and which is almost constant right across the equator, increasing somewhat in regions of heavy rainfall and decreasing in dry regions :—

Latitude	45 N.	17 N.	10 N.-20 S.	40 S.
Depression of snow-line (feet) . . .	2,300	3,000	1,300-2,000	3,300

The character of the ice-sheets and glaciers varied. In Northern Europe, North America and presumably also in Greenland and the Antarctic they were several thousand feet thick and spread out actively from various centres. Siberia was for long considered to have had only minor mountain glaciers, but recent work summarised, *e.g.*, by R. F. Flint and H. G. Dorsey (2) shows that there was at one time a large ice-sheet in north-west Siberia extending over the Arctic shelf, though the ice was thin and inactive. Farther east and south there were extensive piedmont glaciers. In South America the glaciers were also mainly of the piedmont type.

The ice-age was divided into glacial and interglacial periods by a series of large-scale advances and recessions of the ice. These are best known from Europe and North America, and appear to run closely parallel in the two continents. The succession is summarised by F. E. Zeuner (3), K. Bryan (4) and others as follows (the youngest at the top) :—

Alps.	N. German Plain.	Continental U.S.A.
Wurm	Weichsel (including Warthe)	Wisconsin (including Iowan)
Riss	Saale	Illinoian
Mindel	Elster	Kansan
Gunz	?	Nebraskan (Jerseyan)

The Mindel and Riss glaciations were the most extensive. The Mindel-Riss interglacial (Yarmouth Interglacial in America) was very long, of the order of 240,000 years, and was generally mild, but was interrupted by at least one colder period which did not reach glacial intensity. The Riss glaciation had a double maximum in Europe at least. The Riss-Wurm interglacial was short and mild but was interrupted near the middle by a period of sub-arctic conditions in Jutland.

The Wurm glaciation was less extensive than the Riss ; it comprised three maxima, each of less intensity than the preceding one, separated by sub-arctic or even cold-temperate conditions. The recession after the third peak was interrupted by several halts or slight readvances.

In other parts of the world the succession is less complete. In Kashmir and neighbouring territories F. Loewe (5) recognises three glaciations, which he correlates with Mindel, Riss and Wurm, decreasing in intensity, and there are probably traces of the same three in East Africa; in both cases, however, the correlation is not certain. In nearly all other parts of the world the remains of only two glaciations are found, presumably representing the Riss and Wurm, the former always being much the more extensive. In Siberia for example there was no Wurm ice-sheet, only mountain and piedmont glaciers. It is not yet certain that the earlier glaciations did not occur ; their moraines may have been destroyed by later advances.

In discussing the cause of the Quaternary Ice-Age, it is necessary to distinguish between the Ice-Age as a whole, and the succession of glacial and interglacial periods. In Chapter XII. it was shown that the most probable cause for Ice-Ages was elevation and mountain building in extensive high continents, limited accession of warm ocean currents to high latitudes, and probably much volcanic dust in the atmosphere, all of which factors were present at the beginning of the Quaternary. A decrease in the amount of CO_2 in the atmosphere (Chapter VI.) may have been a contributory factor. With the exception of volcanic dust, however, these are all stable factors, which would not change sufficiently rapidly to account for the succession of glacial and interglacial periods. For the latter therefore some other explanation must be found.

The fact that the last two glaciations at least began and ended more or less together in all parts of the world is highly significant. In a minor degree it is paralleled by the recession of the glaciers everywhere within the past hundred years. It shows that glaciations in different regions do not depend only on local conditions, but are mainly controlled by some world-wide factor such as the temperature of the oceans, the heat received from the sun, or the circulation of the atmosphere, or by some combination of them. The late beginning of glaciation in the tropics and Southern Hemisphere, if confirmed, suggests that ocean temperature may be important, because the lag in cooling the oceans would be greater the farther removed the region is from the sources of cold water in the Arctic and North Atlantic.

The first question concerns the lag between the occurrence of mountain formation and the beginning of glaciation, which was discussed in Chapter X. There are four possible reasons :—

1. The slow cooling of the oceans.
2. The erosion of the mountains.
3. The occurrence of a period of explosive volcanic activity.
4. The occurrence of favourable astronomical conditions.

An important factor in fixing the actual beginning of the Quaternary glaciation over the land must have been the general temperature of the sea. At the close of a long warm period the sea is warm throughout its whole depth ; there is none of the very cold bottom water which exists at present. This must be so, for the temperature of the sea depths cannot long remain lower than the temperature of the coldest part of the surface. Now the beginning of the Quaternary glaciation was a period of great elevation in most parts of the north temperate belt. The gap between Greenland and Norway, which at present conducts the Gulf Stream into the Arctic Ocean, was greatly narrowed if not completely closed. Bering Strait probably differed little from its present condition, and there may have been an open channel to the west of Greenland. Now there are some interesting peculiarities in the development of the Quaternary glaciation which may have a bearing on this question of the cooling of the seas. The first glaciation of Europe was most extensively developed in Scandinavia and North Russia ; the British Isles were probably not glaciated until later. The corresponding glaciation in America was developed in the Rocky Mountains of British Columbia and in Labrador, but not in the central parts. The glaciation of British Columbia was apparently an enormous development of valley and piedmont glaciation due to the great height of the mountains, but the North European and Labradorean centres developed true inland ice-sheets. If we suppose that the elevation of the Wyville Thomson ridge between Greenland

and Scotland above its present level shut out the Gulf Stream from the coast of Norway, the Arctic Ocean would lose almost all the supply of heat formerly carried into it by ocean currents and its temperature would begin to fall. The ocean south of the Wyville Thomson ridge would still be very warm, however, and the winds must have brought a considerable amount of heat across the land barrier. It is difficult to estimate the time which would be required under these conditions for the thorough cooling of the Arctic Ocean. Most of the ice formed in the Arctic at present begins with the freezing of a surface layer of relatively fresh water brought down by the great rivers which enter the basin, and which, owing to its smaller density, floats on the main mass of warmer but more saline water. Probably this water must freeze fairly near the coast, otherwise the storm winds would break it up and mix it with the underlying salt water. By analogy with what happens at present at the junction of the Labrador Current and the Gulf Stream, we can say that the fresh layer would become salt more quickly than it would warm up (6). The resulting mixture would be heavier than both the upper and lower layers, and would therefore sink. But while the main oceans were still warm, it seems probable that the heat transferred by southerly winds would suffice to keep this layer of fresh water liquid long enough for the mixing process to destroy it. Thus we conclude that the formation of a cover of floating ice probably did not follow immediately on the elevation of the Wyville Thomson ridge, but had to wait until the cooling of the main oceans had progressed some way.

At first sight it might seem that the accumulation of cold bottom water could not possibly affect the atmospheric processes which go on above the surface of the oceans. Such an influence does take place, however, especially off the western coasts of the continents, where cold bottom water wells up to replace the surface water driven away by easterly winds. Investigations into the effect of the Trade winds on the surface temperature of the North Atlantic have shown that the North-east Trade, blowing off the coast of West Africa, does actually bring up a large amount of cold water from the underlying layers. This cold water has the effect of lowering the temperature of the Gulf Stream, and ultimately the surface temperature of the North Atlantic between the United

States and Ireland, probably by several degrees. If the depths of the oceans were much warmer than at present, this cooling influence would not exist. The meteorological effects of upwelling cold water on the western coasts of South America, South Africa, and Australia are extraordinarily marked, being largely responsible for the desert character of those coasts. It is probable, however, that this cold water comes, not from the greatest depths, but from some intermediate layer, and that a certain accumulation of cold water could take place without affecting surface conditions.

We have no means of knowing how long it took to cool the main body of the oceans, but it was certainly a very long time. As an example of the quantities involved, if we suppose that all the thaw water of the ice, both land and sea ice, which melts each summer in both hemispheres, sank to the bottom of the oceans and spread out there, it would take between ten and twenty thousand years to fill the oceans with cold water. Immediately after the formation of the Wyville Thomson ridge, the annual supply of cooled water was probably not so great as the present annual melting of ice, and at first it was not ice-cold. When we take into account also the cold water which wells up in the tropics and becomes warmed there, so that it has to be cooled again, we see that the thorough cooling of the oceans must have taken several times, perhaps many times, ten thousand years. It is unlikely, however, that this effect could have caused a lag of millions of years.

The second stage in the oncoming of the Quaternary Ice-Age would occur when the general temperature of the oceans had fallen low enough for a covering of floating ice to develop over the Arctic Ocean. This would result in a great cooling of the lands washed by that ocean—Greenland, Norway, and Northern Russia. The Labrador Current may have been in existence before, but now it would carry great quantities of floating ice, and there would be a great lowering of temperature in Labrador and Newfoundland. The decrease in the summer temperature would be greater than the decrease in the winter temperature, and there would also be a marked increase in the storminess and snowfall. All these regions would develop glaciers, which would speedily become ice-sheets. The glaciation of the mountains of British Columbia may have commenced earlier, but probably increased rapidly about this time, while the ice-sheet of the Antarctic probably reached the sea. This may have been the first or Gunzian glaciation.

The second cause of lag is the reaction of the elevated land areas to erosion. The action of frost and running water removed great quantities of rock, much of which found its way into the sea. The lightening of the load caused further uplift, but the topography now being irregular, the higher peaks were at a greater height than before, while the valley-heads formed suitable gathering grounds for the accumulation of snow drifts. It is in such hollows that snow drifts persist longest in Scotland.

Finally, when all other factors were favourable, it is possible that either a period of plentiful volcanic dust or a period of decreased radiation in summer due to astronomical causes, by keeping down the summer temperature, was the actual immediate cause of the beginning of glaciation.

Once the ice-sheets had formed, by raising the effective height still further (both by adding ice to the land and sub-tracting water from the sea), by reflecting solar radiation and cooling the area around them, and by shedding ice into the sea and so depressing ocean temperatures still further, they would tend to maintain themselves and spread, until for some reason they became unstable. We must now examine the possible causes of the break-up of the ice-sheets. These are :—

1. A lowering of the level of the land.

2. A general rise of temperature.

3. A decrease in snowfall.

Since ice weighs about one-third as much as the average rock, the accumulation of 3,000 feet of ice is equivalent to adding the weight of 1,000 feet of rock to the land. This additional weight gradually depressed the land surface, though with a considerable lag, and brought the margins of the ice-sheets under the action of the sea, causing for example, floating ice-barriers which broke away as icebergs. The area of the ice-sheets and consequently their cooling power diminished, initiating an amelioration of climate. The process might be carried far enough for the ice to disappear more or less completely. After the load was removed the land would begin to rise again, causing a return of glaciation. This process might account for the division of a glacial period into two or three peaks, and possibly, though this is more doubtful, for the shorter interglacial periods. Moreover, since each glaciation would wear down the high ground and deposit the material round it in the form of moraines, we should expect each glacial recurrence to be less severe than the preceding one until the topography became unsuitable for glaciation. It cannot account for the long Mindel-Riss interglacial, but during the latter there was a great deal of earth-movement and volcanism in many parts of the world, which would eventually cause a return of glaciation. It is known that after the Wurm glaciation the land in the centres of greatest ice accumulation continued to sink and in late Glacial time the central shores of the Gulf of Bothnia were depressed about 900 feet below their present level ; the recovery is still in progress. Similar subsidence and recovery should have followed each glacial advance but the evidence for interglacial oscillations of the same type was swept away by subsequent advances of the ice. Also, the oscillations of level may have been superposed on a steady sinking of the land, so that each rise was less than the preceding fall. This would account for the gradual decrease in intensity of successive glacial peaks.

A world-wide rise of temperature could be due to the cessation of volcanic activity, to an increase of solar radiation (Chapter IV.) or to astronomical causes (Chapter V.). The first two are rather speculative ; moreover it is very doubtful whether the slight increase of radiation in middle and high latitudes which would result from a cessation of volcanic activity would have much effect on a full-grown ice-sheet. Astronomical effects also seem rather slight, but as the effect of increased warmth in summer would be reinforced by the greater cold of winter which would probably result in a decrease of snowfall, they cannot be ruled out. The good accord between Milankovitch's astronomical scheme and the succession worked out by F. E. Zeuner (Chapter V.) supports the idea that these small astronomical causes may actually have been the controlling factor in the glaciation of the Northern Hemisphere. These large ice-sheets would exercise a dominant effect on the ocean temperatures and atmospheric circulation, and so might well control the glaciation of other parts of the world. There is good reason to believe that the Pluvial periods of low latitudes were in fact controlled by the atmospheric circulation. In this connexion it is interesting to note that according to H. Mortensen (7) there was no pluvial period in the coastal desert of northern Chile. This desert exists because of the upwelling cold water of the Humboldt Current off the coast, which in turn is due to the south-east trade winds blowing off the coast. A strengthening of these trade winds would therefore maintain the desert conditions.

We come finally to the question of precipitation. The supply of precipitation would of course follow variations of solar radiation (Chapter IV.) but it is now generally recognised that the development of ice-sheets would itself cause changes in their supply of moisture. This was first suggested by V. Paschinger (8).

In Chapter IX. we saw that in mountainous country, as we go upwards the total amount of precipitation increases to a certain level, above which it again decreases. With increasing height, also, the proportion of total precipitation which falls as snow becomes steadily greater. Hence we can distinguish a level of maximum rainfall, and above that a level of maximum snowfall. The latter is often very sharply marked ; it depends on the winter conditions, especially the general winter temperature of the lowlands, the vertical temperature gradient, and the relative humidity. The snow-line, on the other hand, depends mainly on the summer temperature. At present in the Alps the snow-line is about 2,000 feet above the level of maximum snowfall. Suppose now the summer temperature decreases while the winter temperature remains unchanged. The snow-line will descend, and if the decrease of summer temperature reaches 6° F., the snow-line will coincide with the level of maximum snow-fall. The supply of snow available for glaciers will now be greatly increased, and this stage will see a great development of glaciers. Even if the cooling is uniform throughout the year, the snow-line will descend more rapidly than the zone of greatest snowfall.

At present in polar regions the snow-line is below the

zone of maximum snowfall, and these regions are widely glaciated. In the Tertiary period, the snow-line must have been above the snowfall maximum even in polar regions. Paschinger considers that the cooling of the temperate regions spread out from the poles, probably in the form of repeated cold waves (*i.e.*, outbreaks of the polar front). Owing to the conservation of heat in the oceans, whence most of the moisture is evaporated, the total precipitation is not diminished at first, while the proportion which falls as snow is increased. Glaciers spread until they reach the sea or some warm lowland where ablation is rapid. Then as the seas cool, the snowfall diminishes, while the lowering of temperature due to the ice itself depresses the zone of maximum snowfall. At the same time, the development of glacial anticyclones cuts off the supply of snow in the interior, so that the snow-line rises, until it is again above the zone of greatest snowfall. The ice-sheets and glaciers now retreat. When the retreat has proceeded far enough, the secondary cooling due to the ice ceases to be effective, the level of maximum snowfall rises to the snow-line again, and the whole process recommences. This is Paschinger's conception of the meteorological cycle of a glacial period ; granted an initial cause, such as elevation, glacial and interglacial (or " intraglacial ") stages will repeat themselves regularly until the immense denudation effected by the ice lowers the mountains—or at least the corries and depressions where snow can gather—below the snow-line. He thinks that this stage has not yet been reached in Europe and that another glaciation is to be expected in due course.

Paschinger points out that the relationship between the level of maximum snowfall and the snow-line accounts for many peculiarities of the Quaternary Ice-Age. In the continental mountain regions of Asia, with very cold winters and hot summers, the two levels are many thousand feet apart, and the glacial cooling was not, as a rule, sufficient to bring the snow-line down to the zone of heavy snowfall. Hence the development of glaciers and ice-sheets was less extensive than in Europe or North America. In equatorial regions, on the other hand, while both snow-line and maximum snowfall are at a great height, the former lies only a short distance above the latter, owing to the absence of seasons. A comparatively slight increase in the snowfall would bring them together, and cause a considerable extension of the mountain glaciers.

This view of the sequence of events in an ice-age undoubtedly contains many elements of truth, and may well account to some extent for the alternation of glacial stages with what I have termed above " intraglacial " stages. The Mindel-Riss interglacial stands in a different category, and cannot be accounted for on any purely meteorological cycle ; it necessarily involves a cessation or great weakening and a subsequent renewal of the ice-forming factors.

As was pointed out in Chapter II., the development of ice-sheets would cause changes in the atmospheric circulation and tracks of depressions, which would react on the supply of moisture. Besides the work of Flint and Dorsey, referred to in that chapter, there have been several other studies of American glaciation on these lines. Thus E. Antevs (9) considers that the Keewatin and Cordilleran ice-sheets in the west and centre developed first. The Keewatin and Scandinavian ice-sheets caused a southward displacement of the Icelandic low which caused frequent north-east winds in Labrador. Once started, the Labrador ice-sheet was fed by cyclonic snowfall on its southern border. Ultimately the area of ice grew so large that the supply of snowfall in the central regions was insufficient to maintain it, and the ice-sheet began to decay. At this stage, however, depressions were still deflected southward, and Antevs thinks that the mountain glaciers and lakes south of the main ice mass may have reached their maximum during the earlier stages of the retreat. Later, however, the storm tracks shifted north again and brought about a rejuvenation of the ice-sheets and a repetition of the series of events. This process might account for short intraglacial periods but not for recurrences after long interglacial periods as warm as or warmer than the present.

There is no doubt that changes in the atmospheric circulation must have brought about changes in the centres of the ice-sheets and it is highly probable that their growth must eventually have resulted in starvation at the centre, but it seems unlikely that this would have resulted in their disappearance. It is also doubtful whether the North American sequence followed the lines of Antevs's argument ; K. Bryan (10) for example, states that American geologists believe in a progressive shift from east to west of the main ice-centre throughout the last (Wisconsin) glaciation, and R. F. Flint (11) thinks that the Labradorean and Keewatin areas were both parts of a single Laurentide ice-sheet fed by maritime air from the south and south-east and expanding southward and westward.

Summing up, we find that for the occurrence of the ice-age as a whole the " geographical " theory seems to be the only adequate one, with possibly some help from CO_2. The actual commencement of glaciation may, however, have been determined by some minor factor such as the astronomical situation or changes of solar radiation. The interglacial periods present the main difficulty, because of their close parallelism in different parts of the world. Astronomical causes seem to come nearest to filling the necessary conditions, but alternating depression and elevation, due to the accumulation and removal of the ice-load, are also probable, while cycles of solar radiation cannot be ruled out. The recrudescence of glaciation after the Mindel-Riss Interglacial was due, at least in part, to renewed mountain building. Finally, the " intraglacial " oscillations were most probably caused by reactions between the ice-sheets and the circulation of the atmosphere.

We must now briefly consider the climate outside the main areas of glaciation.

The part of Central Europe sandwiched between the Scandinavian ice-sheet to the north and the Alpine glaciers to the south must have suffered from a severe climate, which has been studied by P. Kessler (12). He has three lines of evidence—the climate in the neighbourhood of the present ice-sheets of Spitsbergen, Greenland, and the Antarctic, the flora and fauna, and the geological phenomena—and all three present the same picture. The mean annual temperature is below freezing point, and although the summer may have a few short spells of warmth, the winters are very cold. On the margins of the Antarctic continent the summer climate is especially unpleasant. Although the temperature during a relatively warm summer month may average above freezing point, and may go as high as 40° or even 45° F. for a few hours, yet the persistently overcast sky, the frequent storms of snow and sleet, and the general unpleasantness of the weather, are worse than the cold of the interior.

The conditions in these high latitudes, between ice-sheets and the sea, however, cannot be regarded as typical of those in Central Europe far from the Atlantic, especially if, as seems probable, there was a considerable area of land west of France.

The study of the flora and fauna gives results of great interest. The similarity of the plants at high levels in the Alps to Arctic forms suggests that during the maximum extension of the ice these cold-loving species inhabited the low unglaciated ground north of the Alps, and after the ice-age they followed the retreating glaciers upwards to high levels. The general picture shows a region of tundra vegetation, inhabited by the reindeer, the woolly rhinoceros, and the mammoth. The geological phenomena—earth-flows, block-trains, and mounds, ridges or terraces of angular and sub-angular material—point to frost action on a huge scale, the earth and rocks moving down the valleys under the action of repeated freezing and thawing. The general climate of the region appears to have been highly abnormal ; the prevailing winds were probably dry glacial winds from the north-east, but these winds were shallow and were overlain at a small height by moist winds from the Atlantic, which sometimes descended to the level of the ground. The snow-line lay at 3,000 feet in the west and at 5,000 feet in the east, and in the hills the accumulations of snow carved out cirques or corries at these levels. These are mainly on the north-eastern side of the crests, and since Enquist has shown that the greatest accumulation of snowfall takes place on the lee-side, they indicate that the snow-bearing winds at a height of 3,000 to 5,000 feet came from the south-west.

The annual precipitation was small at low levels, but occasionally rain fell in torrential downpours. The evaporation was great, and one of the greatest peculiarities of the cold periglacial climate was that it could ape the formations of the hottest deserts. An important deposit was the loess, an accumulation of the finest wind-blown dust, and there were even small salt lakes in which layers of salt were formed. It is noteworthy that similar saline deposits are forming at present in restricted areas in Spitsbergen and Greenland, a fact which has some bearing on the evidence for Wegener's theory of polar movements.

Outside the limits of the ice-sheets and of the peripheral zone of ice-winds, the weather was probably much as we know it to-day, but more stormy. This applies especially to the Mediterranean region, which must have had a heavy rainfall distributed more or less evenly throughout the year, instead of a moderate or scanty rainfall limited to the winter months as at present. These regions probably had the weather now found on the north-western coasts of Europe. Wandering storms penetrated into the Sahara, which was then one of the most genial regions on the globe, and this region, now a desert, appears to have been one of the main centres in which the human race rose to a dominant position in the world. H. v. Ficker (13) calculated that at the time of the maximum glaciation of the north-west Pamir the rainfall was four or five times as great as at present.

The equatorial regions in general also had a greater rainfall than at present, though with local exceptions. Over the oceans the Trade winds, stronger in consequence of the greater temperature difference between the equatorial and polar regions, brought in more warm moist air than at present. The volume of air ascending in the equatorial belt of low pressure was therefore greater, and the rainfall in the Doldrums and over the eastern equatorial parts of the continents was heavier. The succession of glacial and interglacial periods in the northern continents was paralleled by a succession of pluvial and interpluvial periods in tropical Africa, and by advances and retreats of the mountain glaciers. The exact correlation is not yet determined, but may be as follows :—

A very early lake, the deposits of which have been described by E. J. Wayland (14) as Kafuan, may correspond with Gunz and Mindel. Wayland thinks it had two maxima separated by a period of earth movements. After a long dry interval a large lake (Lake Kamasia) formed from the junction of several existing lakes. E. Nilsson (15) calls this the Great Pluvial and equates it to the Riss. Lake Kamasia then dried up completely and the mountain glaciers disappeared. This interpluvial was followed by the Gamblian period of renewed lake-formation in each of the separate basins. Nilsson distinguishes four successive lake systems, the first three representing the three maxima of the Wurm and the fourth a late Glacial halt or re-advance. Between

Lakes I. and II. and II. and III. there were lower lake levels, between III. and IV. the lakes dried completely.

A similar succession can be traced over a large part of East Africa from the Nile Valley to Rhodesia though the stages of the last Pluvial have not been distinguished. The Upper Nile Valley, however, became desert early and K. S. Sandford (16) considers that in that region there were no changes sufficiently great to be called " Pluvial " and " Interpluvial." It is in fact likely that owing to the preponderance of ice in the Northern Hemisphere the whole system of climatic belts was shifted southwards and that the increase of rainfall was much greater south than north of the equator.

The retreat of the ice-sheets shows a number of halts or re-advances marked by a series of terminal moraines. These present a similar appearance in North America and Europe. There were also a series of fluctuations of lake-levels in East Africa, which most probably represent the pluvial equivalents.

The variations of lake levels do not necessarily represent very great changes of rainfall. In the Nakuru catchment area the present rainfall is about $37\frac{1}{2}$ inches a year. R. E. Moreau (17) from botanical evidence considers that the average rainfall in the last of the Wurmian pluvial stages (Makalian) was about 44-50 inches, while during the arid Post-Makalian period, when the lakes dried completely, it cannot have been as low as 27 inches.

REFERENCES

(1) ANTEVS, E. " The last glaciation." New York, Amer. Geogr. Soc., *Research Series*, no. 17, 1928.

(2) FLINT, R. F., and H. G. DORSEY. " Iowan and Tazewell drifts and the North American ice-sheet." *Amer. J. Sci.*, 243, 1945, p. 627.

(3) ZEUNER, F. E. " The Pleistocene period ; its climate, chronology and faunal successions." London, *Ray Soc.*, 1945.

(4) BRYAN, K., and L. L. RAY. " Geologic antiquity of the Lindenmeier site in Colorado." Washington, *Smithson. Misc. Coll.*, 99, no. 2, 1940.

(5) LOEWE, F. " Die Eiszeit in Kaschmir, Baltistan und Ladakh." Berlin, *Zs. Ges. Erdkunde*, 1924, p. 42.

(6) SMITH, E. H. " The international ice patrol." *Meteor. Mag.*, London, 60, 1925, p. 229.

(7) MORTENSEN, H. " Uber den Abfluss in abflusslosen Gebieten und das Klima der Eiszeit in der nordchilenischen Kordillera." *Naturwiss*, Berlin, 16, 1929, p. 245.

(8) PASCHINGER, V. " Die Eiszeit ein meteorologische Zyklus." *Zs. Gletscherk.*, 13, 1923, p. 29.

(9) ANTEVS, E. "Correlation of Wisconsin glacial maxima." *Amer. J. Sci.*, 243A, 1945, p. 1.

(10) BRYAN, K., and R. C. CADY. "The Pleistocene climate of Bermuda." *Amer. J. Sci.*, 27, 1934, p. 241.

(11) FLINT, R. F. "Growth of North American ice-sheet during the Wisconsin age." New York, *Bull. geol. Soc. Amer.*, 54, 1943, p. 325.

(12) KESSLER, P. "Das eiszeitliche Klima und seine geologischen Wirkungen im nicht vereisten Gebiet." Stuttgart, 1925.

(13) FICKER, H. v. "Die eiszeitliche Vergletscherung der nordwestlichen Pamirgebiete." Berlin, *SitzBer. Preuss. Akad. Wiss.*, 1933, 2, p. 61.

(14) WAYLAND, E. J. "Rifts, rivers, rains and early man in Uganda." London, *J. R. Anthrop. Inst.*, 64, 1934, p. 333.

(15) NILSSON, E. "Quaternary glaciations and pluvial lakes in British East Africa." *Geogr. Ann.*, Stockholm, 13, 1931, p. 249.

(16) SANDFORD, K. S., and W. J. ARKELL. "Palæolithic man and the Nile valley in Nubia and Upper Egypt." Chicago Univ., Oriental Inst., *Publ.*, vol. 17. Prehistoric survey of Egypt and Western Asia, Vol. 2, Chicago (1933).

(17) MOREAU, R. E. "Pleistocene climatic changes and the distribution of life in East Africa." London, *J. Ecol.*, 21, 1933, p. 415.

PART III

THE CLIMATES OF THE HISTORICAL PAST

CHAPTER XVII

The Nature of the Evidence

IT is not many years since it was generally believed that variations of climate came to an end with the Quaternary Ice-Age, a period moreover which was placed hundreds of thousands of years ago. The post-glacial or " Recent " period was supposed to show merely a more or less rapid warming up to the present level, followed by a long period in which the climates of the different parts of the world were exactly as we now find them. It was the International Geological Congress at Stockholm in 1910 which first made the majority of geologists familiar with the existence of a warm period intercalated between the ice-age and the present. About the same time, a number of investigations in different countries combined to prove that the ice-age itself was not so remote as it had seemed to be, and that in fact the post-glacial " geology " of Europe was partly contemporaneous with the " history " of Egypt. But since the geological deposits undoubtedly point to changes of climate, slight indeed in comparison with the preceding ice-age, but still marked enough to leave their traces permanently written on the face of the earth, the unvarying climate of history is evidently a myth. The beginning of the " period of unchanging climate " has advanced later and later before the attacks of geologists, and now, in the minds of most of the authors who concern themselves with the subject, it apparently stands only a few centuries before Christ. But meanwhile a different, and more logical, view has arisen, namely, that the present does not differ from the past, that variations of climate are still in progress, which are similar in kind, though not in extent, to the climatic vicissitudes of the ice-age.

There is, however, one point in which the " historical " period may be said to differ from the " geological " periods ; during the historical period the distribution of land and sea, the heights of the mountains, and the positions of the poles have changed only to a very slight extent. Hence we may regard the geographical factors of climate as practically constant during this period, and any climatic changes which we can discover and confirm must be attributed to non-geographical factors, and most probably to variations in solar radiation. Hence it is in the historical period that we are most likely to be able to trace the effect of solar radiation on climatic changes. Of course this difference between the " historical " and the " geological " periods is more apparent than real ; the length of the historical period is a few thousand years, while the length of even the subdivisions of the geological periods is to be expressed in hundreds of thousands or in millions of years. Nevertheless, we do seem to be living at present in a period of quietude relative to the Quaternary period ; the change from the *Ancylus* to the *Litorina* stages in the Baltic, for instance, represents a greater geographical variation than anything which has happened since.

The interpretation of the term " historical period " adopted in this section is a somewhat liberal one ; it is essentially the period during which the vicissitudes of human life are known and dated to within a few centuries. Archæologists are continually pushing back the boundaries of history, while astronomers, geologists, and others from time to time supply new fixed points or new chronologies. At the present time we have a more or less complete record of human history in South-western Asia since about 5200 B.C., and that date has been taken as the point of origin. For a study of the climatic changes during this period of 7,000 years, we have a variety of material. Instrumental records are of course of the greatest value, but reliable meteorological observations go back a mere three centuries, and for the greater part of the period we have to make the best of less direct evidence. The various lines of attack may be summed up as follows :—

1. Instrumental records and old weather journals.

2. Literary records (accounts of floods, droughts, severe winters, and great storms).

3. Traditions, such as that of the Deluge, which can sometimes be correlated with other data.

4. Fluctuations of lakes and rivers, glaciers and other natural indices of climate, which can often be connected with historical events or dated by laminated clays.

5. Arguments from the migrations of peoples, for which climatic reasons may be assigned with some show of probability. To this we may perhaps add the waxing and waning of civilisations.

6. The rate of growth of trees, as shown by the annual rings of tree-growth, which can be correlated with the annual rainfall.

7. Geological evidence—great advances or retreats of glaciers, growth of peat-bogs, succession of floras, etc., which can sometimes be dated approximately.

The first and second sources of data, meteorological and literary records, and the seventh, geological evidence, are mainly exemplified in Europe, while the fourth and fifth sources provide the main mass of information for Asia and the fourth for Africa ; while the sixth, growth of trees, gives the only exact chronology for North America, where, however, it is highly developed.

Instrumental meteorological records even in Europe date back for only about three centuries, in North America for two centuries, while in other continents they are practically confined to the last hundred and fifty years. Moreover, while old observations are of great interest in discussing variations of weather from one year to another, they are of less value in determining changes of climate extending over a long period. The accuracy of the early instruments is not always above reproach ; some of the early types of rain gauge, for example, do not make adequate provision against the re-evaporation of the fallen water. Defects of exposure may be a serious source of error. It was a common practice among early observers to expose their rain gauges on the roofs of houses, but gauges so exposed do not catch so much water as gauges exposed on the ground in open sites. Even when placed on the ground, they may have been too near to buildings or trees. Most of the sources of error tend to give a rainfall which is too small rather than too great, so that if the early instrumental records appear to indicate that the rainfall was smaller than at present, they must be regarded with suspicion unless they can be confirmed in some way. The rainfall minimum in England indicated by Symons in the eighteenth century was suspected for this reason ; the way in which it was confirmed is described in the next chapter, where, in addition, long rainfall records are discussed from other countries.

There is a curious exception to the comparative modernity of instrumental meteorological records, namely, the measurements of rain in Palestine in the first century A.D. Hellmann (1) states that " the amount of rainfall then considered as normal for a good crop corresponds pretty closely with that deduced from the modern observations of Mr Thomas Chaplin at Jerusalem, whence it can be inferred that the climate of Palestine has not changed."

There are a number of old meteorological journals in which the wind and weather are given, but no instrumental readings. The best known of these are the journal kept by the Rev. William Merle at Oxford from 1337 to 1344 (2), and that of Tycho Brahe (3) at Uranienborg on the Island of Hveen in the Sund from 1582 to 1597. Merle's journal presents a picture of the weather which would not differ greatly from that given by a similar journal at the present day. The winters were certainly not invariably rigorous, for example, 1342 : " It is also to be noted that there was spring-like weather for the whole time between September and the end of December, except on those days to which frost is ascribed, so much so that in certain places the leeks burst forth into seed, and in certain places the cabbages blossomed." Unfortunately, the journal is not a day-to-day record so much as a weekly or monthly summary of the weather, so that it is not easy to extract numerical data like the frequency of rain-days which can be compared with similar figures at the present day. An attempt to count up the rain-days for the two most complete years, 1341 and 1342, omitting only the " extremely light " or " very light " rains, gave totals of 152 and 153 respectively, compared with a present normal of 168, but the difference is of no significance.

The observations of Tycho Brahe seem to be exceptionally favourable for determining a difference of climate between the sixteenth and the nineteenth centuries, because the site could be accurately identified, and a further series of observations was made at the same spot from 1881 to 1898. P. la Cour (3) also has made a careful comparison between Tycho Brahe's observations and the mean results at fourteen stations in Denmark. The most important difference is that the prevailing wind, which is at present from south-west throughout the year, was in the sixteenth century from south-east, especially in winter. In winter, south-east winds are cold and dry, whereas south-west winds are mild and moist, leading to the inference that the winters were more severe in the latter half of the sixteenth century than they are at present. The number of rain-days recorded by Tycho Brahe is about thirty per cent. below the present mean in winter, whereas in summer the two figures are nearly the same. There is, however, the possibility that Tycho Brahe missed some rain-days in winter, when he would have been out of doors less than in summer. The number of days with snow is greater than at present, confirming the view that the winters were colder. H. H. Hildebrandsson, however, pointed out (4) that the period 1582 to 1597 appears to have had severer winters than the remaining parts of the sixteenth century ; Tycho Brahe's observations happened to coincide with a cold spell and were therefore not representative of the century. This conclusion from the observations and Hildebrandsson's commentary will be fitted into their place in the sum total of evidence concerning climatic changes in Europe in the next chapter.

Observations of wind direction are probably the most valuable of all the records of old weather diaries, since they can often be compared directly with present-day records. An analysis of old wind records in the British Isles, made by C. E. P. Brooks and T. M. Hunt (5), presented several results of interest (see Chapter XVIII.). If similar studies were made for other parts of the world, our knowledge of climatic changes would be greatly extended.

Some weather journals, apparently from Alexandria, dating from the early part of the Christian era, described by G. Hellmann, are referred to in Chapter XX. These journals would be of the very greatest importance in demonstrating a change of climate, if it were *absolutely* certain that they were made at Alexandria, and not in Greece. That is the chief difficulty in dealing with early meteorological observations, whether instrumental or not ; there is generally an element of doubt somewhere.

Still less satisfactory are inferences drawn from early descriptions of the climate and physical nature of various countries. The Roman writers described Britain as damp and cloudy, but so would an Italian of the present day, and we are left in doubt as to whether it was any damper or cloudier at the beginning of the Christian era than it is to-day. The general analysis of the literature of the Mediterranean countries initiated by Arago (6), and continued by a large number of meteorologists and antiquarians, has shown that in these countries during the first century of the Christian era the nature of the vegetation and crops, the dates of sowing and reaping, and the animal life, all suggest that the climate differed little from the present. Arago's remarks about the date and the vine have been quoted by every opponent of climatic change for the last ninety years—the date cannot ripen its fruit in a mean annual temperature below 21° C., the vine cannot abide a temperature above 22° C. ; since both date and vine flourished in ancient Palestine, the mean annual temperature must have been 21° C., which is also its present value. It seems doubtful, however, whether the solution can be quite so simple as that ; differences of exposure must come in, and the annual range of temperature from summer to winter. Even if Arago's strict limits of temperature be accepted, in a country of such varied relief as Palestine, the area over which the mean annual temperature at ground-level (as opposed to mean sea-level) lies between the limits of 21° and 22° C. must be quite a small proportion of the whole. The effect of a slight change of climate would be nullified by moving the plantations to a site with a different exposure or at a different level.

There are two curious features of this mass of anti-variation literature started by Arago. The first is that it is almost entirely directed against the idea of a *progressive* change of climate, and not against climatic fluctuations. The old theory of progressive desiccation has been dead for many years, and all this reiteration is merely killing the slain, for to prove that the climate of the first century A.D. resembled that of the present does not prove that the climate of the seventh century A.D. also resembled the present. The reason probably is that the progressive theory offers the opportunity for a definite negation, while the theory of fluctuations does not. The weather of one year differs from that of another year, the weather of one decade from that of another decade ; why should not the *climate* of one century differ from that of another century ? The question is one, not of fact, but of degree, which is much less satisfactory. The second point is that practically the whole of the literature is directed against the idea that the climate of Europe, Asia, Africa, has become *drier*. No one has attempted to prove that the climate has not become wetter, because the fact is so obvious that no proof is needed.

The discussion of the literary records of weather follows a different line of argument. There have at all times been annalists, who wrote down accounts of the striking events of their time. They were not concerned particularly with the weather, but if a great flood or drought, frost or storm, occurred, they wrote it down. These weather notes have been extracted by various commentators, who have often been at great pains to verify the dates and eliminate the errors introduced by copyists, so that a large amount of fairly reliable material is now available. It seems a reasonable argument that if a considerable number of droughts were recorded in one century, the rainfall of that century was abnormally low ; similarly, a large number of floods and storms suggest a heavy rainfall. There are, however, several difficulties to be overcome. The first is that the completeness of the record changes from one century to another. Thus the number of records of droughts may be six in the seventh century and ten in the thirteenth century, but this does not necessarily mean that the latter was the drier. The records of storms and floods may number two in the seventh century and twenty in the thirteenth. The correct way of stating the evidence would be that of the total number of records of raininess in the seventh century, 25 per cent. indicate a high rainfall ; of those in the thirteenth century, 67 per cent., so that the latter century was the wetter. This gives us a satisfactory method of dealing with records of raininess, but, unfortunately, records of temperature cannot be dealt with in the same way, for we have practically only records of severe winters or hot summers, the mild winters and cool summers being less often recorded.

The second difficulty concerns the psychology of the annalists. Vanderlinden (7), in the preface to his " Meteorological Chronicle of Belgium," divides records of this type into three stages. In the earliest stage, the authors are concise ; they merely state " cold winter," " dry year," etc., without any subjective remarks. Later, the records become longer and more fanciful ; often the chronicler breaks into verse. In the third stage, there is a certain amount of manipulation of facts, under the influence of religious or superstitious ideas, and it is not until the end of the eighteenth century that the reports again assume a concise and scientific character. Finally, we have the difficulty that all these annotations tend to be comparative. Suppose that after a long period of dry climate there is a change in the direction of greater rainfall, which accomplishes itself in a period of fifty years, after which the climate continues at its new level. Obviously the period of increasing rainfall and the early subsequent years would suffer by comparison with the dry period which preceded them, while after the rainier climate had prevailed for one or two generations it would be accepted as the normal order of events and would escape comment. Thus, from the records, the period during and immediately following the change would actually appear as a rainfall maximum. I think the maximum rainfall indicated for Europe in the eleventh century is due in this way to the abrupt change from the dry conditions of the tenth, while the maximum of the thirteenth century, which was probably equally if not more pronounced, hardly appears. The oscillation in the ninth and tenth centuries also is probably exaggerated from this cause.

There is some discrepancy between the records of the Classical period and those of the Middle Ages, because the centre of civilisation moved northward in the interval from the Mediterranean to Western and Central Europe, that is, from a generally drier to a more humid climate. The meteorological events which are considered worthy of record by the annalists are those which strike them as most unusual ; in the Mediterranean, a dry summer is taken for granted, while a wet summer is an event to be recorded ; in North-west Europe, where the rainfall is usually sufficient at all seasons, a drought is the more noticeable event. From the records, one might suppose that the Tiber was more often in flood than the Thames ; this is because the floods of the Tiber are short-lived capricious affairs due to sudden heavy storms in the mountains ; they were considered worthy of record, while in the Thames the water rises more gradually but also more regularly, and a certain amount of flooding occurs almost every winter. For these and similar reasons the early records give an appearance of wetness. Hence the climatic curves derived from the literary records have to be " calibrated " by reference to the records of lake levels, advance of glaciers, etc.

Difficulties of chronology may also cause compilations of historical records to give a false impression of the variations of climate. When the exact date of say a drought is uncertain, different annalists may assign it to different years. The compiler collects all these dates, and quotes a drought for each of them. In this way a few months of dry weather may become a drought lasting three, four or five years. Even worse, the original drought may not have any real existence. Thus C. E. Britton (8), in his model meteorological chronology of Britain, considers that the famous three-year drought ended by St Wilfrid is a later invention to glorify the Saint. For these reasons, in this revised edition I have attached much less importance to these early literary records than in the first edition and I have omitted the diagrams of the frequency of different phenomena as more misleading than useful.

When we go back beyond the written annals, we come to the period of tradition. The traditional meteorological event which will spring at once to the mind is the Noachian deluge, which finds its parallel in the legends of many other nations besides the Jews. The Biblical flood was closely similar in its details to a Chaldæan legend recorded by Berosus, but most peoples of the Near East, including the Greeks and Persians, had similar traditions, all of which were probably derived, in part at least, from the Chaldæan. Curiously, the Chinese have a flood legend which is remarkably like that of the Bible. In the Indian version the saving vessel finally landed on the loftiest summit of the Himalayas. There is also a flood legend among the Aztecs of Mexico. The meaning of this widespread tradition is not clear ; if it refers to a single event in Mesopotamia it must be very old, probably earlier than 4500 B.C.

Another meteorological legend is that of the twilight of the Norse gods, when frost and snow ruled the land for generations. This can reasonably be attributed to a great change of climate for the worse which occurred about 500 B.C. But since these traditional meteorological events can only be interpreted in terms of climatic changes with which we are already acquainted through other evidence, they are of little or no help in elucidating the actual climatic variations ; at most, they can serve as a confirmation of other evidence.

The fluctuations of lakes and rivers form in general the most satisfactory evidence for determining changes of climate. The levels of the Central European lakes during the period of lake-dwellings are the most reliable source of information for the long pre-Classical period in Europe. In Western Asia the variations of the Caspian form a useful index of the rainfall during the past 1,500 years or so, eked out by scattered data from other salt lakes. Lakes without outlet are the most satisfactory because they respond readily to changes of rainfall. In both Europe and Asia the fluctuations can generally be dated by archæological or historical correlations ; the pronounced variations of level in the salt lakes of Western North America are of less value because they cannot be dated in this way. In Europe the variations of rainfall in the basins of several lakes are recorded in the thickness of annual layers of sediment, and one of these records goes back to 2300 B.C. In Africa there is a unique series of actual measurements of the levels of the Nile, which are dated to within a year. There is also a large body of evidence about the long-period variations of level of the Central African lakes, which point to large fluctuations of rainfall, but unfortunately these can only be dated very roughly by archæological means. Returning to Central Europe, there is a large body of information as to the fluctuations of the Alpine glaciers, including some pre-Classical fluctuations which can be traced and approximately dated by archæological evidence. The laws which govern the movements of glaciers are not yet fully understood, but a succession of snowy winters and cool wet summers seem to be most favourable for advance and hot dry summers for retreat.

The evidence afforded by racial migrations as to climatic changes depends on the principle that during a period of increased rainfall there is a movement of peoples from regions which are naturally moist to regions which are naturally dry, while during the drier periods the direction of movement is reversed, the naturally moist regions being occupied and the dry regions more or less abandoned. E. Brückner (9) showed, for example, that emigration from Europe to the United

States depended on the rainfall. In order that this principle may be used to determine the course of climatic variations, certain conditions are necessary. First, there must be large areas which are on the borderline between aridity and complete desert ; these areas must be mostly too dry for extensive agriculture, but with sufficient resources to support under average conditions a large nomadic population, while a succession of dry years renders them almost uninhabitable. In close proximity to this arid region there must be a fertile well-watered plain, with a long and accurately dated history. During dry periods the nomads are driven from their homes by lack of water, but they find little difficulty in moving from point to point, and the sedentary agriculturists of the neighbouring plain generally find them irresistible. It is only in Asia that these conditions are fulfilled in perfection, the rich plains of the Tigris and Euphrates, the site of a long succession of civilised states, having on the one side the semi-deserts of Arabia and Syria, on the other side a great dry region extending eastwards and north-eastwards as far as China. We should expect a period of decreased rainfall to initiate a series of great migrations spreading out from the dry regions and recorded in the history of the Mesopotamian states as the invasions of barbarians. The history of Egypt does not give anything like so complete a record, because the desert on either side of the Nile valley is too dry, even under favourable conditions, to support a nomadic population sufficiently large to have made any impression on the might of ancient Egypt. The Hyksos conquest of about 1800 B.C. is the main exception, but the Hyksos themselves probably came out of Asia. The invasions of China from the west provide some evidence of climatic fluctuations in the east of Asia.

The principle that tribal movements were mainly due to drought is insisted on by H. J. E. Peake in his study of the migrations of the Aryans (Wiros, as he prefers to call them). Thus he writes (10, p. 157) :—" We have seen reason for believing that a period of drought, occurring some centuries before 3000 B.C., drove some of them towards the Baltic. . . . But the great dispersal was about 2200 B.C. On this occasion the drought seems to have been more excessive or more prolonged, for it is believed that the steppe was left for a time almost uninhabited." This evidence, however, is circumstantial and needs to be used with care. Consider, for example, the four great outbursts from Arabia, the first of which occurred during the fourth millennium B.C., the second, or Amorite, about 2000 B.C., the third, or Aramæan, from 1500 to 1000 B.C. (according to Peake, mainly 1350 to 1300 B.C.), while the fourth, or Arabian, culminated in the Islamitic expansion of the seventh century A.D. No reasonable cause other than drought can be assigned to the first three of these migrations, but the fourth might be attributed to the influence of the Moslem religion, were it not that it began some time before the birth of Mohammed. The Arabs of the region east of Southern Palestine relate that shortly before the days of Mohammed, or somewhere about A.D. 600, a terrible and prolonged drought caused untold havoc, and the greater part of the tribe migrated to the African coast near Tunis (11). We have also to distinguish between *migrations*, in which large numbers of people (men, women, and children) moved away from one region and occupied another, and *conquests*, in which the ruler of a strong country imposed his government on his weaker neighbours, and established an empire. No one would attribute the conquests of Napoleon Bonaparte, for example, to unfavourable climatic conditions in France—in fact, the reverse conclusion could be argued more plausibly, namely, that owing to favourable conditions the state became powerful enough to dominate its neighbours less favourably situated. The latter is in fact the type of argument adopted by Ellsworth Huntington in his historical studies. In compiling a list of migrations, therefore, we must be careful to omit mere military conquests and raids.

Huntington's contention, as set out in " Civilisation and Climate " (12), is that a certain type of climate, now found mainly in Britain, France and neighbouring parts of Europe, and in the Eastern United States, is favourable to a high level of civilisation. This climate is characterised by a moderate temperature, and by the passage of frequent baro-metric depressions, which give a sufficient rainfall and changeable stimulating weather. Now it is well known that the great centres of civilisation in the past lay in more southerly latitudes than those of to-day, beginning in Egypt, Mesopo-tamia, and the Eastern Mediterranean, and then passing to

Greece and Rome. Huntington attributes these changes in the centres of civilisation to climatic changes associated with the northward shifting of the belt of cyclonic activity. In another volume (13), he gives a detailed comparison of the history of Rome during the Classical period with the climatic changes deduced from the growth of the Sequoias in California (see Chapter XXI.), which he regards as corresponding very closely with the rainfall of the Mediterranean area. If this principle can be maintained, it obviously affords a powerful weapon for deducing the existence of climatic fluctuations during the historical period in other parts of the world, but S. F. Markham (14) has given an alternative explanation, namely that the northward migration of the centres of civilisation has followed improvements in the methods of heating houses, so that civilised activities became possible throughout the year in regions with cold winters, and had no relation to changes of climate.

The width of the annual rings of growth of trees in dry regions is closely correlated with the rainfall during the preceding few years, so that old trees offer valuable evidence as to variations of rainfall during their lifetime. The Sequoias of Western U.S.A. at present provide the only accurately dated evidence of climatic fluctuations in that country previous to the settlement by Europeans. The further description of the way in which the records are interpreted is postponed to Chapter XXI. So far, the method has not been applied to any very old trees outside the United States, but there seems no reason why it should not be almost equally effective in other continents.

Geological evidence by itself plays only a very small part in elucidating climatic changes during the historical period, because it is only rarely that geological deposits can be dated with sufficient accuracy. There are considerable possibilities in the fine seasonally banded clays which have been forming in lakes and quiet fiords in the glaciated regions. These clays and the associated annual moraines have yielded valuable information concerning the rate of retreat of the ice-sheets at different stages, and in the post-glacial period they have served to date the peat-bogs and raised beaches which have supplied abundant evidence of post-glacial climatic changes in Scandinavia, as described in " The Evolution of Climate."

The great deterioration of climate which marked the beginning of the sub-Atlantic period, about 500 B.C., is dated by archæological evidence.

From all this it will be seen that our knowledge of the climatic changes of the historical period has to be drawn from a great variety of sources. While this renders the task of reconstruction more difficult, it has the advantage of offering frequent opportunities for testing the results by comparing the conclusions derived from quite different and independent sets of data. An example of this occurs in the climatic changes of Europe and Asia. The former are deduced almost entirely from geological and archæological data and from the literary records, the latter from migrations of peoples and from the levels of the Caspian. The rainfall curves obtained for these two continents, however, resemble each other so closely that the fluctuations portrayed are obviously real.

REFERENCES

(1) HELLMANN, G. " The dawn of meteorology." London, *Q. J. R. Meteor. Soc.*, 34, 1908, p. 221.
(2) MERLE, REV. W. " Consideraciones temperici pro 7 annis." The earliest known journal of the weather . . . 1337-1344. Reproduced and translated under the supervision of G. J. Symons. London, 1891.
(3) LA COUR, PAUL. " Tyge Brahes meteorologiske dagbok, holdt paa Uranienborg for aarene 1582-1597." Appendix til Collectanea Meteorologica. Kjøbenhavn, 1876.
(4) HILDEBRANDSSON, H. H. " Sur le prétendu changement du climat européen en temps historique." Upsala, 1915.
(5) BROOKS, C. E. P., and T. M. HUNT. " Variations of wind direction in the British Isles since 1341." London, *Q. J. R. Meteor. Soc.*, 59, 1933, p. 375.
(6) ARAGO. " Œuvres complètes." T. 8. Paris, 1858.
(7) VANDERLINDEN, E. " Chronique des événements météorologiques en Belgique jusqu'en 1834." Bruxelles, 1924.
(8) BRITTON, C. E. " A meteorological chronology to A.D. 1450." London, Meteor. Off., *Geoph. Mem.*, 8, no. 70, 1937.
(9) BRÜCKNER, E. " Klimaschwankungen und Völkerwanderungen." Wien, 1912.
(10) PEAKE, H. J. E. " The Bronze Age and the Celtic world." London, 1922.
(11) HUNTINGTON, ELLSWORTH. " The burial of Olympia." London, *Geogr. J.*, 36, 1910, p. 657.
(12) HUNTINGTON, E. " Civilisation and climate." 3rd ed. New Haven, 1924.
(13) HUNTINGTON, E. " World power and evolution." New Haven, 1919.
(14) MARKHAM, S. F. " Climate and the energy of nations." London (Oxford Univ. Press), 1942.

CHAPTER XVIII

EUROPE

WE have no historical records for Europe which go back much more than half-way to the year 5000 B.C. On the other hand, the geological evidence has been studied in great detail, and the chronology of the whole period since the glaciers commenced their final retreat is rapidly being placed on an exact basis. Since the beginning of the Christian era, there is a rich European literature which provides a wealth of material. The " official " end of the Glacial period in Sweden, according to the Scandinavian geologists, is now dated about 6500 B.C. (1), but by this time the climate of Central and Western Europe had become definitely temperate, and the latest glacial period really ended much earlier. The Fenno-Scandian end moraine, dated about 8300 B.C., which encloses most of Scandinavia and Finland, would give in some ways a more appropriate date. After this the rapidly disintegrating remains of the ice-sheet can have had little effect on the climate of Europe. The " post-glacial " stages are set out in Table 20, with their archæological equivalents in Western Europe. The latter is only approximate, since cultures do not appear simultaneously over the whole area.

The dating of the various archæological stages is determined by the known historical sequence in the Eastern Mediterranean, Egypt, and Mesopotamia, and especially the two latter. In Mesopotamia, a fixed point is provided by the total solar eclipse of 15th June 763 B.C., which was recorded in the annals. In Egypt, the dates can be approximately fixed by the heliacal risings of Sirius and by certain new-moon festivals, but 1580 B.C. is the earliest date in Egyptian history which can be regarded as certain within a few years. As regards the dating of earlier periods, finality is still far from being attained even in Egypt or Mesopotamia, and this doubt is added to in dealing with events in Europe which can only be dated by associating them with some event in the East.

Date.	Climatic stages.	Culture.	Vegetation.
6000	Boreal Dry, becoming warmer	Maglemose	Alder, Oak, Elm
5000	Atlantic Climatic	Late Tardenoisean	*Peat*
4000	Optimum Humid	Campignean	Lime
3000	Sub-boreal	Neolithic	Oak giving place
2000	becoming cooler,	Lake-dwellings	to Pine
1000	drier		
B.C.		Bronze	Yew
	Sub-Atlantic	Early Iron	*Peat*
0	Cool, wet	Romano-British	Beech in Central Europe
A.D.			
1000	Near present		

Table 20.—Post-glacial succession, Western Europe.

The broad early stages of the climatic succession are now very well known from innumerable studies of peat-bogs in Northern, Central and Western Europe. The successive stages can be placed in sequence by the microscopical analysis of the pollen grains which are found in peat-bogs and lake deposits. Certain pollen grains of some trees and plants are almost indestructible, and since they are produced in large numbers and scattered by the wind, they are very widespread. The technique of their study has been highly developed by G. Erdtman in a large number of papers, and has enabled a very close comparison to be made of the forest succession in different parts of Europe. The absolute dating is given by the prehistoric objects which are found in the bogs.

Generally speaking, the ground laid bare by the retreat of the ice was a maze of depressions and ridges. The hollows were occupied by lakes and ponds, and the ridges first by an arctic flora, which soon gave place to birch, followed by pine. By about 7000 B.C. the climate was dry and sufficiently warm in summer for the rapid spread of hazel. The rise of temperature continued, and with some increase of moisture, by 6000 B.C. all the western half of Europe was occupied by a rich forest of oak, alder and elm, the alder being favoured by the increasing rainfall. This was the beginning of the " Climatic Optimum," with temperatures up to 5° F. higher than the present, permitting forests to grow much higher up the mountain sides than is possible now. The heavy rainfall, however, favoured the growth of peat, and about 5000 B.C. large areas of forest were killed and buried by peat-bogs. This phase continued until about 2500 B.C., with gradually decreasing temperature, when there occurred a rather puzzling change. The peat ceased to grow and in many places the surface of the bogs was occupied by forests of pine and yew. This was the Sub-boreal, which was formerly supposed to be rather warm and very dry.

It seems certain that at some stage in the Sub-boreal there was a prolonged drought. Lakes decreased in area and in a few places trees grew on the floor of dried-up lake basins below the level of the outlet ; providing definite evidence that at these places the rainfall was less than the evaporation, and this enables us to estimate the actual rainfall. The following table gives estimates for four such lakes in places for which I have been able to obtain statistics or estimates of the present rainfall and evaporation. In a drier climate evaporation was presumably more active than at present and I have accordingly increased the evaporation figures by one-third.

	Estimated present evaporation.	Present rainfall.	Estimated sub-boreal rainfall.	Per cent. of present.
	inches.	inches.	inches.	
Donegal . . .	15·6	50	21	42
Connaught . .	15·6	40	21	52
Sager Lake nr. Bremen .	13·3	27	18	67
Seehof, Lunz Austria . . .	20·0	56	27	48

The average of the last four figures in this table indicates a rainfall of only about half the present amount in Western and Central Europe. This is less than the rainfall of the dry year 1921 and points to a real and very marked change of climate in Europe.

Botanists, however, have questioned the possibility of a dry period of such intensity lasting for over a thousand years. G. Erdtman (3) points out that there was a steady development of forests throughout the Sub-boreal, and he considers that during this period there was a gradual change towards the cool moist Sub-Atlantic type, but at the very end there was a relatively short period of perhaps 200 years which might be described as a dry heat wave, giving place abruptly to much cooler and moister conditions. Similarly H. Godwin and A. G. Tansley (4) write of southern England :—

"The best opinion of archæologists and pre-historians generally is beginning to question the validity, for these islands at least, of the clear-cut conception of a wet Atlantic and a dry Sub-boreal period. Evidence of a major climatic Atlantic-Sub-boreal transition in Britain, comparable with the thoroughly well established Boreal-Atlantic and Sub-boreal-Sub-Atlantic changes, is often lacking and there may perhaps have been several important alterations of climate between say 5500 and 500 B.C. But it is certain that part at least of the Bronze Age in England was relatively dry. In contrast with Neolithic settlements on chalk summits, chalk uplands seem to have been practically uninhabited during the Bronze Age, plausibly because of shortage of water.

In the middle of 1st millennium B.C. climate became cool and wet ; of the reality of that transition there is no doubt."

H. Godwin (5) considers that chalk and limestone soils, now mostly occupied by natural beechwoods, were too dry to carry close woodland during the Sub-boreal and raised bog surfaces were much drier than in the Sub-Atlantic. But a bog-surface is very sensitive to climatic change, and in some British bogs there are several layers indicating a cessation and resumption of growth. E. M. Hardy (6) found five dry phases in the bogs of Shropshire, which he places in the upper part of the full Boreal, in the Atlantic, in the Sub-boreal, at the top of the Sub-boreal and about midway in the Sub-Atlantic. These would be about 6000 B.C., 5000 B.C., 1200-800 B.C., 700-500 B.C., and at the beginning of the Christian era.

E. Granlund (7) from an intensive study of Swedish peat bogs, also found evidence of five dry layers, which ended about 2300, 1200, 600 B.C. and A.D. 400 and 1200, that about 600 B.C. being the best developed. This probably represents the "Grenzhorizont" of the Sub-boreal which is widespread in Europe.

About 500 B.C. there was a very rapid large-scale climatic change. Over large areas the Sub-boreal forests were killed by a rapid growth of peat, which was certainly caused by a great increase of rainfall and probably also a fall of temperature. As many of the bogs formed at this time are now drying up, the rainfall was probably much greater than at present. From this peak it has gradually declined, but with a number of oscillations of smaller amplitude. The general results of these peat-bog investigations forms the main basis of the top three curves of Fig. 31.

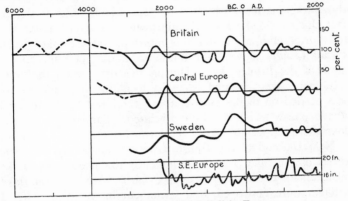

Fig. 31.—Variations of rainfall in Europe.

The records of peat-bogs are borne out by the evidence of land mollusca as summarised by A. S. Kennard and other writers in England. These point to a very wet climate in the Atlantic period. The beginning of the Sub-boreal was still humid but the rainfall was decreasing and by 1500 B.C. the wet period had largely passed. About 1000 B.C. the climate resembled the present.

A detailed study of the changes in level in the Central European lakes has been made by H. Gams and R. Nordhagen (2). The earliest part of the Neolithic, while warm, was decidedly moist, but it appears that the greater part of the lake-dwelling period was one of low water and of relatively high temperature. It is supposed, in fact, that the lake-dwellings were established, not in the lakes themselves, but on peat-bogs which are now covered by the waters of the lakes. The succession of events has been made out most completely in the Feder See basin, but other lakes confirm it. During the Neolithic period the lake was smaller than now, indicating a dry period which seems to have culminated about 2200-2000 B.C. This was followed by a period of somewhat greater rainfall, but drier than the present, and somewhere in this period was a " high-water catastrophe "—a brief régime of floods which destroyed many of the lake-dwellings. This flood period cannot be far distant in time from the great eruption of Bronze Age peoples from the Hungarian plain, which probably occurred soon after 1300 B.C., and carried the Phrygians into Asia (see p. 320). It is quite likely, in fact, that the flood was the stimulus which caused the migration to a drier climate. After this moist period the lakes again shrank rapidly to a second minimum, about 1000 B.C., at the close of the Bronze Age. During this dry period the chief settlements were located in moist localities, and agriculture was carried on in places now above the forest level and even above passes which are now glaciated. Sand dunes on the lake shores appear to have been formed mainly in this period.

Near the end of the Hallstatt period, about 500 B.C., the levels of the lakes rose suddenly ; in the Boden See (Lake Constance) the rise exceeded 30 feet. Most of the lake villages were destroyed, and settlement in the Alps reached a minimum, while the occasional remains are concentrated in the warmest valleys. The Alpine mountain settlements, including even those where mining for metalliferous ores and salt was carried on, were abandoned, and Gams and Nordhagen remark that this climatic fluctuation had the appearance of a catastrophe. From this peak the rainfall gradually declined and by the Roman period was very little above the present, since in the first century A.D. the Boden See was near its present level, and roads were made across the bogs. From about A.D. 180 to 350, settlements were again concentrated in the driest localities, indicating a return of moist conditions. In the fourth and fifth centuries A.D. many German settlements were established on low ground, now swampy, and this dry period probably continued until the end of the tenth or the middle of the eleventh century, interrupted in the eighth century by a moist interlude indicated by a rise of lake levels.

Traffic across the Alpine passes, as shown by the transmission of culture, became important about 1800 B.C. (8) when the Brenner Pass first became traversable, and reached a maximum at the end of the Bronze Age and in the Early Hallstatt period, or about 1200-900 B.C. The valley settlements of the Late Hallstatt period developed independently apparently in complete isolation, and traffic across the passes was at a minimum. There was a slight revival at the end of the La Tène period and in the early Roman Empire (200 B.C. to A.D. 0), but it was not until between A.D. 700 and 1000 tha this traffic again developed on a considerable scale. There was a re-advance of the glaciers in the western Alps about A.D. 1300, followed by a retreat to a minimum extent in the fifteenth century. Near the end of the sixteenth century the glaciers advanced rapidly and about 1605 they overran settlements which had been occupied since the beginning of history. About the same time the glaciers advanced in the Eastern Alps, Iceland, where they almost reached the moraines of the Late Glacial stages, and probably in other parts of the world ; and the period from 1600 to 1850 has been termed the " Little Ice-Age." There were minor maxima of glaciation about 1820 and 1850 ; since then the glaciers and ice-sheets have been in rapid retreat in all parts of the world. A useful summary of the advances and retreats of glaciers is given by F. E. Matthes (9).

From Russia we have a remarkable series of measurements by W. B. Schostakowitsch (10) of annual layers in the mud deposits of Lake Saki, a salt lake on the west coast of the Crimea in 45° 7' N., 33° 33' E., separated from the sea by a strip of sandy beach. The total thickness of the deposit is several metres, most of the individual layers measuring only a few millimetres. The measurements, in tenths of a millimetre, were made partly on photographs and partly on the original sections ; the earliest layer is dated 2294 B.C., but in some parts of the sections it was difficult to distinguish the lines separating the annual layers and there may have been some errors in calculating the age of parts of the sections.

This series is a valuable climatic record, comparable in importance with the tree-rings of Western America. The variations in the thickness of the layers are almost certainly due to variations of rainfall, and in particular of heavy rainstorms which would cause rapid run-off. The measurements show a number of isolated years with very thick layers five or ten times as thick as neighbouring layers, which is consistent with this suggestion. A possible connexion with rainfall was further examined by comparing overlapping five-year means with the variations in level of the Caspian Sea. Although the curves show many differences of detail, there can be no doubt that their general course, from maximum to minimum and back to a second maximum, is similar. Since rainfall is presumably an important factor in determining the fluctuations of the Caspian, it seems probable that the deposits of Lake Saki are also a rough measure of rainfall. The measurements, smoothed over 50 years, are shown in the lowest curve of Fig. 31. This shows the high maximum just before 2000 B.C., but the remaining fluctuations do not resemble those to the north and west very closely. We may complete the picture for South-east Europe with some data for Lake Ostrovo in West Macedonia collected by M. Hasluck (11). From historical evidence he concludes that the lake was high in Byzantine days and in the eleventh and thirteenth centuries, after which it was low from at least 1400 to the early nineteenth century. These details fit in excellently with Schostakowitsch's data.

When we turn to North-west Europe we find fewer records to which exact dates can be assigned, but the evidence fits in well with that from Central Europe. About 1200 to 1000 B.C. there is evidence of considerable traffic between Scandinavia and Ireland, probably indicating a minimum of storminess. There is other evidence that the climate of Ireland during the Bronze Age was dry and favourable, and that a high civilisation developed both there and in Scandinavia. The Iron Age was a time of great peat-formation, and peat-beds on the Frisian dunes between two layers of blown sand are dated 100 B.C. Some peat-bogs in Northern France were formed during the Roman period. This was a time of eclipse for the northern peoples, who were unable to maintain their high culture. In about 120 to 114 B.C., a period of especially great storminess in the North Sea (Cimbrian flood) caused the wanderings of the Cimbri and Teutons, and gave the coast of Jutland and North-west Germany its present form (18). About A.D. 500 there began a second period of high culture, centred on Tara, in which Irish learning was greatly esteemed in Europe, and a further great outburst of maritime activity.

The thirteenth century was very stormy in the North Sea, and many inroads of the sea were reported in the annals. C. E. Britton (12) gives the following frequencies of " marine floods " in Britain :—

A.D.	1001-1050	1051-1100	1101-1150	1151-1200	1201-1250	1251-1300	1301-1350	1351-1400	1401-1450
	1	1	1	4	3	11	2	2	1

The worst years seem to have been 1176 in Lincolnshire, 1250, when much of Winchelsea was destroyed, and 1287-8 when there were three inundations. There was a corresponding maximum on the coast of Holland. Some of these storms did great damage. In the winter of 1218-19 the coastal defences of Holland and Frisia were broken through and large areas inundated.

A study of the history of water mills in East Kent, by G. M. Meyer (13) points to a period of heavy rainfall in South-east England, which began some time before 1087 (Domesday Book) and ended during the thirteenth century. The rainfall about 1303 was much less than in 1217, and is probably still less to-day.

I have already alluded to the remarkable absence of mention of ice in the early reports of Norse voyages to Iceland and Greenland (Chapter VIII.), and this subject is discussed further in Chapter XXII. It seems probable from the descriptions that previous to A.D. 1000 the climate of Iceland was more genial than to-day ; considerable areas cultivated in the tenth century are now covered by ice. Some remarkable evidence concerning climatic change in Greenland is discussed in Chapter XXI. The Iceland glaciers probably reached their maximum extent of the Christian era in the first half of the fourteenth century. In the fifteenth and sixteenth centuries they retreated, to advance again in the seventeenth century, several farms being destroyed about 1640 or 1650. Since then there has been a slight retreat.

The records of ice in Danish waters were collected by C. I. H. Speerschneider (14), who concluded that they show no evidence of essentially different conditions in the past. The first record of a severe ice-year which he could discover was for the year 1048. For later centuries he gives the following table :—

Century.	Severe Ice.	Ice-free.
12th	..	3
13th	3	..
14th	11	..
15th	9	5
16th	19	9
17th	26	12
18th	27	33
19th	42	58

From the very few occasions when the duration of the ice could be determined, he forms the following summary :—

Period.	Cases.	Duration (Days).
1296-1546	4	49
1583-1595	5	49
1619-1674	6	76
1700-1750	6	51

In opposition to Speerschneider, I think that both these tables point to a period of cold winters in the seventeenth century, and the first also suggests that the fourteenth century was cold. The detailed records place this minimum of temperature in the first half of this century.

Before proceeding to a further discussion of the literary records, it will be well to summarise the results already gained (Table 21). In this table the records are collected in chronological order.

The literary records of Europe provide a great mass of information which has been described by various authors. The British records were collected by E. J. Lowe (15) and have been exhaustively analysed up to A.D. 1450 by C. E. Britton (12). Records of British droughts have also been collected by G. J. Symons (16). For Belgium an extensive compilation has been made by E. Vanderlinden (17). In addition, I have made use of a compilation by R. Hennig (18) which relates to the whole of Europe, but may be rather uncritical. For Britain up to 1450, I have used Britton's work exclusively, omitting years with both wet and dry periods and for droughts after that date I have relied exclusively on Symons's work.

His rather uncritical assiduity makes his numbers somewhat excessive, and there is a discontinuity after 1450. From these sources I have summarised the number of years in each half-century with

(a) Great storms, floods, heavy rain, or wet summers.

(b) Hot dry summers or droughts.

These are shown in Table 22.

Under (a) an attempt was made to eliminate floods or great rains associated with isolated thunderstorms in summer.

It will be observed that all the records increase in frequency as they approach the present day ; this is, of course, simply because the volume of literature becomes greater. In addition to this, certain periods stand out because the records include an abnormal percentage of storms and rain, or of drought. Thus the period from about 600 to 800 appears to have been rather dry and, at the beginning, mild, while 800 to 950 and 1050 to 1350 were generally rainy. The period 1701 to 1750 comes out as unusually dry. In order to obtain the figures of " raininess " of Europe, the three records were

B.C.	
5400	Moist and warm.
5000	Drier and cooler.
4500	Moist, rather warm.
4000-3000	Becoming cooler and drier.
2200	Very dry especially in Central Europe.
2000	Rainy period.
(1275 ?)	Short maximum of rainfall (lake villages destroyed).
1200-1000	Dry and warm, sea traffic.
850	Somewhat moister and cooler.
700- 500	Dry and warm.
500	Sudden increase of rainfall, much cooler. Beginning of Sub-Atlantic.
A.D.	
0	Climate similar to present.
100	Drier and warmer.
180- 350	Wetter.
600- 700	Dry and warm. Alpine traffic.
800-1200	Little ice. Rainfall heavy in Central Europe.
1200-1300	Great storminess, mild winters, probably rainy.
1600	Beginning of general advance of glaciers.
1677-1750	Dry period generally, mild winters.
1850	Beginning of general recession of glaciers.

Table 21.—Fluctuations of climate in Europe.

Period.	Europe (General).		Britain.		Belgium.		Total.		" Raininess."
	(a)	(b)	(a)	(b)	(a)	(b)	(a)	(b)	
100-51 B.C.	2	0	..	..	..	..	2	0	
50 B.C.-0	5	1	..	..	..	..	5	1	
A.D.									
0-50	3	0	..	..	..	..	3	0	
51-100	7	2	..	..	..	..	7	2	
101-150	4	0	..	..	..	..	4	0	
151-200	5	0	..	..	..	..	5	0	
201-250	1	0	..	..	..	..	1	0	
251-300	3	1	..	..	..	..	3	1	
301-350	1	1	..	..	..	..	1	1	
351-400	3	1	..	..	..	..	3	1	73
401-450	6	0	..	..	..	..	6	0	75
451-500	2	1	..	..	..	..	2	1	82
501-550	5	1	..	..	..	..	5	1	80
551-600	16	3	..	..	..	..	16	3	72
601-650	4	2	..	..	..	..	4	2	67
651-700	6	3	..	1	..	..	6	4	60
701-750	3	1	0	2	..	..	3	3	50
751-800	4	5	1	0	..	..	5	5	50
801-850	8	1	0	0	3	1	11	2	85
851-900	10	5	2	0	6	2	18	7	72
901-950	6	1	4	1	0	0	10	2	83
951-1000	5	8	1	0	3	1	9	9	50
1001-1050	21	8	2	0	5	2	28	10	74
1051-1100	18	3	5	3	11	3	34	9	79
1101-1150	22	7	6	4	13	3	41	14	75
1151-1200	22	9	11	6	22	8	55	23	71
1201-1250	28	7	16	6	19	8	63	21	75
1251-1300	25	5	13	7	8	3	46	15	75
1301-1350	24	4	11	5	8	1	43	10	81
1351-1400	17	16	7	4	18	8	42	28	60
1401-1450	15	7	11	1	16	8	42	16	72
1451-1500	8	4	3	6	12	8	23	18	56
1501-1550	..	..	2	6	11	12	13	18	62
1551-1600	..	..	10	12	22	7	32	19	63
1601-1650	..	..	8	15	19	12	27	27	50
1651-1700	..	..	20	22	17	8	37	30	55
1701-1750	..	..	17	29	23	18	40	47	46

Table 22.—Numbers of (a) storms and floods, (b) droughts, in Europe.

combined in the following way. First of all the numbers under the heading (a) for Britain, Belgium, and Europe in each half-century were all added together, irrespective of whether or no some of those from different sources referred to the same year. It was considered that if a particular year appeared as stormy in more than one record, greater weight should be attached to it, and this process of simple addition answered the purpose. The same process was then followed with the records under (b). The number of records under (a), i.e., storms, floods, and great rains, entered to each half-century was then expressed as a percentage of the total number under both (a) and (b). The numbers up to A.D. 600 were then smoothed by adding together the records for five successive half-centuries and allocating the mean to the middle period of the five.

On the whole the figures for " raininess " obtained from the literary records agree with the data from other sources, but add little to the latter. The apparently rainy character of the years 451 to 650 is due entirely to a large number of records of floods in the Tiber collected by Hennig for the last half of the sixth century, and the true figure for raininess should probably be much lower, comparable with the period 651 to 800. The last part of Table 22 is over-weighted by Symons's droughts but there seems little doubt that the period round 1700 actually was abnormally dry (see p. 309).

Actual rainfall records date from the seventeenth century. In England a record at Townley, near Burnley in Lancashire, commenced in 1677 and continued with intermissions until 1704. Meanwhile, in 1697, a record was commenced in Upminster, Essex, which continued until 1716. From this date there was a gap of nine years, after which records began at Southwick, near Oundle, in 1726, and at Plymouth in 1727. Both these are overlapped by a long record from Lyndon, Rutland, commencing in 1737, and from that date there is no difficulty in carrying on the story. The records from 1726 onwards were collated by G. J. Symons who published a well-known table of the annual values in *British Rainfall* for 1870, but owing to the gap the earlier records could not be compared directly with his series, while the Plymouth record came to light after Symons had completed his calculations. The method employed by Symons was to calculate for each station the average for the period over which its records extended, and to express the rainfall for each year as a percentage of that average. The various series were then reduced to a common basis by means of corrections deduced from the years during which pairs of records overlapped. This process is sound if the records are truly homogeneous, and if the stations are close together and overlap by a sufficient number of years, but it becomes hazardous when it is applied several times to isolated records and the overlapping periods are short. For example, the reduction from the average at Southwick to that at Lyndon is based on an overlap of only three years, of which 1737 was much wetter at Lyndon than at Southwick, while in 1738 and 1739 Southwick was slightly wetter than Lyndon. Hence some sort of a check seems to be necessary, and this was provided by expressing the early figures as percentages of the normal for the period 1881 to 1915, and comparing these figures with the percentages obtained by Symons. The results show very close agreement :—

		Percentage given by Symons.	Percentage of 1881-1915.
1726-1736.	Southwick . . .	99	97
1740-1760.	Lyndon	78	78
1727-1752.	Plymouth . . .	83	84

This table is a remarkable testimony to the accuracy and judgment of the reductions carried out by Symons. It is reassuring on another point also. We know nothing as to the construction and exposure of these old gauges, but if they had recorded less than the correct amounts, the figures in the second column of percentages would have been lower than those in the first. The figures may be accepted as reasonably accurate, and this encourages us to accept also the figures for Townley and Upminster. These cannot be reduced to the present-day normal by Symons's method, so they have been compared directly with estimated normals for the period 1881 to 1915.

In this way we obtain the following figures of rainfall in percentage of present normal :—

		Per cent. of normal.
1677-1686.	Townley	89
1689-1693.	Townley	90
1697-1703.	Townley and Upminster . . .	91
1704-1716.	Upminster	93
1726-1730	. . . .	100

The rainfalls for the different decades, revised and brought up to date by J. Glasspoole, are given in the second column of Table 23, expressed as a percentage of the long period average for 1726 to 1940. Figures in brackets indicate an incomplete decade. The third column gives values for Paris

Years.	England.	Paris.	Holland.	Mean.	Years.	England.	Paris.	Holland.	Mean.
1677-1686	89	—	—	89	1821-1830	105	97	103	102
1689-1703	90	108	—	99	1831-1840	100	99	95	98
1704-1720	93	98	—	95	1841-1850	103	103	107	104
1721-1730	(101)	79	—	86	1851-1860	99	101	97	99
1731-1740	92	86	(101)	91	1861-1870	97	95	104	99
1741-1750	87	88	90	88	1871-1880	112	104	110	109
1751-1760	99	—	109	104	1881-1890	101	90	103	98
1761-1770	102	—	113	107	1891-1900	98	93	104	98
1771-1780	97	105	95	99	1901-1910	98	101	99	99
1781-1790	96	98	103	99	1911-1920	107	106	102	105
1791-1800	99	79	90	89	1921-1930	107	117	101	108
1811-1820	99	96	97	97	1931-1940	103	109	97	103
					1941-1947	98	—	—	—

Table 23.—Rainfall means by decades, percentages of long period average, Western Europe.

collected from various sources and expressed as percentages of the average for 1771 to 1940. The fourth column gives a long series for Zwanenburg in Holland corrected to Hoofddorp. The final column of this table gives the means of the three series, as far as they go.

Two long records are also available for Sweden, namely, Uppsala, near Stockholm (1741-1940) and Lund in the extreme south (1748-1910), but these differ so greatly in the early years that they cannot both be correct. They agree in showing a low rainfall before 1760, but whereas at Uppsala the period 1761 to 1820 is recorded as very dry, at Lund the years 1771 to 1810 are excessively wet. These two series have therefore been omitted.

Table 24 gives figures for Padua and Milan. The record for Padua runs from 1725 to 1900, that for Milan begins in 1764. The Padua series was corrected to Milan to make one record. The percentages for Paris, Hoofddorp and Milan were calculated by Miss N. Carruthers.

Years.	Per cent.	Years.	Per cent.	Years.	Per cent.	Years.	Per cent.
1725-1730	(111)	1781-1790	91	1841-1850	118	1901-1910	101
1731-1740	89	1791-1800	99	1851-1860	106	1911-1920	108
1741-1750	102	1801-1810	104	1861-1870	92	1921-1930	79
1751-1760	112	1811-1820	103	1871-1880	100	1931-1936	102
1761-1770	113	1821-1830	99	1881-1890	108		
1771-1780	87	1831-1840	105	1891-1900	105		

Table 24.—Rainfall means by decades, Milan, Italy.

The rainfall minimum in the first half of the eighteenth century appears to have been widespread. Table 25 gives the values for places with a record covering at least ten years of the period, as percentages of recent averages.

Place.	Period.	Per cent. of normal.	Place.	Period.	Per cent. of normal.
Uppsala . .	1721-1731	87	Berlin . . .	1729-1739	91
	1739-1749				
Lyndon . .	1740-1749	77	Paris . . .	1701-1750	86
Southwick .	1726-1736	97	Bordeaux .	1714-1750	87
Upminster .	1701-1716	84	Padua . .	1725-1750	103
Plymouth .	1727-1750	94	Charleston .	1738-1750	100
			(South Carolina)		

Table 25.—Rainfall of first half of eighteenth century.

The figures for Western and Central Europe are remarkably consistent, and point clearly to a persistent dry period during these years.

There is one other interesting feature about Table 23. Rainfall maxima occur in 1761-1770, 1821-1830, 1871-1880, 1921-1930, and possibly 1689-1703, i.e., at intervals of between 50 and 60 years. The Milan series shows maxima in 1725-1730, 1761-1770, 1801-1810, 1841-1850, 1881-1890, and 1911-1920, i.e., an average interval of nearly 40 years. W. H. Bradley (see p. 108) found a cycle of about 50 years persisting for a period of several million years in the Eocene of the West-central United States. The cause is unknown, but if this cycle is verified from other records it may be of great economic importance.

Fig. 32 aims at giving a reconstruction of the variations of temperature in Western Europe. The uppermost curve is based mainly on the estimates of botanists from plant remains, especially tree pollen (see p. 296) and is necessarily generalised. The lower curve is constructed from the estimates of the character of each winter given by C. Easton (19). Up to 1200 records are scanty ; from A.D. 401 to 800 only 4 to 8 per century and from 801 to 1200 only 7 to 17 per half century. These refer mainly to severe winters and the direct average of Easton's figures is therefore too low. I applied a rough correction by plotting the mean of the years given against the number of years and drawing a smooth curve through them, then reading off the difference between this curve and the actual mean. The results show cold periods about 900, 1100 and 1551 to 1700, and warm periods about 1175, 1300 and 1725. This curve is repeated in the right hand part of the upper curve of Fig. 32.

Reliable instrumental records of temperature begin about

the middle of the eighteenth century. Series for Lancashire were discussed by G. Manley (20) and compared with those for Edinburgh, Oxford, Durham and Stockholm. A. Labrijn (21) gives a series for De Bilt, Holland, from 1741 to 1940. All these show a series of oscillations of the order of 30 to 40 years, while the winter temperatures, but not those for other seasons, have in addition a steady rise which persisted from about 1810 until it was interrupted by the series of cold winters in the present decade. This rise of winter temperature has affected the greater part if not the whole of the Northern Hemisphere and is an important climatic fact.

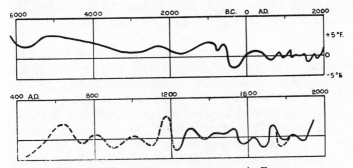

Fig. 32.—Variations of temperature in Europe.
Upper curve, general.   Lower curve, severity of winters (Easton).

A good deal of information about variations of wind direction in Britain is available from various sources. An interesting paper by Leonard S. Higgins (22) gives some inferences as to the prevailing wind directions in South Wales since about 400 B.C. The sand dunes have formed where the coast faces west or south-west, and have since moved inland at intervals. The archæological and historical evidence, which is plentiful, is discussed in detail, with the following results :—

1. Blown sand was present, and had recently been increasing, about 400 to 200 B.C.

2. After that date the area appears to have become stable, and the dunes fixed by vegetation, until the end of the thirteenth century.

3. Soon after 1300, references become frequent to moving sand obliterating roads and pastures and burying

buildings. By 1553 an Act of Parliament " touching
the sea sands of Glamorgan " had become necessary.

4. After about 1550, conditions appear to have gradually
   become more stable, and the dunes became increasingly
   fixed by a plant cover.

The inroads of the sand depend on the conjunction of
several circumstances, especially a period of abnormally high
tides and a period of stormy west or south-west winds. The
results of the investigation, therefore, suggest periods of
stormy winds from west or south-west before 200 B.C. and again
from about 1300 to 1550, while during the intervening and
succeeding periods these winds were less frequent or less
strong.

From 1341 to 1343 there are a number of entries of wind
direction in the weather diary kept by the Rev. W. Merle.
These were analysed by C. E. P. Brooks and T. M. Hunt (23)
and show a dominant wind from west-south-west.

In the latter half of the sixteenth century, however, the winds
appear to have been more easterly. G. M. Meyer wrote to
me : " The following quotation is taken from ' A Restitution
of Decayed Intelligence,' by Richard Verstegan, first published
at Antwerp in 1605 ; it appears on p. 109 of a copy dated
London, 1634 : ". . . old shippers of the Netherlands
affirming, that they have often noted the voyage from
Holland to Spaine, to be shorter by a day and halfe sayling
than the voyage from Spaine to Holland."

The natural inference to be drawn from this record is
that, during the sixteenth century, easterly winds were more
prevalent in the English Channel than westerly winds—
contrary to modern experience. The following quotation
from a book published in 1579 though ambiguous is to
the same effect : " And the winds in Winter blow about the
morning, but in the Sommer about the evening, and in
the Winter out of the East, and also in the Sommer, out
of the West."

The observations of Tycho Brahe at Uraniborg in Denmark
from 1582 to 1597 also point to conditions over North-west
Europe differing from those prevailing at present. According
to the discussion of these observations by D. La Cour (24) all
easterly winds were more frequent than at present, and the

prevailing direction was actually south-east instead of south-west.

About the middle of the seventeenth century the winds appear to have been rather variable in England. Sir Francis Bacon (25, p. 35) wrote : " In Europe these are the chief stayed winds, North windes from the Solstice, and they are both forerunners and followers of the Dog-starre, West windes from the Equinoctiall in Autumne, East windes from the Spring Equinoctiall ; as for the Winter Solstice, there is little heed to be taken of it, by reason of the varieties." On page 39, however, there is an indication that the wind was mainly westerly. Probably the winds were more variable than at present but with some predominance from the west.

From 1667 onwards (23) there are a sufficient number of old weather diaries to give a fairly complete picture. The direction and " constancy " (per cent.) for each half-century are as follows :—

	1667-1700	1701-1750	1751-1800	1801-1850	1851-1900	1901-1930
London						
Direction . .	237	226	190	221	235	239
Constancy . .	19	15	35	21	19	21
Edinburgh						
Direction . .	—	243	245	253	257	241
Constancy . .	—	31	15	25	26	30
Dublin						
Direction . .	—	241	248	258	262	256
Constancy . .	—	19	22	27	25	17

The resultant directions are shown in degrees from north through east ; 180° is a south wind, 225° a wind from south-west and 270° from west. Constancy is the ratio between the total movement of the air from the resultant direction, assuming that each wind has unit velocity, divided by the number of winds and multiplied by 100. A constancy of 33 means that two-thirds of the winds blew more or less from the prevailing direction and the remainder from opposing directions. The figures show for London a steady swing from nearly W.S.W. to nearly south, and back again. The most remarkable periods at London were from 1740 to 1747 and from 1794 to 1810, both of which show a predominance of easterly winds. The Dublin series also shows mainly easterly winds in 1740 to 1748, but observations are missing for the second period. The decade 1740 to 1749 was the driest in England since observations began. During the period 1794 to 1810 the winters in England were exceptionally severe. The years 1901 to 1930 were remarkable for the great steadiness of the W.S.W. winds, which is no doubt related to the rise of winter temperatures.

We are now in a better position to examine Huntington's contention, set out in the preceding chapter, that the level of civilisation in Rome fluctuated in accordance with the average rainfall. The vigorous Roman life of the early Republic, based on intensive agriculture, was maintained during the period from 450 to 250 B.C. Towards 250 B.C. the spirit of discipline and rural simplicity began to decay, from 225 to 200 B.C. was a period of economic stress, and the second century B.C. witnessed a great decline of agriculture. During this period malaria became endemic, according to E. Huntington (26), because the decreasing rainfall was no longer sufficient to maintain flowing water in the rivers throughout the long hot summers, so that stagnant pools and marshes were formed which provided favourable breeding-grounds for mosquitoes. After the land law of Spurious Thorius in 111 B.C., however, agricultural disturbances declined and the price of land rose rapidly, but the vine and olive replaced grain as the main agricultural product. By 90 B.C. there was a marked increase in general luxury and comfort which reached a high level from 75 B.C. to about A.D. 50. From A.D. 80 onwards, however, there was a gradual decline, and A.D. 180 to 190 were years of famine and pestilence. From A.D. 193 to 210 there was a slight increase in prosperity, but then began in full force the long " decline and fall of the Roman Empire."

When we compare these variations in the level of prosperity with the estimates of rainfall we obtain the following result :—

	450-250 B.C.	250-200 B.C.	200-100 B.C.	100 B.C.-A.D. 50	A.D. 80-200	A.D. 200 onwards.
Civilisation.	Very high.	Falling.	Low.	High.	Mainly low.	Very low.
Rainfall.	Very heavy until 300, then decreasing.	Decreasing.	Light until 150 B.C., then increasing.	Heavy at first, then decreasing.	Light.	Light and generally decreasing.

The agreement is surprisingly good, but the climatic changes seem to precede the changes of civilisation by about fifty years. This is actually rather more probable *a priori* than a direct concordance.

If this principle is sound and if the rainfall curve is correct, we should expect to find a recrudescence of civilisation in the Mediterranean from the middle of the eleventh to the middle of the thirteenth centuries. At that time a large part of this region was in the hands of the Moslems. It was seen in the last chapter that the Moslem outbreak from the seventh century onwards was the fourth of a great series of waves of emigration from Arabia which are attributed to dry periods. The Moslem armies rapidly overran North Africa and Spain, and at first their achievements were largely military and religious. About the eleventh century, however, they began to develop a high civilisation, and Egypt for a time took almost its old place as a leader of thought. As an example of the high level of Moslem culture in Spain, we have the Alhambra at Granada (thirteenth century). Italy also reached a high level during this period, when the city states, especially Venice and Genoa, became famous—for example, the cathedral of St Mark in Venice was built in the eleventh and twelfth centuries. The agreement seems to afford additional support to the rainfall maximum of this period in Europe, and it justifies us in using, with due caution, variations in the level of civilisation as indications of climatic change in other regions also.

It is with regard to ancient Greece that the discussion of Huntington's theory of civilisation and climate has been most vigorous. In " The Burial of Olympia " (27), Huntington first put forward for discussion the theory that up to about 400 B.C. Greece had been well watered and forested, with perennial streams unsuited to the development of mosquitoes, but that after that date the rainfall diminished greatly. The streams were reduced in summer to stagnant pools and swamps, with the result that malaria became endemic and undermined the vitality of the population. The driest period began about the seventh century A.D., and resulted in the accumulation above the ruins of Olympia of about fifteen feet of silt. At present the river Lodon is again cutting a channel through this silt. Unfortunately, the hydrographical system of this river is so peculiar that it is doubtful whether any significance can be attached to this deposit of silt.

The evidence of the Classical writers is very conflicting, but E. G. Mariolopoulos (28) will not admit that there has been the slightest change in the climate of Greece since Classical times, basing his arguments chiefly on the descriptions of the fertility of the country, the nature of the streams and rivers, and the dates of sowing and of harvest. He is able to make out a strong case against climatic change since at least 350 or 400 B.C., and perhaps the best verdict is one of " not proven." Mariolopoulos gives one interesting quotation from Plato which shows that questions of climatic change are not new, but agitated the scientific circles of ancient Greece as well as those of to-day ; it reads exactly like the report of the recent Drought Investigation Committee of South Africa, the decrease in fertility being attributed to the washing away of the soil.

REFERENCES

(1) ANTEVS, E. " Swedish late-Quaternary geochronologies." New York, *Geogr. Rev.*, 15, 1925, p. 280.

(2) GAMS, H., and R. NORDHAGEN. " Postglaziale Klimaänderungen und Erdkrustenbewegungen in Mitteleuropa." München, *Geogr. Gesellsch. Landesk. Forschungen*, H. 25, 1923.

(3) ERDTMAN, G. " Some aspects of the post-glacial history of British forests." London, *J. Ecol.*, 17, 1929, p. 42.

(4) GODWIN, H., and A. G. TANSLEY. " Prehistoric charcoals as evidence of former vegetation, soil and climate." London, *J. Ecol.*, 29, 1941, p. 117.

(5) GODWIN, H. " Pollen analysis and forest history of England and Wales." Cambridge, *New Phytologist*, 39, 1940, p. 370.

(6) HARDY, E. M. " Studies of the post-glacial history of British vegetation." V. " The Shropshire and Flint Maelor mosses." Cambridge, *New Phytologist*, 38, 1939, p. 364.

(7) GRANLUND, E. " Den Svenska hogmossarnasgeologi." Stockholm, *Sverig. geol. Unders.*, Afh.C., 26, 1932, no. 1.

(8) CHILDE, V. G. " The Danube thoroughfare and the beginnings of civilisation in Europe." *Antiquity*, 1, 1927, p. 79.

(9) WASHINGTON, NATIONAL RESEARCH COUNCIL. " Physics of the air." Vol. IX., " Hydrology," Chapter 5.

(10) SCHOSTAKOWITSCH, W. B. " Bodenablagerungen der Seen und periodische Schwankungen der Naturerscheinungen." Repr. from Leningrad, *Mem. Hydr. Inst.* See London, *Meteor. Mag.*, 70, 1935, p. 134.

(11) HASLUCK, M. " A historical sketch of the fluctuations of Lake Ostrovo in West Macedonia." London, *Geogr. J.*, 87, 1936, p. 338.

(12) BRITTON, C. E. " A meteorological chronology to A.D. 1450." London, Meteor. Office, *Geoph. Mem.*, 8, No. 70, 1937.

(13) MEYER, G. M. " Early water-mills in relation to changes in the rainfall of East Kent." London, *Q. J. R. Meteor. Soc.*, 53, 1927, p. 407.

(14) COPENHAGEN, DANSK METEOROLOGISK INSTITUT. Medd., No. 2. " Om isforholdene i Danske farvande i ældre og nyere tid, aarene 690-1860." Af C. I. H. SPEERSCHNEIDER. Kjøbenhavn, 1915.

(15) LOWE, E. J. "Natural phenomena and chronology of the seasons." London, 1870.

(16) [SYMONS, G. J.] "Historic droughts." *British Rainfall*, 1887, p. 22.

(17) VANDERLINDEN, E. "Chronique des événements météorologiques en Belgique jusqu'en 1834." Bruxelles, 1924.

(18) HENNIG, R. "Katalog bemerkenswerter Witterungsereignisse von den ältesten Zeiten bis zum Jahre 1800." Berlin, *Abh. K. Preuss. Meteor. Inst.*, Bd. 2, No. 4, 1904.

(19) EASTON, C. "Les hivers dans l'Europe occidentale." Leyde, 1928.

(20) MANLEY, G. "Temperature trend in Lancashire, 1753-1945." London, *Q. J. R. Meteor. Soc.*, 72, 1946, p. 1.

(21) LABRIJN, A. "200 jaar temperatuurwaarnemingen in Nederland." *Hemel en Dampkr.*, Groningen, 40, 1942, p. 41.

(22) HIGGINS, L. S. "An investigation into the problem of the sand dune areas on the South Wales coast." *Archaeologia Cambrensis*, June, 1933.

(23) BROOKS, C. E. P., and T. M. HUNT. "Variations of wind direction in the British Isles since 1341." London, *Q. J. R. Meteor. Soc.*, 59, 1933, p. 375.

(24) LA COUR, D. "Tyge Brahes meteorologiske dagbok holdt paa Uraniborg for aarene 1582-1597." Appendix til Collectanea Meteorologica. Kjøbenhavn, 1876.

(25) BACON, FRANCIS. "The naturall and experimentall history of winds." 1653.

(26) HUNTINGTON, E. "The pulse of progress." New York and London, 1926.

(27) ——. "The burial of Olympia." London, *Geogr. J.*, 36, 1910, p. 657.

(28) MARIOLOPOULOS, E. G. "Étude sur le climat de la Grèce. Précipitation. Stabilité du climat depuis les temps historiques." Paris, 1925.

CHAPTER XIX

ASIA

THE interior of the great continent of Asia has for the last twenty centuries or more been occupied by nomadic tribes, and we have no great body of dated literary records such as that which facilitated our study of the climate of Europe during the Christian era. On the other hand, south-western Asia includes the sites of some of the oldest known civilisations of the globe, and we have a rich mine of historical material from which to draw conclusions. In Chapter XVIII. we found some evidence that even in such a climatically favoured continent as Europe, variations in the rainfall from one century to another strongly influenced the level of civilisation, and to some extent determined the wanderings of peoples. During a period of increased rainfall there is a movement from regions which are naturally moist to regions which are naturally dry ; this movement is noticeable both in the locations of Alpine settlements and in the migrations of whole tribes and nations. During the drier periods the direction of movement is reversed ; the naturally moist regions are occupied and the dry regions are more or less abandoned. In the great Eurasian continent there is a progressive diminution of rainfall from west to east, which extends to the eastern boundary of the region of monsoon rainfall in China, and large parts of the interior of Asia are on the borderline between aridity and complete desert. Hence, it is in Asia that we should expect to find droughts recorded most vividly in history, in accordance with the principles set out in Chapter XVII.

Evidence from the earlier periods is given by the semi-arid settlement at Anau on the northern margin of Persia. This site was occupied from time to time and abandoned during the intervening periods, and since there is no evidence of conquest, while the periods of abandonment are represented by desert formations, it is generally accepted that the interruptions were due to drought (1). The dating of the earlier settlements is by means of the relative thickness of deposits formed above them, the latest is partly historical. The first settlement began about 9000 B.C. ; the second, which immediately succeeded it, about 6000 B.C. The last part of the first settlement and the whole of the second show evidence of gradually increasing drought, and the settlement was abandoned soon after 6000 B.C. The site was reoccupied, after an interval of desert conditions, about 5200 B.C. This third settlement continued until about 2200 B.C. with a short interruption, probably due to drought, about 3000 B.C. In 2200 B.C. there began a period of intense drought, and about this time not only Anau, but other settlements, such as Susa and Tripolje, were abandoned (1). The site was not reoccupied until the Iron Age, probably not much earlier than Persian times.

The evidence for climatic changes in Asia since 100 B.C. was summarised by Ellsworth Huntington in his last book (2). He gives curves of caravan travel in Syria after C. P. Grant (3), his own reconstruction of the evidence of lakes and ruins in Asia, and a tabulation of the frequency of migrations to and from the dry areas since 400 B.C. from A. J. Toynbee (4), all of which agree in their main features. They point to rainy periods about A.D. 0-200, 400-500, 700-1100, 1250, and 1500-1700, and dry periods about 300, 500, 1100, and 1400.

The Syrian desert lies across the main land routes between Asia and Europe-North Africa, and in its present state offers a formidable obstacle even to motor transport. It would now be almost impassable by camel caravans, and in dry periods these went by a circuitous route. At other times, however, they struck boldly across it, which would have been impossible without a much heavier rainfall. Huntington notes that about 550 B.C. Nabonidus set up his headquarters at Terma in north-west Arabia, now a small village, and his son sent him couriers and food supplies regularly by camel across the desert.

In order to carry the curve of migrations farther back, Table 26 was compiled from three main sources (1, 5, 6), and was carried from 5200 B.C. down to the year A.D. 50. It will be observed that the migrations are almost always from the drier to the wetter regions ; the exceptions are the

reoccupations of Anau about 5200 and in the first millennium
B.C., and a wave of migration which began in Central Europe
about 1275 B.C. and penetrated as far as the east of Asia
Minor, where it formed the Armenians.

	B.C.	
	5200.	Anau reoccupied.
Before	5000.	Sumerians and some Semites occupied Mesopotamia.
4000-3000.		First Semitic wave from Arabia.
	3000.	Aryans to Baltic.
	2650.	Sumer overrun from the north.
	2600.	Amorites invade Egypt.
	2450.	Kassites.
	2360.	Kassites.
2300-2050.		Great dispersal of Aryans (or Wiros).
	2225.	Steppe-folk invade Tripolje area.
	2200.	Evacuation of Susa and Anau.
		Semites in Mesopotamia.
	2170.	Kassites.
	2072.	Kassites.
	2045.	Kassites.
	2000.	Amorites. Tumulus folk.
	1926.	Kassites.
	1800.	Hyksos conquer Egypt.
	1750.	Eruption from Iran.
	1746.	Kassites.
	1700.	Aryans into Punjab.
	1500.	Aryans.
1500-1000.		Great unrest in Western and Central Asia.
1350-1300.		Aramæan wave from Arabia.
	1275.	*Migration eastward to Armenia.*
	1180.	Elamites.
	(1050.	Dorians.)
ca.	600.	Anau reoccupied.
	500.	Arabs.
300-250.		Break up of Hellenised States in West Asia.
	160.	Saka.
A.D.	50.	Jafnite wave from Yemen.

Table 26.—List of migrations.

In order to construct a climatic curve from this table, a
value was assigned to each century (or other convenient
period according to the detail of the record) based on
the " wet-ward " component of migration. The numbers
assigned ranged from —5 for the period about 600 (reoccupa-
tion of Anau, apparently complete cessation of migration

from the dry regions) to $+5$ for the great drought about
2200 B.C. From these numbers the curve shown in Fig. 33
was constructed up to the beginning of the Christian era.
The later portion of that curve is derived mainly from the two
curves given by Huntington (2) and Toynbee's diagram of
migrations, controlled by the levels of the Caspian (see below).

The climate of Western and Central Asia during the
Christian era is best determined from a study of the fluctuations
of the Caspian and other salt lakes without outlet. The
level of such lakes is determined by the rainfall and evaporation.
If the rainfall increases, the level of the lake rises and it overflows its shores until it offers a sufficiently increased surface
for evaporation to balance the greater rainfall over the basin.

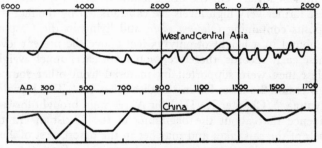

Fig. 33.—Variations of rainfall in Asia.

If the increase of rainfall is very great, the lake may rise until
it finds outlet to the sea. With decreasing rainfall the level
sinks and the area decreases until the evaporation is again
only sufficient to balance the rainfall. A decrease in the rate
of evaporation per unit area would have the same effect
as an increase of rainfall. Only one of these natural rain-
gauges, the Caspian Sea, lies sufficiently near to the world of
antiquity for its variations of level to be brought into our
chronology, but the variations of the remaining lakes have
evidently been similar, and we may infer that the fluctuations
of rainfall indicated by the Caspian represent with fair
accuracy those of the whole of Central and Western Asia, in
spite of the complications introduced by variations in the
course of the Oxus River. The fluctuations of the Caspian
were carefully studied by E. Brückner (7), and have been
further discussed by Ellsworth Huntington (8).

The first definite reference to the Caspian is given by Herodotus, about 438 B.C. Huntington interprets Herodotus' description as implying that the length of the Caspian from north to south was about six times the breadth from west to east. At present its length is only between three and four times its breadth, but if it became deeper, it would expand very little in an east-west direction and very greatly to the north. He also considers it probable from the description that the Sea of Aral was united with the Caspian. For these reasons Huntington believes that when Herodotus wrote, the Caspian stood about 150 feet higher than now. Strabo, in A.D. 20, gave descriptions from which Khanikof has estimated that the Caspian stood at that time 85 feet above its present level. On the other hand, L. Berg (9) states that these former very high levels are contradicted by the fact that deposits containing *Cardium edule*, still living in the Caspian, are found to a height of only 75 feet above the present level. These early records, therefore, would not carry much weight, unless they were supported by material from other sources. This is not the case, for in the discussion on " The Burial of Olympia " (Chapter XVIII.) Sir Aurel Stein brought forward evidence that about the beginning of the Christian era the levels of the salt lakes and marshes in the desert west of Tunhuang were about the same as to-day. These form part of a defensive line which was completed by a Chinese wall, built about 100 B.C. and abandoned early in the first century A.D. Wherever this wall abuts on any of the lakes or marshes it can clearly be traced down to within a few feet of the actual water level in the spring of 1907.

Nothing further is known as to the level of the Caspian until the middle of the fifth century, but J. W. Gregory (10) states that in A.D. 333 the Dead Sea stood at its present level. Between A.D. 459 and 484 the " Red Wall " was built as a barrier against the Huns. At this time the level of the Caspian must have been very low, for the wall extends below water 18 miles from the shore, and a caravanserai at the old port of Aboskun is now under water ; there are other submerged houses and cities of unknown date in different parts of the basin. The level was at least 15 feet below the present.

Istakhri, an Arab geographer, in A.D. 920, records that the old wall at Derbent projected into the sea so far that six of its towers stood in the water, and from this Brückner concludes that the level was 29 feet above the present. There is also evidence that Lake Seistan, in Persia, was high about A.D. 900.

The caravanserai at Baku, which, according to Brückner, was built in the first half of the twelfth century, indicates a level 14 feet below the present. In 1306 to 1307 the level was 37 feet above the present ; this may be partly due to the fact that the Oxus entered the Caspian instead of the Sea of Aral at that time, but it is significant that almost in the same year Dragon Town, on the shores of Lop Nor, was destroyed by the rising water. In 1325 the level was still high.

Early in the fifteenth century the Caspian swallowed up a part of the city of Baku (level 16 feet above present). The level was still high in 1559 and 1562.

For the seventeenth to nineteenth centuries Brückner gives the following levels :—

1638.	15 feet above present level.	1830.	1 foot above.
		1843-1846.	2 feet below.
1715-1720.	1 foot above.	1847.	1 foot above.
1730-1814.	Relatively high.	1851-1860.	1 foot below.
1815.	At least 8 feet above present.	1861-1878.	1 to 3 feet above.

The changes of level in the Caspian are shown in Fig. 34.

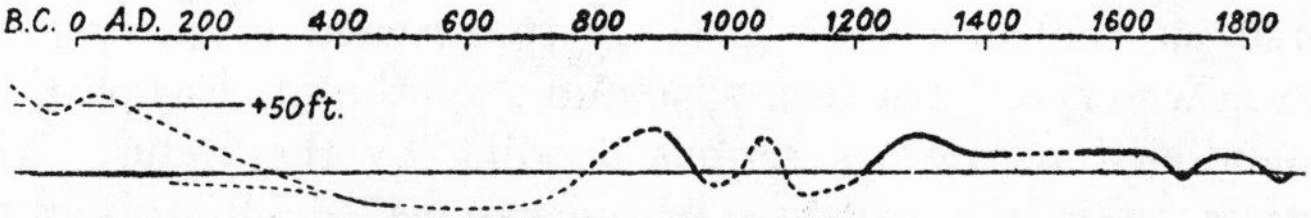

Fig. 34.—Variations in the level of the Caspian.

The curve of rainfall in Asia since 5200 B.C. agrees well with those for Europe (Fig. 31). Both rise to a maximum between 5000 and 4000 B.C., falling steadily to a minimum about 2200 B.C. The small maximum at 1275 B.C. is dated by evidence of migration from Europe to Asia ; in Europe it is supported by other evidence which refers to about this period but cannot be dated accurately. The great maximum of rainfall about 500 B.C. in Europe appears to have come somewhat earlier in Western Asia, and after this there is some conflict of detail which may be due to lack or misinterpretation of data, or to errors of dating. The oscillations in Western Asia *appear* to have been more pronounced than in Europe, but the scales of the diagrams are only relative, and arid areas are much more sensitive to small variations of rainfall than more humid regions. The dry period about A.D. 700 on which Huntington lays so much stress is also found in Europe, and a general excess of rainfall about A.D. 1300 is common to both. Considering that the two curves are based on entirely different and independent data, the measure of agreement seems to prove that at least the major climatic oscillations are real and widespread.

It should be noted, however, that R. C. F. Schomberg (11) does not admit the reality of these variations of rainfall in Central Asia, considering that the apparent evidence is due to other causes, especially changes in river courses, which easily erode the soft sandy soil. In the discussion of the last of these papers several speakers questioned this verdict. It is, of course, against all climatological experience that rainfall could have maintained a dead level for several thousand years, but it is not unlikely that Huntington has magnified the range of the oscillations.

There is some evidence of former moister conditions in Northern India. E. J. H. Mackay (12), describing excavations at Mohenjo-Daru, near the west bank of the Indus about 270 miles above Karachi in a very dry region, states that about 2750 B.C. culverts were specially constructed to carry away storm water, and between 2750 and 2500 the site was partially abandoned because of serious flooding by the Indus. This agrees with the paucity of migrations about this time. Reference may also be made to some speculations by V. Unakar (13) on the interpretation of the RG-VEDA. He finds evidence of three types of climate : first a period of cool weather with rains fairly uniformly distributed through the year and few thunderstorms ; second a stormy period when winter depressions gave copious rains ; and finally a period of increasing drought. The date is uncertain, but Unakar provisionally places the sequence as possibly covering the period 5000 to 2000 B.C.

A few references given by C. W. Bishop (14) to events

in China throw some light on the climatic changes in that country. Thus we have :—

1766 B.C. First dynasty overthrown by a popular revolt following seven years of drought. Fig. 33 shows a secondary minimum of rainfall in Western Asia about this date, but not comparable with the minimum of 2200 B.C.

1122 B.C. Second dynasty overthrown by popular revolt and invasion from the west. This disturbance clearly coincides with a minimum in Western Asia.

842-771 B.C. Period of turmoil and invasion from the west. Great drought, accompanied by disturbances, recorded for about 822 B.C. This is in remarkable agreement with Erdtman's " dry heat-wave " in the closing centuries of the Sub-boreal in Europe.

Four hundred years of anarchy and confusion began in the third century A.D., and this again coincides with a dry period in Europe and especially in Central Asia, and a shorter period of disintegration in the tenth century also falls in a period of drought.

Co-Ching Chu (15) published an analysis of the Chinese archives since A.D. 100 on the same lines as that for Europe described in the preceding chapter. His results for all China, tabulated by centuries, are as follows :—

Century. A.D.	Floods.	Droughts.	"Raini- ness."	Century. A.D.	Floods.	Droughts.	"Raini- ness."
2	18	35	34	10	36	64	36
3	15	24	38	11	41	69	37
4	5	41	11	12	56	58	49
5	18	37	33	13	43	77	36
6	10	41	20	14	57	60	49
7	13	43	23	15	24	54	31
8	31	41	43	16	43	84	34
9	24	43	36	17	67	82	43

Table 27.—Floods and droughts in China.

There is a general tendency for the number of floods to increase relatively to the number of droughts in the later centuries ; when this is allowed for, the fourth, sixth, and seventh centuries and, later, the fifteenth and sixteenth centuries stand out as predominantly dry ; the second and third, eighth, twelfth, and fourteenth centuries as wet. The general agreement with the results of the similar tabulation for Europe is very good. The curve of raininess in China, based on these data, is shown below the curve for Western Asia in Fig. 33.

Co-Ching Chu also remarks that " In a recent bulletin published by the U.S. Department of Labour, Ta Chen has found that Chinese migration can be grouped into three periods : those of the seventh, fifteenth, and nineteenth centuries. During the first period, Chinese migrated to the Pescadores and Formosa ; in the second period, to Malaysia ; and in the third, about 1860 . . . with destinations in Hawaii, North America, and South Africa. Mr Chen found that the most significant causes of emigration are pressure of population and droughts and famines ; while during the last century Chinese emigration was much accelerated by the ease of communication and by the demand for labour to open up new lands. Such, however, cannot be said of the seventh or fifteenth centuries." The dryness of these two periods shown in Table 27 is confirmed by these results.

Co-Ching Chu also gives some records bearing on the variations of temperature. The number of severe winters per century during the sixth to sixteenth centuries falls to a minimum between A.D. 600 and 800, rises to a maximum between A.D. 1100 and 1400, with a well-marked crest in the fourteenth century, and subsequently decreases again. The variations of frequency, like those of raininess, run closely parallel with the variations in Europe. The author also examines the dates of the latest spring snowfall in each decade at Hangchow during the period 1131-1260, and finds that the average date, 9th April, is nearly a month later than the date of the latest spring snowfall during the period 1905-1914, suggesting that the climate was colder and stormier in the twelfth and thirteenth centuries than at present, and confirming the evidence of the severe winters. It is also interesting to remark that the author finds a parallelism between the occurrence of a severe climate and the frequency of sunspots.

K. A. Wittfogel (16) finds some slight evidence that in North China about 1600 to 1100 B.C. the winter was warmer than at present, interest in crops and agricultural rainfall starting very early in the year. He thinks that the summer rainfall may have been slightly greater than now.

Finally, some reference is necessary to the ruins of Angkor

(17), a lost city, formerly the centre of the powerful and highly civilised Khmer Empire, which developed in the steaming jungle of French Cambodia, between the Mekong River and the frontier of Siam, in about 14° N. Angkor flourished between A.D. 600 and 1200, and the empire seems to have reached its highest point about A.D. 1000. Climatically, the region is very similar to Yucatan, which also had a high civilisation, the relics of which are now buried in thick tropical jungle (see Chapter XXII.), though it lies nearer to the equator than does Yucatan, and belongs definitely to an extension of the equatorial rain-forest belt. The climatic conditions of Angkor are very unfavourable for the development of a high civilisation at the present day, and the founding of a great city (the population is estimated at a million) in such a site suggests a much drier climate about A.D. 600. The causes of the break up of the empire and the abandonment of the city are not known, but it is an interesting possibility that about A.D. 1000 or 1050 the climate became moister in conformity with the changes in other parts of the world, and that the inhabitants fought a losing fight against the advancing tide of tropical vegetation for nearly two centuries before they finally gave up the struggle and migrated to more open country.

REFERENCES

(1) PEAKE, H. J. E. "The Bronze Age and the Celtic world." London, 1922.
(2) HUNTINGTON, E. "Mainsprings of civilisation." New York and London, 1945.
(3) GRANT, C. P. "The Syrian desert." New York, 1938.
(4) TOYNBEE, A. J. "A study of history." Vol. III. Oxford Univ. Press, 1934.
(5) "THE CAMBRIDGE ANCIENT HISTORY," vol. i. Cambridge, 1923.
(6) HADDON, A. C. "The wanderings of peoples." Cambridge, 1919.
(7) BRÜCKNER, E. "Klimaschwankungen seit 1700 . . ." Vienna, 1890.
(8) HUNTINGTON, E. "The pulse of Asia." Boston and New York, 1907.
(9) BERG, L. "Das Problem der Klimaänderung in geschichtlicher Zeit." *Geogr. Abh. hrsg. von. A. Penck in Berlin*, 10, H. 2, 1914.
(10) GREGORY, J. W. "Is the earth drying up?" London, *Geogr. J.*, 43, 1914, p. 154.
(11) SCHOMBERG, R. C. F. "The aridity of the Turfan area." London, *Geogr. J.*, 72, 1928, p. 357.
——. "The climatic conditions of the Tarim Basin." *idem*, 75, 1930, p. 313.
——. "Alleged changes in the climate of southern Turkestan." *idem*, 80, 1932, p. 132.
(12) MACKAY, E. J. H. "Further excavations at Mohenjo-Daro." London, *J. R. Soc. Arts*, 82, 1934, p. 206.

(13) UNAKAR, V. "Meteorology in the RG-VEDA." *J. Asiat. Soc.*, *Bombay Branch*, 9, 1933, p. 53 ; 10, 1934, p. 38.
(14) BISHOP, C. W. "The geographical factor in the development of Chinese civilisation." New York, N.Y., *Geogr. Rev.*, 12, 1922, p. 31.
(15) CHU, CO-CHING. "Climate pulsations during historic time in China." New York, N.Y., *Geogr. Rev.*, 16, 1926, p. 274.
(16) WITTFOGEL, K. A. "Meteorological records from the divination inscriptions of Shang." New York, N.Y., *Geogr. Rev.*, 30, 1940, p. 110.
(17) CANDEE, H. CHURCHILL. "Angkor the magnificent." (Witherby), 1925.

CHAPTER XX

AFRICA

THE most important source of information as to the variations of rainfall in Africa is provided by the levels of the River Nile. As is well known, the Nile commences in Lake Victoria, in Central Africa, and flows to Lake Albert as the Victoria Nile. From here it continues as the Bahr-el-Jebel, becoming known as the White Nile after the junction of the Sobat River. At Khartoum it receives the Blue Nile, and near Berber the Atbara River, both of which originate in the mountains of Abyssinia. From the junction of the Blue Nile to the Mediterranean, a distance of 1,800 miles, it receives no appreciable accession of water. The level of the Nile passes through an extremely regular annual variation ; the water is at its lowest in April and May, it rises slowly and irregularly in June and the first half of July, but rapidly and steadily in the latter half of July and the first half of August, remaining high during September and commencing to fall rapidly in October. The regular annual flood is the source of the fertility of Egypt ; without it the whole land would be a barren desert, and hence the levels of the flood have been recorded annually, probably for some thousands of years. The lowest level reached at the stage of low water has been recorded less regularly. Many of these records have been lost, but enough remain to form a very valuable series, which has been collected and published by Prince Omar Toussoun (1). The values of maxima and minima at the Roda gauge, Cairo, smoothed by forming fifty-year means commencing at successive intervals of ten years, are shown in Fig. 35.

It is necessary to understand exactly the significance of both the high and low levels. The White Nile drains a large area of equatorial Africa which has a considerable annual rainfall distributed fairly evenly throughout the year ; moreover, it passes through two large lakes, Victoria and Albert, which further regulate the flow. Hence the White Nile above its junction with the Sobat River discharges an almost constant volume of water throughout the year (2). The Blue Nile, the Atbara, and the Sobat River, on the other hand, originate in Abyssinia, which receives the greater part of its rainfall in the summer months. At Addis Ababa the mean annual rainfall is 46 inches, and of this amount 35 inches fall in June, July, August, and September. Hence it is these rivers, and especially the Blue Nile, which supply the waters of the annual flood, in which the White Nile plays very little part. The Abyssinian rainfall is monsoonal ; Sir Henry Lyons (3) showed that it is closely related to the pressure in

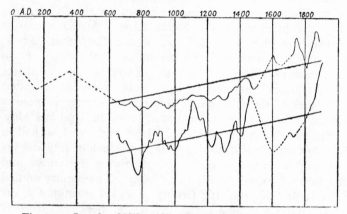

Fig. 35.—Levels of Nile. Flood stage and low-level stage.

the neighbourhood of Cairo, high pressure preceding a low Nile flood, and low pressure a high flood. Variations of pressure at Cairo are representative of those found over a wide area, extending from Beirut to Mauritius, and from Cairo to Hong Kong. Pressure variations in this area are generally opposite to those in South America and adjoining regions, and the fluctuations of the Nile flood therefore represent the " see-saw " of pressure between the old and the new worlds. The minimum level, on the other hand, depends chiefly on the rainfall in the equatorial belt of low pressure, which is very closely connected with the intensity of the general circulation of the atmosphere. For this reason the level of the Nile during the stage of low water is the better guide to the general rainfall of equatorial Africa, while the flood levels represent the monsoon rainfall of the eastern highlands. We should expect the former to show a closer relation to the rainfall of Europe than the latter.

It will be noticed that both maxima and minima show a general upward trend. This is due to the deposit of silt, which has been steadily raising the level of the Nile bed for thousands of years, at the rate of about 0·1 metre per century. This is represented by the straight sloping lines in Fig. 35, but it is probable that the rate has varied from time to time. The steep rise of the minima in the latest years is due to artificial control of the water, especially by the Delta Barrage, and the data for these years are useless for our purpose.

The records are made up as follows : there is an almost continuous series of records of both high and low levels extending from A.D. 641 to 1480. From 1480 to 1830 there is a broken record of the flood levels, and the data of low levels are very scanty. From 1830 onwards the series are again complete, but their value is greatly lessened by the irrigation works. There are no connected series of records known earlier than A.D. 641, but Prince Omar Toussoun gives determinations of the levels which constituted " weak floods," " good floods," and " strong floods " in the fifth century B.C., and in the first, second, and fourth centuries A.D., as well as in the later centuries. From the seventh to the nineteenth centuries the mean of these three levels averages 0·48 metre below the mean annual level for the century, but this difference is very variable, with some tendency for a secular increase. Hence the means obtained from these weak, good, and strong floods can be regarded merely as indications, which point to generally good floods in the fifth century B.C., and in the first and fourth centuries A.D., and rather poorer floods in the second century A.D.

The maximum about 500 B.C. finds some support in Herodotus' "History" (4). Prince Omar Toussoun quotes a passage to the effect that unless the flood rose to a level of 15 or 16 coudées (" cubits," 8·0 or 8·5 metres) it did not overflow the fields ; he therefore takes this as representing an average good flood. He thinks that these figures refer to the " effective flood," that is, the height to which the flood rose above the level of low water, for he points out that " the levels given by Herodotus, four centuries B.C., are the same (actually they are rather higher) as those cited by Ammien Marcellin (Ammianus Marcellinus), four centuries A.D. Now it is impossible, with the continual elevation of the soil, that after an interval of eight centuries these levels should have remained the same, if they had for base the same zero. Hence it is necessary to consider all the levels mentioned by these authors as being effective levels."

I do not agree with this argument, for the irrigation value of a flood depends on its gauge level, and not the range from low water, and I think the gauge level would be more likely to be reported. The figures quoted would agree equally well with the assumption that in the fifth century B.C. the floods were generally good, while in the fourth century A.D. they were generally poor. Even if the figures do refer to " effective floods," however, they still point to relatively high floods in the fifth century B.C. From the seventh to the nineteenth century A.D. the average " effective height " of a good flood is 7·4 metres, and the figures quoted by Herodotus exceed this by half a metre or more. This piece of evidence cannot, however, be regarded as more than an indication.

Herodotus has another passage (Book II., Chapter 97) which may be interpreted in the same sense ; " When the Nile overflows, the country is converted into a sea, and nothing appears but the cities, which look like islands in the Aegean." This happens now only when the flood is very high.

A passage from Pliny, quoted by Prince Omar Toussoun, states that the regular flood of the Nile in the first century A.D. was 16 coudées (8·5 metres). Whether this refers to the actual or " effective " flood, it represents a rather good supply of flood water, and consequently points to good rains in Abyssinia.

During the period of complete records from the seventh to the fifteenth centuries there is a fairly good agreement between the flood levels and the low-water stage, although the fluctuations of the latter are the more violent. Both show a minimum about 775, a maximum about 870, a minimum about 960, a maximum at 1110, and a double minimum at 1220 and 1300. After 1370 the curves become divergent and generally opposed ; it seems as if the figures recorded were mainly the highest maxima of the floods and lowest minima of the low-level stage. The true rainfall during this period is probably to be obtained by a judicious blend of both sets of data.

There are no other sources of information for Africa which can compare in detail with these Nile flood records. The climate of the Mediterranean provinces of Africa has been exhaustively examined by H. Leiter (5) on the basis of the literary references, mainly in the Roman writings. Leiter finds no evidence that there has been any appreciable variation of rainfall since Roman times, but he thinks there may have been a slight rise of temperature during the historic period. C. Negri (6) similarly examined the evidence for changes of climate in Cyrenaica, and concluded that there had been no appreciable change of climate. Both these authors were concerned more with the question of progressive desiccation than with fluctuations of climate ; they prove fairly conclusively that the rainfall of this northernmost belt of Africa was not much greater during the early centuries of the Christian era than it is at present, but they pay comparatively little attention to the possibility of prolonged wet or dry periods in the intervening centuries. Nevertheless, their careful work does seem to show that in this part of the earth's surface the climatic fluctuations were probably of smaller amplitude than in the regions to the north and to the south, *i.e.*, in Europe and at the sources of the Nile. On the other hand, a meteorological register kept at Alexandria by Claudius Ptolemaius in the first century of the Christian era, and described by G. Hellmann (7), strongly suggests a considerable change of the summer climate in Northern Egypt since that date. Hellmann examines the register in detail, and concludes that there is no direct evidence that the observations were not actually made in or near Alexandria. The record appears to state quite clearly the name of the observer and the place where the observations were made. Fortunately, there is a long series of good recent observations in Alexandria with which these early observations can be compared.

This register was re-examined for me by Miss L. D. Sawyer (8), who concluded that the disagreement between Ptolemaius' register and recent observations is not so great as is represented by Hellmann. The following table is based on her figures, converted to percentages :—

Winds from	N.	NE.	E.	SE.	S.	SW.	W.	NW.	Variable.
First Century									
Hellmann .	7	0	2	2	28	10	31	20	—
Sawyer . .	11	0	2	2	30	8	25	14	8
									Calm.
Present . . .	35	11	7	4	5	5	5	25	3

Table 28.—Frequency of winds in Alexandria.

Miss Sawyer points out that northerly winds are most frequently referred to in winter, when they are now least regular ; in summer there are references to the beginning and end of the Etesian winds in Egypt over 40 days apart, during which period other winds are mentioned occasionally but no reference is made to actual north winds on any of the days in that period. This indicates that it is the unusual rather than the more commonplace weather conditions that are referred to. Hence, while the register indicates that the winds in the first century differed appreciably from the present, the difference is not so impossibly great as Hellmann makes out.

Miss Sawyer did not discuss the descriptions of weather phenomena, but the same conclusion seems to hold.

The number of storms is rather high, but such observations are of a relative nature. Observations of rainfall occur under various designations, most of which are clear. The significance of one term (*takas*) is not quite clear, but it probably means " fine rain." The frequency of rain, as shown in Table 29, does not differ greatly from the present frequency (though, if " fine rain " is included, the number is rather high), but the annual variation, with its entries in summer, is quite

Days	Jan.	Feb.	Mar.	Apr.	May	June	July	Aug.	Sep.	Oct.	Nov.	Dec.
Rain—												
1st Century	4	3	0	5	3	1	2	0	3	4	3	2
1st Century, " Fine rain "	1	0	1	3	4	5	0	0	2	0	2	2
Present	11	6	5	1	1	0	0	0	0	1	7	10
Thunder—												
1st Century	1	0	1	1	2	2	1	1	1	0	0	0
Present	0·7	0·5	0·3	0·2	0·3	0·1	0	0	0	0·4	0·7	1·5
Great Heat—												
1st Century	0	0	0	0	0	3	8	6	1	0	0	0
Present	0	0	0	2	6	12	0	1	3	2	0	0
" Weather Changes "—												
1st Century	4	0	3	2	4	0	2	1	4	1	0	1

Table 29.—Frequencies of meteorological phenomena in Egypt.

different from the present. Summer in Alexandria is now
completely rainless. The distribution of thunder also differs
from the present, and the annual total is high. The occurrences of " great heat " in the old register also fail to agree
with the present climate. The constant strong winds from
north or north-west in summer temper the heat and give this
season fewer very hot days than the seasons immediately
before and after the midsummer months. The column for
" Present " under hot days shows the number of occasions
during the period 1873 to 1896 on which the hottest day of
the year occurred in the month in question.

Very remarkable also are the frequencies of " weather
changes." At present the period from May to September
is one of almost uninterrupted fine dry weather. The
Calendar of Antiochus, dated about A.D. 200, which also
refers to Egypt, is still more remarkable in this respect, since
out of fifty-one records of " weather changes," nineteen
occur in the period May to September.

These observations point to a climate very different from
the present summer climate of Alexandria, and resembling
much more that of Northern Greece. The divergence is,
in fact, so striking that in spite of the apparent trustworthiness
of the record, Hellmann considers that there must be something wrong, since a climatic change of this degree would imply
that meteorological conditions differed from the present,
not alone over Northern Egypt, but over a very wide region,
if not over the whole earth. The records agree, however,
with our somewhat scanty records of the Nile flood for that
period, and with the evidence from Kharga Oasis described
in the next paragraph, while in Chapter XXII. we shall see
that the climatic changes did in fact extend over a large part
of the Northern Hemisphere in a manner agreeing with the
requirements of meteorology.

H. J. L. Beadnell (9) has given us an interesting study of
the probable changes in the water supply of the oasis of
Kharga, which lies 100 miles west of the Nile valley and 400
miles south of the Mediterranean, and further details have
been added by Miss Caton-Thompson and Miss E. W.
Gardner (10). The water in this oasis is provided by wells,
the rainfall being extremely small and erratic.

In Pleistocene times the Kharga depression was supplied

with water by springs which formed mounds, but it is not now accepted that the floor of the depression was ever occupied by an extensive lake. In the Neolithic (7000-5000 B.C.) these springs ceased to flow, and holes were dug in the tops of the mounds to obtain water. The dead or dying springs were covered by dune sands. From Early Egyptian to Persian times (5000-525 B.C.) the oasis was practically uninhabitable owing to lack of water. The Persians replenished the supply of water by sinking deep artesian wells. In Kharga itself the oldest ruins belong to the time of Darius, about 500 B.C., which was a period of great prosperity.

The oasis continued to be of importance until the beginning of the seventh century A.D., but there was a temporary decline in the third and early fourth centuries. In the seventh century the oasis decayed. No further evidence is available until the twelfth century, when the oasis was found to be almost depopulated. In A.D. 1225 Kharga appears to have been more prosperous than it was about 1150, and by A.D. 1300 there appears to have been a still further improvement. After this we have no information.

At first sight, this history of the Kharga Oasis seems to be very complete evidence of changes of climate in this part of Africa, but Beadnell expresses a doubt. The water supply is entirely derived from wells in two layers of water-bearing sandstone, and the greater part of the supply is artesian, rising to the surface under considerable pressure and forming flowing wells. These water-bearing sandstones underlie a great area in this part of Africa, but it is only in depressions that they are at a small enough depth to be tapped. The origin of the water is doubtful ; there are three possible sources —the Nubian reaches of the Nile, the great swamp regions of the Sudán, and the rains of Abyssinia or Darfur. Beadnell thinks that the Nile is the most probable source. The total amount of water in the sandstone is very large, and may represent the accumulation of hundreds or even thousands of years. During Roman times a large number of wells were put down, and these gradually drained the water-bearing sandstones, so that the water supply fell off. After the Romans left, wells which became choked were not cleaned out, and the water supply decreased still further.

This suggested explanation shows that the variations of

water supply in Kharga Oasis do not necessarily represent synchronous variations of rainfall ; at least the evidence is not so convincing as it appears at first sight. Nevertheless, the variations in the water supply agree so well with the evidence from other parts of Northern Africa that they must be due in part to climatic causes, if not in the region of the oasis, at least in the region where the water originates. The evidence may be set out in parallel columns as shown in Table 30, the last column being in part an anticipation of later paragraphs.

Date.	Kharga Oasis.	Nile.	Other localities.
500 B.C.	Prosperous.	Floods high.	
A.D. 100	Extensive well boring.	Floods good.	Alexandria rainier in summer.
A.D. 200	Temporary decline of prosperity.	Floods poorer.	
A.D. 400	Improvement.	Floods good.	Mandingan Empire, A.D. 320-680 (see below).
A.D. 700	Great decline.	Minimum level, A.D. 700-1000.	Mandingan Empire broke up.
A.D. 1150	Almost depopulated.	Rise of low-level stage ca. A.D. 1100.	
A.D. 1225	More prosperous.	Level low.	Traffic in now waterless Eastern Desert.
A.D. 1300	Still further improvement.	Level low.	Prosperous Sudanese States.

Table 30.—Variations of climate in Northern Africa.

The chief disagreement is between the Kharga Oasis and the levels of the Nile from A.D. 1100 to 1300. If the evidence derived from the Kharga Oasis for this period stood alone, one would have no hesitation in rejecting it. It has, however, a certain amount of support from other regions. The figures of raininess in Europe, after a temporary maximum about A.D. 1075 fell to a minimum in 1175, rising again to a maximum about 1325. Similarly, in Asia the Caspian (Fig. 34) appears to have reached a low level about 1150, rising again to a maximum in 1300. Mr G. W. Murray informs me that there is evidence of considerable pilgrim traffic across the Red Sea to Jeddah in the thirteenth century from a port afterwards abandoned for lack of drinking water. At about the same time we have the prosperous Mossi States of the Sudan (see below). All this seems to confirm the Kharga evidence. It is possible that at this time the fluctuations in the equatorial regions were following a different régime from those in the north temperate rainfall belt. The levels of the Nile in Fig. 35 show that while the fluctuations of the Nile flood from A.D. 800 to 1300 were generally similar to those of the low-water stage, they were on a very much smaller scale. It has been pointed out that the low-level stage represents mainly the fluctuations of the equatorial rainfall, while the flood level represents the rainfall of Abyssinia, and this difference of scale may possibly imply that the equatorial fluctuations at that time died out northwards, and that the northern part of the continent came under the influence of the north temperate fluctuations. It is evident, however, that further data will be required before the climatic fluctuations in this part of the world can be set out in reliable detail.

The Sahara itself is at present occupied only by a few wandering tribes. There is, however, some historical evidence that during at least one, and possibly two historical periods the level of civilisation in the desert was appreciably higher than at present (11). The first period is that of the Mandingan Empire. The Mandinke are Sudanese negroes, who, according to native tradition, maintained a Saharan Empire from about A.D. 320 to 680. After this date the empire broke up, and it was not until the thirteenth century that the second and more authentic cultural period began. Early in the fourteenth century " the greatest Sudanese State of which there was any authentic record " was centred at Mali, in French Guinea ; about the same time the Mossi formed a powerful state in the great bend of the Niger, and the Housa State was developed at Kano in Nigeria. The Moslem Empire of the Central Sudan spread over a large part of the Sahara in the thirteenth century, and increased rapidly in importance until it reached its highest pitch about 1526-1545.

The shore-lines of the enclosed Central African Lake basins of Nakuru-Elementeita and Naivasha, according to

E. Nilsson (12) point to a number of post-glacial fluctuations.
At the end of the last main pluviation (Lake III.) the lakes
dried completely. The first post-pluvial lake (IV.) reached
almost to the depth of Lake III. but persisted for a shorter
time. The lakes again dried completely ; after they re-
formed (Lake V.) there were minor oscillations, superposed
on a gradual fall, but no complete drying out. Lake V. is
associated with Leakey's "Gumban A and B" cultures
(Nakuran) (13). A Gumban A site has been found in the
deposits of Lake V. The Gumban B burial was near the
lake and contains fish bones, and Leakey considers that it
was of about the same date. It yielded a bead which cannot
be earlier than 3000 B.C. and was probably later. In 1931 I
thought the Nakuran lake must represent the Sub-Atlantic,

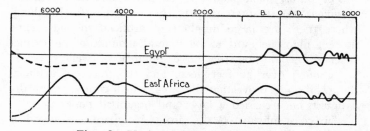

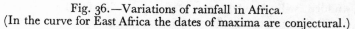

Fig. 36.—Variations of rainfall in Africa.
(In the curve for East Africa the dates of maxima are conjectural.)

which was then placed about 850 B.C. (but is now dated about
500 B.C.) and this correlation was accepted by the archæologists. Working backwards, the dry stage between lakes
IV. and V. then represents the "Climatic Optimum."
Nilsson, however, correlates Lake IV. with the Neolithic
of Egypt, about 6000 to 5000 B.C., i.e., roughly the Atlantic
stage of Europe. The Gumban A and B are not certainly
contemporaneous and the latter may possibly represent a
high-level stage of the lake subsequent to the highest level.
If we date the Nakuran as late as 500 B.C. there is a long
interval after the Late-glacial which is not accounted for.
In the lower curve of Fig. 36 I have adopted Nilsson's correlation but extended the rainy period to about 1000 B.C.,
and indicated the main arid period as ending about 5500 B.C.
This leaves the Sub-boreal and Sub-Atlantic of Europe

represented only by minor oscillations of the lake levels. This is quite possible, since the Sub-boreal is not now regarded as very dry and the Sub-Atlantic was not associated with a marked advance of the glaciers and ice-sheets. The dating of this curve is to be regarded as conjectural.

On this interpretation, Lake VI., which appears to have been quite important, represents the Sub-Atlantic, and Lake VII. may date from the thirteenth and fourteenth centuries. There is some evidence (14) that Lake Tanganyika was at a much lower level less than 1,300 years ago, when it consisted of two lakes separated by an isthmus. The natives have a legend of the submergence of this isthmus and the joining of the lakes.

C. W. Hobley (15) has collected a certain amount of evidence as to climatic changes in East Africa. Much of this is purely geological, and no dated historical evidence is given. Most interesting is the reference to Jubaland, where there are large numbers of artificial mounds, some 30 feet high, believed to be funeral mounds of an extinct race. "In addition the large number of well-excavated wells, often over 40 feet deep, and the traces of artificial dams, all go to prove that this area, which is now practically a desert, once carried a large and organised population."

The coast of East Africa from Mombasa northward is studded with ruined towns of the Mahommedan period, but their climatic significance is not obvious.

Farther south we have the ruined cities of Mashonaland, of which the best known is Zimbabwe. These are now attributed to native construction in the fourteenth century, but they may be a reflection of outside influences rather than a product of high local culture due to favourable climatic conditions.

REFERENCES

(1) TOUSSOUN, Prince OMAR. "Mémoire sur l'histoire du Nil." Le Caire, *Mem. Inst. Egypt*, vol. ix.

(2) LYONS, H. G. "The physiography of the River Nile and its basin." Cairo, 1906.

(3) LYONS, H. G. "On the relation between variations of atmospheric pressure in North-east Africa and the Nile flood." London, *Proc. R. Soc.*, A. 76, 1905, p. 66.

(4) HERODOTUS, The History of. Transl. by GEORGE RAWLINSON, 2 Vols., Everyman.

(5) LEITER, H. "Die Frage der Klimaänderung während geschichtlicher Zeit in Nordafrika." Wien, *Abh. K. K. Geogr. Gesellsch.*, 8, 1909, p. 1.

(6) NEGRI, C. "Sul clima della Libia attraverso i tempi storici." Roma, *Mem. Acc. Nuovi Lincei*, ser. 2, vol. 1.

(7) HELLMANN, G. "Über die Ägyptischen Witterungsangaben im Kalender von Claudius Ptolemæus." Berlin, *Sitzungsber. preuss. Akad. Wiss.*, 13, 1916, p. 332.

(8) SAWYER, L. D. "Note on Egyptian winds in Ptolemy's 'Prognostics' and Hellmann's criticism of them." London, *Q. J. R. Meteor. Soc.*, 57, 1931, p. 26.

(9) BEADNELL, H. J. LLEWELLYN. "An Egyptian oasis. An account of the oasis of Kharga in the Libyan desert, with special reference to its history, physical geography, and water supply." London, 1909.

(10) CATON-THOMPSON, G., and E. W. GARDNER. "The prehistoric geography of Kharga Oasis." London, *Geogr. J.*, 80, 1932, p. 371.

(11) KEANE, A. H. "Man, past and present." Rev. ed. Cambridge, 1920.

(12) NILSSON, E. "Quaternary glaciations and pluvial lakes in British East Africa." *Geogr. Ann.*, Stockholm, 13, 1931, p. 249.

(13) LEAKEY, L. S. B. "The stone age cultures of Kenya Colony." Cambridge, Univ. Press, 1931.

(14) THEEUWS, R. "Le lac Tanganyika." *Mouvement Géogr.*, 33, 1920, col. 625; 34, 1921, col. 49.

(15) HOBLEY, C. W. "The alleged desiccation of East Africa." London, *Geogr. J.*, 44, 1914, p. 467.

CHAPTER XXI

America and Greenland

OUR chief source of information about the climatic changes in North America is the rate of growth of the "Big Trees" or *Sequoias* of California. Some of these trees are of astounding age, and carry our records back long before the Christian era. There have been difficulties, of course, and the close comparison and averaging of the records of a large number of trees have been required to give a reliable record of the rate of growth. Douglass has used these records very effectively in investigating rainfall periodicities in California, but for the discussion of long-period climatic fluctuations certain corrections are necessary, and these are difficult to determine. Huntington (1) believes that the course of climatic variation is the same in California as in Central Asia, and he accordingly employs the levels of the Caspian for the final calibration of his curve of tree growth. This method, however, is fraught with danger ; if one wishes to compare variations of climate in America with those in Asia, the American curve should, if possible, be derived entirely from local evidence. The material for such a discussion on American evidence only is provided by a valuable Carnegie Institution publication entitled " Quaternary Climates " (2), a collection of papers by J. Claude Jones, Ernst Antevs, and Ellsworth Huntington. In this volume Antevs gives the results of a reinvestigation of the tree-growth data, based on all Huntington's measurements (451 trees) corrected for age by intrinsic evidence only.

Huntington's curve, even after correction for age, longevity and " flaring " (see p. 345) shows a much greater variability in the earlier years than in the later, which is most probably due to the much smaller number of trees. I accordingly applied a scale correction which decreased from the beginning to the end of the curve and was designed to reduce the variability about the mean to approximately the same value throughout. This corrected curve is the uppermost in Fig. 37.

The middle curve is that given by Antevs (2) for trees in dry situations, slightly smoothed to bring out the more lasting variations.

Antevs first divided Huntington's material into two groups, trees growing in dry situations and trees growing in wet situations. The width of the rings in each decade was plotted separately for each tree or, in some instances, for small groups

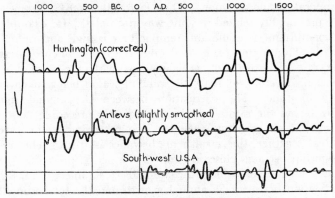

Fig. 37.—Variations of rainfall in U.S.A.

of trees of the same age. A smooth " middle line " was then drawn through the graph, while the maxima and minima were connected by drawing smooth " tangents " on either side of this middle line. The fluctuations were then reduced to the same basis by dividing the distance of any point on the graph from the middle line by the distance between the tangents at that point. Since it proved difficult to draw tangents for the earlier portions of the curve, when the trees were young and growing rapidly, only the middle line was drawn for these parts of the curves, the distances from this line being used without correction. For this part of the curves, before 200 B.C., the fluctuations therefore appear to be greater in magnitude than for the later part.

The results for the different trees or groups of trees were then added together, a correction being applied to allow for the incoming of successive groups.

The curves based on trees growing in dry and moist situations show good agreement after A.D. 800, when they are based on a large amount of material. The chief difference is the retardation of the maxima on the "dry" curve. Previous to A.D. 800 the agreement is not so good, presumably owing to the smaller amount of material. The differences are largely in the minor fluctuations, and when the curves are smoothed a better agreement is obtained.

These curves cannot be regarded as measures of the rainfall only ; they must include other factors of tree growth such as temperature and sunshine. These factors themselves, however, are presumably related in some way to rainfall ; for example, when sunshine is abundant, temperature is high and rainfall small. Huntington (2, p. 162) gives the results of correlating the rate of growth of 112 Sequoias with the rainfall of Sacramento, 1863 to 1910, the rainfall season being taken as July to June. The correlation between the annual tree growth and the rainfall of the immediately preceding season is small ($+0 \cdot 13$), but when the two preceding seasons are added together the coefficient becomes $+0 \cdot 22$. The relationship becomes closer the longer the period preceding the tree growth over which the rainfall is summed, and the correlation between the annual growth and the total rainfall of the ten preceding years is $+0 \cdot 58$. Thus the curves of tree growth evidently reproduce with a fair degree of accuracy the variations of rainfall from one decade to another. In plotting the corrected curves of tree growth in Fig. 37, the values have been assigned to a date five years before the year in which the growth ring was formed, in order to allow for this lag.

The method by which Antevs corrected his curves obviously tends to eliminate all fluctuations of long period, and the curves cannot be expected to show the major fluctuations of rainfall similar to that between the seventh and eleventh centuries in Europe. A better measure of the long-period variations is probably given by the data corrected by Huntington's method (1), but omitting the "Caspian correction factor." Two corrections are applied, for age and for longevity. Trees grow at different rates according to their age, young trees usually growing rapidly and old trees slowly. Trees which are destined to have a long life usually grow more slowly at first than their neighbours which are likely to die much sooner.

The " corrective factor for age " was readily obtained by averaging the rates of growth of a number of trees in the corresponding years of their lives. In this way the climatic and other peculiarities of the individual years (which come at different times in the lives of different trees) are eliminated ; it is found, for example, that the average growth of trees 1 year old is 0·10 inch ; 10 years old, 0·15 inch ; 40 years old, 0·20 inch ; 100 years old, 0·10 inch ; 200 years old, 0·05 inch ; and so on. The growth curves of the individual trees are then corrected for age by dividing the first year's growth by 0·1, the tenth year's growth by 0·15, and so on.

The " correction for longevity " was obtained in a similar manner, the average rate of growth of trees which when felled had lived for 100 years, 200 years, 300 years, and so on, being plotted against the number of years of growth and a smooth curve drawn through the points.

There is another source of error, namely, that due to " flaring " and " buttressing " at the base of the trees. " Flaring " means the spreading out of the base of the tree so that instead of descending nearly vertically, the trunk meets the ground at an angle, like a cone. The correct width of the growth-ring would be that measured at right angles to the surface of the trunk ; since the measurements were actually made on a horizontal surface, the widths found are too great. " Buttressing " means that the cross section of the trunk of an old tree is not circular, but develops protuberances and furrows which add to its strength. The measurements were more easily made across the buttresses than across the furrows, and this again tends to increase the apparent growth in the later portions of the curve.

The later portions of Huntington's curve can be checked by two incidental facts mentioned by him. He states (1) that in moist places there are plenty of young trees of all ages, but on dry mountain slopes, while there are plenty of mature trees 500 or more years old, there are no young trees except an occasional seedling or tree of three or four years' growth. This suggests that the climate has been drier than that of to-day since about A.D. 1400, and that previous to that date it was considerably moister. The other fact is that farther north, on the shores of Mono Lake, the rings of growth of a tree killed by the rising salt water show that it had been growing since 1775. This proves that the level of the lake has not been as high as it is at present since at least 1775. Both these facts fit in with the later portion of curve and confirm the correction for " flaring."

J. Claude Jones (2) has discussed the variations of level of the remnants of Lake Lahontan in the Great Basin of Western U.S.A. This was one of the great lakes formed in Western America during the Quaternary glaciation ; at its maximum it covered a continuous area of about 8,500 square miles. As the level fell it split up into a number of separate lakes, and at present the old basin of Lake Lahontan is occupied by large desert " playas " with several small lakes in depressions —Humboldt, North and South Carson, Pyramid, Winnemucca, Walker, and Honey Lakes. The discussion centres mainly on Pyramid and Winnemucca Lakes. Jones gives a variety of observations of the deposits of calcareous tufa, etc., but his interpretation of them is obviously erroneous, since they lead him to the conclusion that the mastodon and the camel lived on in North America into historic times. He has confused recent phenomena with those belonging to the Quaternary pluvial period, and it is necessary to go over his work and sort out the data from the two periods.

Calcareous deposits or tufas are widely distributed over a large part of the old basin of Lake Lahontan. The tufa is in three forms, lithoid (stony), dendritic, and crystalline. The lithoid and dendritic forms are found at all levels in the basin, and appear to have been formed by the activities of algæ ; the crystalline form occurs as a mineral (thinolite), which seems to have been altered from crystals of aragonite deposited from a saturated solution of calcium carbonate ; it occurs only in the lowest levels of the basin. Evidently, it was not until Lake Lahontan had dwindled almost to its present small remnants that the water became sufficiently saturated for calcium carbonate to be deposited directly without the agency of plant life. There seems no doubt, as E. Antevs points out in the second memoir of the collection, that the mass of the calcareous tufa, including the thinolite, was formed during the shrinkage of Lake Lahontan after the Quaternary expansion ; when the thinolite was deposited, the lake must have been intensely salt, but at present Pyramid and Winnemucca Lakes are only slightly salt, the salinity in 1882 being 0·35 per cent. in Pyramid and 0·36 per cent. in Winnemucca.

If we know the total amount of salt in a lake and the average amount carried in by the rivers in the course of a year, we can calculate the period in years since the lake was fresh. As Pyramid and Winnemucca Lakes are separated only by a low divide, and both receive branches of the same river, the Truckee, a very small expansion would suffice to unite them in a single lake, so that for the purposes of the discussion they can be treated as one. Jones gives four separate determinations of the age of the system. The first two depend on the method described above, of dividing the total amount of saline matter present in the lakes by the annual contribution of the river, but calculating the age from the chlorine and sodium separately (2, pp. 28-29). Jones writes :—

Using a detailed map of Pyramid and Winnemucca Lakes, it is possible to obtain the volumes of the lakes by determining the areas of the sub-lacustrine contours by means of a planimeter and calculating the volumes of the respective sections. Such a determination indicated the amount of water present in Pyramid Lake as 7·787 cubic miles and 1·142 cubic miles in Winnemucca Lake. The Truckee River has an average flow based on measurements at Vista, a station in the Truckee Canyon below all the larger tributaries, made during the years 1899 to 1911 inclusive, of 0·274 cubic mile per year. It would take Truckee River 28·42 years to supply the water at present in Pyramid Lake, and 4·17 years additional to fill Winnemucca Lake. But Pyramid Lake contains 1,455 parts per million chlorine, Winnemucca Lake 2,184 parts per million, and the Truckee only 13 parts per million. . . . It would therefore take the Truckee 3,180 years to supply the chlorine in Pyramid Lake and 701 years additional to furnish that of Winnemucca Lake, or 3,881 years for both. A similar calculation, using sodium instead of chlorine, gave 2,447 years necessary, and the other substances gave still lower results. Of these calculations the first is probably more nearly the truth, as chlorine is the least likely to be removed from solution. No great degree of accuracy can be claimed, for many factors may have influenced the result. While the Truckee River is the only stream of considerable volume that flows into the lakes, yet a considerable amount of water is supplied by the intermittent streams and springs about the borders. . . . The amount of salts carried probably varied somewhat with the increase in flow of the river . . . although . . . the present data indicate no very great change. Of the factors mentioned, only one, the last, would tend to make the period greater, while the others would cause the actual period to be less than the calculated duration. . . .

The question may be approached from an entirely different angle. As it happened in 1913, the level of Pyramid Lake was but 5 inches below the level at the time of Russell's visit. A sample was collected near the locality where he obtained his southern sample, and the chlorine determined. The gain during the 31 years that had elapsed between the two visits was 23 parts per million. As the lake had essentially the same volume in both instances and the samples were taken at the same locality, the variable factors are eliminated as far as possible. Dividing the total chlorine found in 1913 by the gain and multiplying the result by 31, the years that had elapsed, gave 1,956 years as the time necessary for the chlorine to accumulate, providing the present conditions had not been materially changed.

This method is open to the criticism that it depends on but two analyses, and while it is of value as corroborative evidence, yet it cannot be considered as conclusive.

Still another method, one used by Russell, may be employed. Knowing the total amount of chlorine in the two lakes and the rate of evaporation, the length of time necessary to evaporate enough water to supply the chlorine may be determined. Using the recent analyses, Pyramid Lake contains 1,440 parts per million of chlorine. Assuming that the water carried into the lake was as fresh as in the Truckee River, the water before evaporation contained 13 parts per million of chlorine. This would make it necessary to evaporate 111 cubic miles of river water to concentrate 1 cubic mile of Pyramid Lake water, or, since the lake contains 7·787 cubic miles, 864·36 cubic miles have been evaporated since the beginning of Pyramid Lake. Similarly, 180·43 cubic miles additional would be required to furnish the chlorine in Winnemucca Lake.

The loss of water from the surface of a lake by evaporation depends on two factors, the rate of evaporation and the area of the lake. With regard to the rate of evaporation, no actual measurements are available for Pyramid or Winnemucca, but Jones gives determinations obtained in three different ways : the evaporation from open pans on land at Fallon averages 65·14 inches a year, and Bigelow, from observations near Reno, concluded that the evaporation over open water is about five-eighths of that from a pan on land, giving the evaporation from the lakes as 40·7 inches a year. Salton Sea, formed by a break of the Colorado River, but now receiving no appreciable supplies of water, has been falling at the average rate of 55·6 inches a year. Finally, the average inflow of the Truckee River into Pyramid and

Winnemucca, divided by the present areas of these lakes, gives an annual evaporation of 52 inches. From these three determinations, Jones estimates the average evaporation as about 50 inches a year. If the lake had always maintained its present area of about 370 square miles, the concentration of the chlorine would require 4,300 years. Jones thinks, however, that the average level of the lake has been higher than the present level, and since there is a shelf cut in the rock 110 feet above the present surface of the lake, showing that for a long period the lake stood at that level, he adopts 110 feet as the average level. This gives an average area of 550 square miles, and a duration of 2,400 years. It seems certain, however, that the rock shelf dates from an earlier period, probably the close of the Quaternary pluvial period. Antevs (2, p. 102) quotes Gale and Huntington to the effect that there is a well-marked outflow channel through Emerson Pass, 70 feet above Pyramid Lake. Jones (2, p. 40) gives the maximum level of this pass as 78 feet, and states that there is no evidence of overflow, the summit of the pass having a floor of fine clays and silts. Jones realised the importance of this point, and appears to have examined the ground thoroughly. In either event, however, it is obviously impossible for Winnemucca and Pyramid Lakes to have stood more than 78 feet above their present level without overflowing and becoming fresh. Hence the average level of 110 feet above the present and the average area of 550 square miles are too great, and the age determination of 2,400 years too short. This method, therefore, gives the age of the present lakes as probably something less than 4,300 years.

Thus we have four determinations, which are not, however, quite independent of each other. These give respectively 3,880, 2,447, 1,956, and 4,300 years. From these figures it seems probable that Pyramid and Winnemucca Lakes were fresh some time between 2,000 and 4,000 years ago.

A lake may become fresh in one of two ways, either by overflowing into another basin, so that there is a flow of water through it which sweeps away the accumulated salt, or by becoming dry for a period long enough to allow the salts to be buried below subaerial deposits. A short dry period will not suffice ; the overlying deposits must be so thick that when the lake forms again the salts are protected from the water and are not redissolved. The answer to the question in which of these two ways Pyramid and Winnemucca Lakes became dry, depends on the interpretation of the phenomena in Emerson Pass. If Jones is correct, the formation of the present lakes was preceded by a long dry period which ended 2,000 to 4,000 years ago, and since then they have never overflowed. If Gale is correct, the lakes probably overflowed 2,000 to 4,000 years ago, and so became fresh, and there is no evidence of a preceding dry period. Since Jones examined the pass with Gale's work in mind, he is the more likely to be correct.

This conclusion is supported by an investigation of W. van Winkle into the salt contents of Abert and Summer Lakes, in Oregon, which are remnants of the old Quaternary Lake Chewaucan, north of Lake Lahontan, also without outlet. Van Winkle writes (3, p. 123) : " A conservative estimate of the age of Summer and Abert Lakes, based on their concentration and area, the composition of the influent waters, and the rate of evaporation heretofore assumed, is 4,000 years. It is quite possible that the lakes are recent pools, and that the salt and soda deposits of Early Quaternary Chewaucan Lake lie buried beneath them."

It is known from the work of Antevs and de Geer that the major variations in the rate of recession of the ice-sheets at the close of the Quaternary glaciation in North America ran closely parallel with the variations in Scandinavia. In the post-glacial period, the peat-bogs show a succession of wet and dry periods which closely resembles the Scandinavian succession. H. P. Hansen (4, 5) states that in Eastern North America the climatic succession shown by the peat-bogs closely resembles that of North-west Europe :

Eastern N. America.	North-west Europe.
Ice-retreat (Hudsonian)	Late-glacial.
Spruce, fir (cool, moist)	Pre-boreal.
Pine (warmer but still cool)	Boreal.
Oak and hemlock (warm, moist)	Atlantic.
Oak and hickory (warm, dry)	Sub-boreal.
Oak, chestnut, spruce (cooler, moister)	Sub-atlantic.

The pollen profiles reveal consistent and definite evidence for a dry period, which is best developed in east Washington and Oregon. In the North-west Pacific states of U.S.A. the dry period was less developed owing to the proximity of the Pacific. Early in the dry period there was a great explosive eruption of Mount Mazama ; the distribution of the pumice shows that at the time the winds blew from south and west. The climate was cooler and moister in the early post-glacial than at any time since, but the earliest forests differed little from the present. There is evidence of a rather wet period in the Puget Sound region some time after 7000 B.C. when the moisture-loving hemlock expanded. There was another abrupt spread of hemlock about 2000 B.C. (The dates appear to be estimates from the thickness of sedimentary deposits.)

K. Bryan (6) states that the bogs of Eastern Canada generally originated in ponds and lakes, bordered by a forest richer than the present, indicating a warm period. A cooler moister climate was followed by a warm dry period which again changed to the present cool moist climate. The changes in North America cannot be dated by archæo-logical evidence, but there seems no reason to doubt the approximate synchronism of corresponding stages in North America and Scandinavia, especially as the two areas seem to be linked up to some extent by the deposits in Iceland and Greenland. This would give a long dry period during the third and second millennia B.C. in Eastern North America, which fits in excellently with the lake records.

In this connexion there is some interest in a note in *Nature*, 28th November 1925, p. 796, to the effect that the first settlement of the arid coast of Southern California is now dated about 3,000 years ago. The concordance with the lake evidence may be accidental, but it may mean that before that date the supply of fresh water was insufficient for settlement.

The result of this discussion, therefore, seems to be that about 3,000 years ago there was *either* the end of a long dry period *or* a period of relatively heavy rainfall, probably the former, possibly both, and in any event an increase of rainfall. Jones' discussion of the levels indicated by the lake terraces is useless for the climates of the historical period, because they are all above the level of the Emerson Pass, and therefore belong to an earlier period in the history of Lake Lahontan, more than 3,000 years ago.

Further evidence is supplied by the levels of Owens Lake, in Southern California (2, p. 200). There seems to be no doubt that the freshening of this lake occurred through the level rising so high during a wet period that the lake overflowed its basin. This lake is supplied by the Owens River, and, according to H. S. Gale, analyses show that the river, at the point where it was tested, would require 4,200 years to supply the chlorine and 3,500 years to supply the sodium now in Owens Lake. This gives 4,000 years as the maximum period since the freshening, but there are several factors which make this estimate too high. The analyses were taken at a point some way up the valley, and omit the lower third of the basin, which contains old saline clays, and across which the river flows slowly; moreover, no allowance is made for greater rainfall in the past. Hence, Huntington concludes that the most probable length of the period which has elapsed since Owens Lake was fresh is between 2,000 and 2,500 years.

Finally, we have the evidence of Walker Lake (2, p. 46). This is fed by Walker River, and there is no possibility that it was ever freshened by overflowing. Its salt content is only 0·25 per cent., and Jones calculates that the present rate of supply would accumulate this amount in about 1,160 years or possibly less. Jones' theory of the origin of the lake is that it was formed by a change in the course of Walker River at the time of maximum level of Lake Lahontan, but this change, if it took place, must have occurred in Quaternary times. All that the lake shows us is that a minor dry period ended about 1,100 years ago.

Thus we may sum up the evidence afforded by the lakes as follows :—Some time after the great expansion of the lakes associated with the Quaternary glaciation of America, there occurred a long period of desiccation, in which Abert and Summer Lakes, and probably also Pyramid and Winnemucca Lakes, dried up completely. About 3,000 years ago this period of desiccation was brought to an end by an increase of rainfall to a value above its present amount, refilling the basins of Abert and Summer Lakes, but without causing them to overflow, filling the basins of Pyramid and Winnemucca Lakes, and *perhaps* causing them to overflow, and also filling Owens Lake and causing it to overflow. This was the time of greatest rainfall in Western North America during the historical period. Then followed a period of decreased rainfall, not enough to cause the complete disappearance of Abert, Summer, Pyramid, and Winnemucca Lakes, but enough to dry up Walker Lake for a period long enough to bury the salt accumulations. This secondary dry period ended about 1,100 years ago, or A.D. 800. The lakes do not give any indication of the happenings after A.D. 800. There is some historical evidence ; thus Huntington (1) states that when the Aztecs founded the city of Mexico, about A.D. 1325, the level of the lake of Mexico was high, and that another period of high water occurred about 1550.

Let us now see to what extent these lacustrine fluctuations fit in with the curve of tree growth. Unfortunately, the tree-growth curve helps us little during the early most crucial period of the change from dry to moist conditions. Antevs' curves show a maximum at 840 B.C., a minimum at 740 B.C., and a second maximum at 660 B.C. Huntington's data show a maximum at 960 B.C., a minimum at 780 B.C., and a second maximum at 660 B.C. ; the maximum at 840 B.C. on Antevs' curves is barely indicated. These early growth curves are based on relatively few trees, and the corrections are uncertain, so that we cannot say more than that the wet period had definitely begun by 660 B.C., but may have begun two centuries or more earlier. It is interesting to note that the Chinese records also indicate a dry period from 842 to 771 B.C. The rainfall maximum indicated by the overflow of Owens Lake presumably corresponds with the rapid growth of the trees from 480 to 250 B.C., shown on all the curves ; during this interval the rainfall reached its absolute maximum for the whole period since 1000 B.C. The period of drought during which Walker Lake dried up seems to extend from about A.D. 400 to 850 ; it is shown much more definitely on Huntington's than on Antevs' curve, owing to the different methods of correction adopted. The deep minimum shown by all the curves in the fifteenth century was apparently of too brief duration to cause even Walker Lake to dry up completely.

The alternation of dry and wet periods has also been traced by Huntington (1) in the archæological remains of Arizona and New Mexico. In these dry regions, whose crying need is water for agricultural purposes, he distinguishes three periods of maximum occupation or prosperity. In the oldest of these the people, whom he terms the Hohokam, were not limited to the neighbourhood of the water-courses but lived on the open plateau, and apparently depended on rainfall instead of on irrigation. The second people, the Pajaritans, lived partly on the irrigable land, but partly on the Pajaritan plateau. The latest pre-Columbian race was the Pueblo, who depended on irrigation, but lived in valleys where there is not now sufficient water for that purpose. The Pueblo village of Gran Quivera was still populous at the coming of the Spaniards. These three periods evidently correspond with three wet periods ; there is no evidence of continuity, and in some places, *e.g.*, Chaco Valley, the deposits containing remains of the different periods are separated by silts without human remains, formed during dry periods. The last of the three evidently represents the rainfall maximum from about A.D. 1200 to 1400 ; it was the least important of the rainfall maxima. No dates can be assigned to the earlier periods, but the Pajaritan occupation presumably includes the period from 750 B.C. to A.D. 400, when the rainfall was much heavier. The first occupation, by the Hohokam, appears to be much older, and may have occurred during a very early rainfall maximum older than the oldest of the trees, but represented in Eastern North America by the period of peat-formation corresponding with the Atlantic stage in Europe.

The lowest curve in Fig. 37 has been reconstructed from a paper by E. Schulman (7) in the Colorado Plateau, based not only on living trees but also on specimens of timber from ruins of Indian buildings. Apart from the trough and peak in the thirteenth to fourteenth centuries it tends to vary oppositely to Huntington's curve. This is probably due to its low latitude (35-40° N.) and is of interest in connexion with Huntington's theory of the shift of the climatic belts. With regard to the scale of the curves, it may be remarked that the maximum area of Pyramid and Winnemucca Lakes was less than one and one-half times the present area, so that the maximum of rainfall indicated by the curve is less than 50 per cent., and probably not more than 25 per cent., greater than the present rainfall.

Farther south we have the remarkable ancient Mayan civilisation of Yucatan (8, 9). This country is at present covered by almost impenetrable forests, the climate is hot, moist, and enervating, while the inhabitants are idle and uncultured. Buried in the forests are the ruins of great cities, decorated by elaborate carving, and indicating a greater and more progressive population and a high level of civilisation, one of the features of which, as is well known, was the construction of an elaborate calendar. The problems of Mayan history and chronology have not yet been completely solved, but it seems probable that before 400 B.C. there was little forest, the winters being dry and cool. Between 400 B.C. and 100 B.C. the climate became somewhat moister and more uniform ; this is the time of the earliest carvings. The highest level of culture was reached in the period 100 B.C. to A.D. 300, first in the south, later in the north. By A.D. 300 climate had become less favourable. The deterioration continued in A.D. 300-450 ; the forest advanced and civilisation declined in the south. From A.D. 450 to 900 the forest spread over the whole country, especially north Yucatan, and civilisation fell to a low ebb. There was a marked improvement in A.D. 900-1100 accompanied by a great deal of building, but climate deteriorated again in 1100-1300. From 1300 to 1450 there was a climatic improvement but culture did not respond to any extent. From 1450 onwards climate has been continually unfavourable. Sapper (9) thinks that the decline of civilisation was due partly to climatic changes and partly to the introduction of malaria.

Since it takes time for both forests and civilisation to respond to the effects of climatic changes, we may date the latter about 50 years earlier than the changes in the level of civilisation. This gives us the following comparison with Huntington's curve :—

Western U.S.A.		Yucatan.	
Wet	500-250 B.C., 100 B.C.-A.D. 200.	Dry	500 B.C.-A.D. 250.
Dry	A.D. 300- 800.	Wet	A.D. 400- 850.
Wet	A.D. 900-1100.	Dry	A.D. 850-1050.
Dry	A.D. 1100-1300.	Wet	A.D. 1050-1250.
Wet	A.D. 1300-1400.	Dry	A.D. 1250-1400.
Dry	A.D. 1450-1550.	Wet	A.D. 1400-

In the dry regions of Asia and Arizona the periods of high culture were attributed to an increase of rainfall, but Yucatan now suffers from too much rain, and any increase would make the conditions even less favourable than at present. Hence Huntington (8) infers that the great periods of Mayan history were times of decreased rainfall in Yucatan. Since they coincide with rainy periods farther north, we are evidently dealing here with a redistribution of rainfall. The way in which this was probably brought about will be discussed in the next chapter.

The supposed climatic changes in Greenland have been a matter of controversy for many years, but excavations, described by Hovgaard (10), appear to establish their existence beyond doubt. Icelanders settled in Greenland in the tenth century A.D., and two colonies were established, the Eastern Settlement, just west of Cape Farewell, and the Western Settlement, 170 miles up the west coast. The settlers brought with them cattle and sheep, which were successfully reared at first, and they even attempted to grow grain, but before very long the colonies became dependent on supplies from Norway. Norway itself was passing through a time of stress, however, and the visits of ships became fewer and fewer, until some time in the fifteenth century they ceased altogether, and the colonies were lost sight of. For many centuries their fate was unknown, but the history of the Eastern Settlement has now been made out by the excavations of a Danish archæological expedition at Herjolfsnes, near Cape Farewell. The most important evidence is derived from the excavation of the churchyard, in soil which is now frozen solid throughout the year, but which, when the bodies were buried, must have thawed for a time in summer, because the coffins, shrouds, and even the bodies were penetrated by the roots of plants. At first the ground thawed to a considerable depth, for the early coffins were buried comparatively deeply. After a time these early remains were permanently frozen in, and later burials lie nearer and nearer to the surface. Wood became too precious to use for coffins, and the bodies were wrapped in shrouds and laid directly in the soil. Finally, at least five hundred years ago, the ground became permanently frozen, and has remained in that condition ever since, thus preserving the bodies. The remains show a gradual deterioration in the physique of the colonists ; their teeth especially are much worn, indicating that they lived mainly on hard and poorly nourishing vegetable food.

The change of climate indicated by these facts is borne out by the evidence as to the ice conditions. When the colonies were first settled, there were traces of the former existence of the Eskimos, but none then lived so far south. The Eskimos follow the seals, which frequent the edge of the ice, and this indicates that in the tenth century the ice-edge in Baffin Bay lay far to the north. In the thirteenth century the Eskimos reappeared and advanced persistently southward, until by the middle of the fourteenth century they had occupied the Western Settlement, which apparently they destroyed.

The accounts of the early Norse voyages to Greenland are remarkably free from references to ice conditions, and, in fact, as O. Pettersson (11) points out, it is difficult to understand how their protracted explorations could have been carried out if the ice conditions had been anything like those of the present day. Pettersson's chart of the old Norse sailing routes shows a track direct from Iceland to the east coast of Greenland in latitude 66° N., then down the coast to Cape Farewell, and up the west coast. According to the documentary evidence which he adduces, this route—at present almost impossible—was followed until about A.D. 1200, when it was abandoned for a more southerly route. On the other hand, as early as A.D. 998 a shipwrecked party was ice-bound on the east coast of Greenland, probably near or north of Angmagsalik. It is to be noticed that the ship was wrecked on the coast and not on the ice.

The early climatic history of Greenland, therefore, appears to have been somewhat as follows :—When the country was colonised in the tenth century its climate was much more favourable than at present, for herds of sheep and cattle thrived. There was less ice than at present in the East Greenland Current, and it is even possible that at first there was no ice at all ; Baffin Bay seems to have been largely free of ice. But in the second half of this century the climate was already deteriorating, and about A.D. 1000 there came a foretaste of the coming ice. After this, conditions apparently improved slightly, and the colony appears to have prospered during most of the eleventh and twelfth centuries. Towards the close of the twelfth century deterioration again set in, and the ice conditions rapidly became very bad. The summer thaw became shorter and shorter, and about A.D. 1400 the ground became permanently frozen. Communication with the mother-country was broken, life became too hard to bear, and the colonies finally perished.

For South America the only evidence I can find of climatic changes in the historical period is given by E. Taulis (12) who estimated the rainfall of each year from 1535 to 1931 in Chile on a scale of 1-5. From his figures it appears that there were wet periods about 1550 and in 1684 to 1700, and a great drought from 1770 to 1783. This agrees with Antevs' curve of tree growth.

REFERENCES

(1) HUNTINGTON, E. " The climatic factor, as illustrated in arid America." Washington, 1914.

(2) WASHINGTON, CARNEGIE INSTITUTION. Publication No. 352. " Quaternary climates." Papers by J. CLAUDE JONES, ERNST ANTEVS, and ELLSWORTH HUNTINGTON. Washington, July 1925.

(3) WASHINGTON, U.S. GEOLOGICAL SURVEY. Water Supply Paper 363. " Quality of the surface waters of Oregon." By WALTON VAN WINKLE. 1914.

(4) HANSEN, H. P. " Postglacial forest succession and climate in the Oregon Cascades." *Amer. J. Sci.*, 244, 1946, p. 710.

(5) ———. " Postglacial forest succession, climate and chronology in the Pacific North-west." Philadelphia, *Trans. Amer. Phil. Soc.*, 37, Pt. 1, 1947.

(6) BRYAN, K. " Palæoclimatology in North America as a result of the study of peat bogs." *Zs. Gletscherk.*, 20, 1932, p. 76.

(7) SCHULMAN, E. " Nineteen centuries of rainfall history in the Southwest." Milton, Mass., *Bull. Amer. meteor. Soc.*, 19, 1938, p. 311.

(8) HUNTINGTON, E. "Civilisation and climate." 3rd ed. New Haven, 1924.

(9) SAPPER, K. " Klimaänderungen und das alte Mayareich." *Beitr. Geoph.*, Leipzig, 34 (Koppenbd 3), 1931, p. 333.

(10) HOVGAARD, W. " The Norsemen in Greenland. Recent discoveries at Herjolfsnes." New York, N.Y., *Geogr. Rev.*, 15, 1925, p. 605.

(11) PETTERSSON, O. " Climatic variations in historic and prehistoric time." *Svenska Hydrogr.-Biol. Komm. Skr.*, 5. Göteborg, 1914.

(12) TAULIS, E. " De la distribution des pluies au Chili. La periodicité des pluies depuis quatre cents ans." Genève, *Mat. étude calam.*, 9, 1934, p. 3.

CHAPTER XXII

The Interpretation of Climatic Fluctuations in the Historical Period

FROM the preceding four chapters we see that during the historical period there have been several climatic fluctuations of quite appreciable magnitude ; in some parts of the Northern Hemisphere the fluctuations were closely similar over wide areas, while in other parts they were in distinct opposition. We must now try to discover the causes of these variations. Fig. 38 gives general curves of estimated

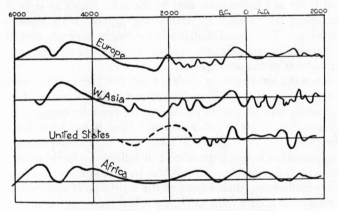

Fig. 38.—Variations of rainfall, world.

rainfall for Europe, Asia, North America and for East Africa from the equator northwards. The curve for Europe was constructed by superposing the first three curves of Fig. 31 and drawing a mean curve through them ; that for U.S.A. was drawn in the same way from Huntington's and Antevs' curves. The first three curves of Fig. 38 show a good deal of resemblance with some discrepancies which may be due to difficulties of precise dating. The discrepancies in the curve for Africa are greater but, as previously stated, the dating of this curve is only conjectural ; the maximum shown at 1250 B.C. may easily be pushed back to 2000 B.C.

The first three curves all refer to the region between about 35° and 65° N., in which the rainfall is mainly brought by barometric depressions (Chapter II.), while the curve for Africa aims at showing the variations in the equatorial belt of low pressure. Between these two lies the sub-tropical high pressure belt where the rainfall is mainly monsoonal, and is greatest when the general circulation of the atmosphere is weakest. Our information about past climates in this region is scanty, but such as it is, it suggests that the variations were in the opposite direction to those farther north. First we have the rainfall maximum at Mohenjo-Daru in India about 2750 to 2500 B.C., which comes in the middle of the long dry period in Europe and Asia. Then the variations in Yucatan are directly opposed to those in the Western United States, and there are indications that in the later stages at least the variations of rainfall in Cambodia agreed with those in Yucatan. This zonal distribution strongly suggests that the variations of rainfall are related to changes in the zonal circulation of the atmosphere.

Since the total amount of air is fixed, the average barometric pressure over the whole surface of the planet must be always the same, and an excess in one region must be compensated by a deficit in some other region. Now it has been found, especially by Sir Gilbert Walker (1), that this process of compensation is not haphazard ; it follows a clearly marked, though not inviolable rule. When, in a region where pressure is normally high, such as one of the sub-tropical anticyclones, it rises even higher than usual, there is a tendency for pressure to be higher than usual in all those parts of the world where it is normally high, and lower than usual in all those parts of the world where it is normally low. That is to say, if pressure is above normal in, for example, the Azores anticyclone, it will tend to be above normal over a belt stretching more or less completely round the globe from Hawaii to the north of Mexico, and across Bermuda and the Azores to North Africa. On the other hand, pressure will tend to be below normal over the belt of storminess which runs from Kamchatka and the Aleutian Islands across Southern Canada and New-foundland to Iceland, the British Isles, Norway, and the North of Asia. At the same time there will also be a tendency for pressure to be below normal near the equator.

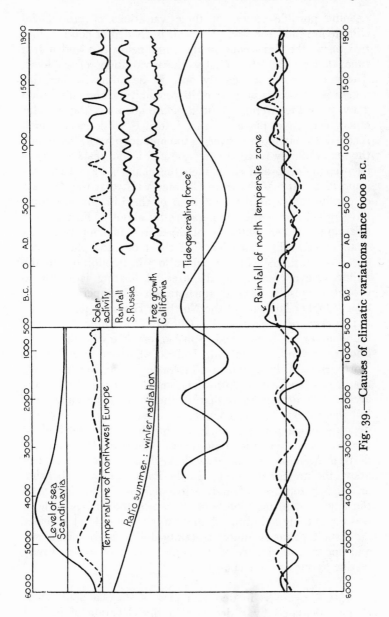

Fig. 39.—Causes of climatic variations since 6000 B.C.

Some possible causes of these variations of post-glacial climate are shown in Fig. 39. This is divided at 500 B.C. into two parts, the time-scale on the right being two and a half times that on the left. First we have variations of sea level. These are known accurately only in Scandinavia, but this region is important because of its proximity to the only broad gateway to the Arctic. A little before 5000 B.C. the Ancylus emergence gave place to the Litorina subsidence, which reached its maximum between 4500 and 4000 B.C., after which the sea gradually receded. By 500 B.C. it had reached nearly its present level, and since then the changes have been unimportant. It was shown in Chapter VIII. that the decrease of continentality at the maximum of the Litorina Sea must have raised the winter temperature by about 5° F., and that this agreed closely with the observed rise of temperature in Scandinavia. The subsidence of the land, and the readier access of southerly winds, would also affect conditions in the Arctic Ocean, decreasing the area of floating ice, and so probably cause a general amelioration of temperature over all the higher latitudes of the Northern Hemisphere.

Evidence of a post-glacial Climatic Optimum has been found in Franz Josef Land, Spitsbergen, Norway, the Baltic Shores, Iceland, Greenland, Ireland, Eastern and Central North America, Patagonia and Tierra del Fuego, New Zealand, Southern and Eastern Australia, South Africa and the Antarctic. We do not know that all these are of the same date, but there is one feature common to a large proportion of the deposits which points strongly in this direction—the bulk of the evidence for the post-glacial Climatic Optimum is derived from or associated with beaches raised a few feet above the present sea-level. In the Baltic the raised beaches are higher and are definitely associated with a subsidence of the land, but in most parts of the world the change of level was remarkably uniform. A uniform change of level at many far distant points is almost certainly due to a rise of the sea and not to a subsidence of the land. A rise of sea-level may be due to one of three causes :—

(a) A decrease in depth of part of the sea floor, compensated by a decrease in the elevation of part of the land area.

(b) An increase in the volume of sea water without change of mass, owing to a decrease in density.

(c) The actual addition of water to the oceans.

The general rise of sea-level during the Climatic Optimum is hard to estimate precisely, owing to the difficulty of obtaining the exact levels of old sea beaches, but it seems to have been of the order of 10 feet. Of the three possible causes of this rise, (a) can be dismissed very shortly. The principal land areas in which there was extensive post-glacial subsidence are Scandinavia and North America north of the Great Lakes, but both these subsidences were largely compensated by elevation of the land to the southward ; moreover, the period of maximum depression was probably over before the height of the Climatic Optimum. Causes (b) and (c) are both possible, but are difficult to estimate.

The mean depth of the oceans is approximately 12,000 feet. Taking the coefficient of expansion of water as ·00015 for one centigrade degree, we find that an increase of temperature by 1° C. or 1·8° F. would raise the mean level of the surface by 1·8 feet. Thus a rise of the mean temperature of the whole mass of the oceans by 5° F. would raise the general level by 5 feet. The increase in the surface temperature of the northern North Atlantic approached 5° F. and in the Arctic and Baffin Bay the increase was probably even greater. The temperature of the lower oceanic layers is determined by that of the polar oceans, but the warming of the whole ocean mass would be very slow and the general rise of temperature is not likely to have been nearly so much.

The only ways in which water can be added to the ocean are by a decrease in the level of enclosed lakes unconnected with the sea, and a decrease in the volume of the ice-sheets and glaciers. The volume of water in enclosed lakes without outlet is so small in comparison with the area of the oceans that it can be neglected. The ice-sheets, however, are on a different scale. The ice-covered area in Greenland and the Antarctic is about six million square miles, and the average thickness of the ice is nearly 5,000 feet. If all this ice were melted, it would raise the general level of the oceans by from 140 to 190 feet. The area occupied by the oceans is about 140 million square miles, or 23 times the area occupied by ice, so that in order to raise the level by ten feet, it would be necessary to melt off 230 feet of ice. Because of the difference of density between glacier ice and water we may put the figure at 250 feet. Now we know that even in the much less intense warm period of the early Middle Ages the boundaries of the Greenland ice-sheet retreated appreciably, which implies a corresponding diminution of thickness, so that in the pro-longed warm period of the Climatic Optimum a lowering of the average level of the ice-sheets by 250 feet is quite possible. These two factors, increase of ocean temperature and increase in the mass of water, appear to be quite competent between them to raise the general level of the oceans by 10 feet, the greater part of this being due to the melting of ice.

The second factor to be considered in our climatic re-construction is the annual range of temperature. As described in Chapter V., the obliquity of the ecliptic appears to have reached a maximum about 8150 B.C., and to have decreased steadily since that date. Also, about 8500 B.C. the earth was farthest from the sun (aphelion) in the northern winter, whereas it is now farthest from the sun in the northern summer. Both these factors would cause an appreciably greater seasonal range of radiation in the ninth millennium B.C. than at present. This change is shown by the second full curve on the left of Fig. 39.

Over most of the world these " astronomical " changes do not affect the total supply of solar radiation appreciably, but in the Arctic, which receives little or no solar radiation in winter, the effect of increased seasonal contrast is to increase the total solar radiation considerably.

The broken curve between these two full curves represents the variation of temperature in North-west Europe, copied from Fig. 32. It is seen that the left-hand part of this curve is a mean between the curve of sea-level and that of summer radiation. The " Climatic Optimum " occurred about 5000 B.C., after which temperature fell gradually until about 3000 B.C. The fluctuations in the years between 3000 B.C. and 500 B.C. will be discussed later.

The climate of the Boreal phase, about 6000 B.C., appears to have been definitely more continental than at any sub-sequent time, with cold winters and hot dry summers in many parts of the Northern Hemisphere. This is probably due to the combination of high obliquity with winter in aphelion ; in North-west Europe the larger land area and the shutting off of the Baltic was also a factor. The present situation, decreased obliquity and winter in perihelion, gives the opposite effect of mild winters and cool summers. The ice-sheets retreated rapidly between 8000 and 7000 B.C., while since the beginning of the Christian era there does not seem to have been any general retreat, only long-period oscillations about a mean position. To this extent the long-period changes of climate since 8000 B.C. support the astronomical theory.

The third factor is the variation of solar activity. Sir Gilbert Walker (2) has pointed out that the contrast between the zones of low and high pressure is apparently controlled to some extent by variations of solar activity. When sunspots become more numerous, pressure increases in the areas where it is already high, and decreases in those where it is already low. In the temperate storm belts, a high sunspot number tends to be associated with low pressure, great storminess, and heavy rainfall. According to Huntington and Visher (3), the belt of storminess in the Northern Hemisphere moves southward and increases in intensity at times of many sunspots, but moves northward and decreases in intensity at times of few sunspots. C. E. P. Brooks (4) found that the annual frequency of thunderstorms shows a fairly close relation to the sunspot number. In many parts of the world, including Siberia, Sweden, Norway and Scotland in the north and the West Indies, South-eastern U.S.A., Southern Asia and the Tropical Pacific in the south, the frequency of thunderstorms is greatest when sunspots are most numerous. Between these two belts is a region including England and Wales, Holland, Germany and the Northern and Western U.S.A., in which the relation is small, but still generally positive. Since in the interior of the continents and in tropical regions a good deal of rain is associated with thunderstorms, this suggests that rainfall maxima should coincide with maxima of sunspots, and there is other evidence that on the whole the total rainfall over the land areas is greatest when sunspots are most numerous. Our next step must, therefore, be to construct a curve which will represent the variations of solar activity over as long a period as possible.

For our knowledge of sunspot frequencies since 1749 we are mainly indebted to the researches of R. Wolf, who has compiled a complete table beginning with that year (5). The earlier data are based mainly on a long but rather fragmentary series of records from China. The first record of a sunspot occurs in the Chinese archives in A.D. 188, and the first aurora in A.D. 194, but records only become frequent in the fourth century, apparently reaching a maximum about 374. Another maximum, both of spots and aurora, occurs in 535-540, the first half of the sixth century giving us 20 records of spots and 13 of auroræ. In the seventh and eighth centuries the number of records is very small, rising to another maximum about 840, in which year there are records of 90 sunspots, while brilliant auroræ occurred in 839 and 840. The records of sunspots again become very few between A.D. 850 and 1070, though there is a secondary maximum of auroræ about 993. There is a great outburst of sunspots in the years 1077 to 1079, and the frequency of both spots and auroræ remains very high until about 1250, with probably a secondary maximum about 1201. The last half of the thirteenth and the first half of the fourteenth centuries again show a falling off in the records, but about 1370-1375 they become very numerous, and Wolfer considers that the absolute maximum of solar activity during the Christian era occurred in 1372. If so, this maximum was of very brief duration, 1or there are no records of sunspots between 1383 and 1511, while the frequency of auroræ also decreases. From about 1676 to 1725 there was an extraordinary dearth of sunspots, followed by maxima about 1778 and 1837.

From these figures of sunspots and auroræ an attempt has been made to construct a curve of solar activity since the occurrence of the first spot in A.D. 188. The early portion of this curve is not reliable, probably depending more on the accidental circumstances which led to the making and preserving of records than on the variations of the phenomena observed, but it seems probable that the maxima of the eleventh and fourteenth centuries, and perhaps also that of the ninth century, are real. From 1750 to 1940, the curve is based on 10-year means of the relative numbers. This curve of solar activity is shown at the top right-hand side of Fig. 39.

The construction of a curve of solar activity during the past few thousand years would be facilitated if we had any definite knowledge as to the cause of the sunspot cycle. Various hypotheses have been put forward, the favourite being the disturbance of the sun's surface by the influence of the planets, especially Jupiter. Jupiter completes his journey round the sun in 11·86 years, but when the influence of the other planets is added to that of Jupiter the result is an irregular recurrence with an average length of slightly less than 11·8 years, which bears some resemblance to the sunspot curve. The divergences are, however, too great for the complete acceptance of this theory. H. H. Turner (6) has devised an interesting alternative, which supposes that sunspots are due to the impact of meteors belonging to a swarm (the " Sunspot Swarm ") which pursues an elliptical orbit round the sun with a period which averages slightly over eleven years, but varies in length owing to interference with the Leonid Swarm. The latter has a period of 33¼ years. According to Turner, the " Sunspot Swarm " originated in A.D. 271 owing to the Leonid Swarm coming into conflict with the rings of Saturn, but so far as I am aware there is very little if any positive evidence for the existence of the Sunspot Swarm of meteorites, and in Chapter IV. we found some evidence for the existence of an eleven-year cycle in meteorological phenomena long before the beginning of the Christian era.

The two curves on the right below the sunspot curve show the variations in the thickness of the annual layers of Lake Saki, South Russia, and the thicknesses of growth rings of trees in Western U.S.A. according to Antevs, slightly smoothed. Previous to about A.D. 800 these do not show much relationship to the curve of solar activity, but as previously stated, the latter curve is not reliable for this period. The minimum about A.D. 250, for example, may be due entirely to a gap in the records. The long minimum of solar activity between 600 and 750 is reflected in the Russian curve, as is the peak about 850. The sunspot maximum of 1077-1079 is reflected in the highest peak of Antevs' curve, though the latter appears to come about 30 years earlier, and the general shape of the Russian curve from 1100 to 1300 is very similar to the solar curve. The great peak about 1372 also appears in both rainfall curves, though again somewhat early in the tree rings. Finally, the long dearth of sunspots from 1676 to 1725 is faithfully reflected in the annual layers ; it is also shown in the actual rainfall observations in Western Europe (Chapter XVIII.).

The short record of the low-level stage of the Nile (Fig. 35) can only be relied upon between A.D. 640 and 1400, but during this period it presents considerable similarity to the sunspot curve. Thus we have :—

| Sunspot maxima A.D. | 620 | 840 | 1077 | 1200 | 1370 |
| Nile, low-level stage . | 645 | 880 | 1100 | 1225 | 1375 |

The maxima of level in the low-water stage of the Nile apparently follow sunspot maxima by intervals of from five to forty years, but the order of importance of the maxima differs greatly in the two curves. In the Nile, the great crest at 1100 completely dominates all the later variations, and the peak at 1375 is insignificant. In this connexion it must be remembered that the Nile curves have been corrected on the assumption that the deposition of alluvium raises the level of the whole valley, including the low-level channel, at a uniform rate. The alluvium is deposited by the flood ; the low stage of the Nile is supplied by water which has been filtered by its passage through a series of lakes. Hence a consistently high level at the time of low water, as happened about 1100, might perform a good deal of erosion, and by cutting out a deep if narrow channel, result in a long series of very low minimum levels in subsequent years. This would be facilitated by the series of weak floods, which apparently occurred at about the same time, and which would bring little alluvium. For these reasons it is possible that the depression between about A.D. 1150 and 1400 may not indicate the true level of equatorial rainfall.

Finally we come to a hypothesis due to O. Pettersson (7) that variations of climate in the historical period have been caused by long-period variations in the circulation of the oceans caused by changes in the " tide-generating force." The latter varies with the declination and proximity of the sun and moon to the earth and, in addition to shorter variations, reaches maxima at variable intervals which average about 1,700 years. Pettersson gives the dates of these maxima as about 3500 B.C., 1900 B.C., 250 B.C., and A.D. 1433, and of the minima as about 2800 B.C., 1200 B.C., and A.D. 550. These variations are shown in the middle curve of Fig. 39.

Pettersson points out that in addition to the surface tides, which would have a greater range at maxima than at minima of the tidal force, there are also internal tides, formed where a relatively light less saline layer rests on a heavier more saline layer, and these submarine tides have actually been measured in the entrances to the Baltic, attaining a range of 80 to 90 feet. Submarine waves also enter the Arctic basin, where they were first traced by Nansen. At periods of maximum tide they are stronger and are able to break up the ice, leading to an increase in the amount of drift ice carried out into the North Atlantic by the polar currents. At tidal minima, on the other hand, the ice is broken up to a much less extent, and so there is little drift ice in the Greenland and Iceland seas.

Drift ice is an important factor in increasing the storminess and consequently the rainfall of temperate latitudes, and in deflecting the storm tracks into lower latitudes, and we should accordingly expect the maxima of tidal force to be maxima of rainfall also. This appears to be the case with the tidal maxima of 1900 B.C., 250 B.C., and A.D. 1433, while the minima in 2800 B.C., 1200 B.C., and A.D. 550 are also clearly shown. Further, with increased drift ice in the Atlantic we should expect lower temperatures in the coastal regions of Western Europe, and the temperature curve shows that on the whole this was so. It seems that there is good support for Pettersson's theory as well as for that of solar activity, and that the actual variations of climate since about 3000 B.C. may have been to a large extent the result of these two agents.

To facilitate comparison, the rainfall curves for Europe, Asia, and the U.S.A. in Fig. 38 have been combined in the lowest full curve of Fig. 39. This was constructed by first superposing the three curves and sketching in a general average. The positions of maxima and minima on all the individual curves of Figs. 31, 33, 35, and 36 were then marked on the curve, which was adjusted to bring the peaks and troughs into accord with the " majority verdict." This was done to eliminate as far as possible the uncertainties of dating. The broken curve is Pettersson's curve, extended backwards by assuming a periodicity of 1,750 years, and modified, after A.D. 100, by superposing on it the curve of solar activity.

The results are of great interest. From A.D. 100 onwards the fit is quite as good as can be expected from the nature of the data, both in the broad swing and in the peaks at intervals of two or three hundred years. From 3000 B.C. to A.D. 0 the long-period variations of rainfall also fit very well. From 2000 B.C. onwards, shorter oscillations of rainfall are superposed on these long waves, and it is a reasonable supposition that these may also be related to variations of solar activity. Before 3000 B.C., however, the two curves are in direct opposition. There are three possible reasons for this :—

1. The dating of the rainfall curve is incorrect. Although the curves for Europe and Asia were constructed independently, their dating before 2500 B.C. ultimately depends mainly on the dating of the cultures of Egypt and the Euphrates valley, which may not yet be quite established.

2. The extrapolation backwards of the tidal curve may be incorrect. The error is unlikely to be sufficiently great to invalidate the apparent opposition.

3. The climatic effect of tidal variations was reversed about 3000 B.C. The possibility of this depends on the ice conditions in the Arctic. The apparent opposition may be accidental, due to the fact that the Litorina subsidence happened to coincide with a minimum of the tidal force. It is of interest, however, to attempt a reconstruction of the history of the Arctic on the assumption that the opposition is real.

During the Climatic Optimum the mean temperature of the Arctic was many degrees higher than now, as shown, for example, by the growth of peat-bogs in Spitsbergen. Owing to the high obliquity and winter in aphelion the winters were cold and the summers very mild, the net result being a gain of solar heat. The Arctic Ocean may have been in a stage at which ice floes tended to form in winter but to break up and melt in summer without much ice finding its way into the polar currents. In such conditions a slight excess or deficit of heat would make a large difference to the development of the ice.

In periods of minimum tidal force the amount of warm water finding its way into the Arctic basin was also a minimum, and an ice-sheet would be likely to form in winter, which, when it broke up in summer, would supply some ice for the polar currents to carry into the Atlantic, though less than at present. These were the rainy periods. At maximum tidal force on the other hand the Arctic received a large amount of warm saline water. Both the high temperature and salinity would act against the freezing of the ocean surface, so that in these periods there may have been either no ice at all or so little that it broke up and melted away in summer without any ice reaching the North Atlantic. In such circumstances depressions would tend to follow northerly tracks into the Arctic Ocean instead of across Europe, giving a dry period in Western Europe. This is not entirely speculative ; much the same happens under present conditions when a long spell of southerly winds between Iceland and Novaya Zemlya drives the ice edge unusually far north, and this is almost invariably followed by a drought in Western Europe (8). Tidal force alone, however, cannot account for the relatively heavy rainfall of the Atlantic period compared with the present. In Europe this might be put down to the larger and warmer Baltic of the Litorina Sea, but if as seems probable the rainy period occurred also in Asia and North America some more general cause must be sought. This may be either the generally higher temperature of the oceans due to the smaller amount of sea and glacier ice, or possibly a period of increased solar activity.

By about 2500 B.C. these favourable conditions had largely passed. Scandinavia had risen almost to its present level, and the contrast between winter and summer had greatly decreased. Hence the Arctic Ocean became cooler and more liable to freeze. The main characteristic of the Sub-boreal in Western Europe seems to have been the instability of its climate, periods of drought and heat alternating with periods when the climate resembled the present, at intervals of perhaps a few hundred years. It is not unlikely that the Arctic ice-cap had now reached the critical stage between non-persistence and persistence—it was difficult for it to become established, but once firmly formed it was difficult to destroy. In such circumstances wide oscillations of climate would be expected.

I think that, paradoxically, persistence would be aided by increased tidal force, which would cool the Gulf Stream Drift by the ice carried into it by the polar currents. It must be remembered that a firm unbroken ice-cap grows more slowly than broken sea ice which allows the sea to freeze between the ice-floes.

The final stage came about 500 B.C. when for some reason the Arctic ice-cap at last became firmly established, apparently very extensively, after a few centuries of heat and drought. The reason for this change is not clear ; it may have been due to a change of solar activity or possibly to explosive volcanic activity. This period of an established Arctic ice-cap and stormy weather apparently lasted for about a thousand years, but the favourable climate of Iceland and Greenland, and the absence of ice in the accounts of the early Norse voyages, suggest that some time after A.D. 500 the ice-cap reverted to the semi-permanent stage, and remained so until nearly 1200.

The latest maximum of tide-generating force is dated by Pettersson as A.D. 1433. Although this force has been represented in Fig. 39 by a smooth curve, this is far from being the real condition. Superposed on the long period are shorter ones of about 90 and 9 years. From Pettersson's diagram the actual maxima apparently occurred in three sharp peaks about 1340, 1430, and 1520. Now the main peak on Antevs' curve of tree-growth comes in the decade 1331-1340, with a secondary peak about 1431-1440. In the Lake Saki varves the peak is about 1450 with secondary peaks at 1390 and 1530. These dates fit in rather better with Pettersson's curve than with the curve of solar activity. We note also that about 1340 the westerly winds in England were more persistent than at present, indicating a more stable Icelandic low.

On the other hand the great "storm floods" of the twelfth to fourteenth centuries, on which Pettersson sets great store, come on the whole before the absolute maximum of Pettersson's curve. The maximum damage occurred in 1170-1178, 1240-1253, 1267-1292, 1374-1377, and 1393-1404, the middle one being the most prolonged and severe ; there was also a great flood in 1421. These fit in better with the sunspot curve than with Pettersson's. Marine inundations require a combination of violent storms and high tides and these disasters of the

twelfth to fourteenth centuries may represent maximum storminess in the North Sea associated with great solar activity, at a
time when the tidal range was approaching its maximum.
There may also have been a slight subsidence of the land about
this time, especially in the fen districts where drainage operations had been carried on. In any case the main period of
great marine inundations would not be likely to continue after
the tidal maximum, for by that time the most vulnerable areas
would have been overflowed and protective measures taken.

The remarkable agreement between Pettersson's rather
obscure " tide-generating force " and the major variations
of climate since 3000 B.C. is surprising, and seems to show that,
under favourable conditions, comparatively small causes may
have disproportionately large effects. The favourable conditions are the stratification of the upper layers of the North
Atlantic, the existence of the submarine Wyville Thomson
ridge at just the right depth between Atlantic and Arctic,
the critical stage of the Arctic ice, just over the border between
non-glacial and glacial, and the maximum range of the tidal
force itself. Such a coincidence is not likely to have recurred
often in geological time, and in spite of the apparent effectiveness of Pettersson's tidal force in recent millennia we may
safely discount it as a permanent agent in climatic changes.

We may conclude this summary of post-glacial history with
brief references to four more recent climatic chapters :—

1. The dry period of the sixteenth century.

2. The great outburst of glaciers about 1600.

3. The dry period of 1701-1750 in Western Europe.

4. The rise of winter temperature in 1850-1940.

Comparatively rapid variations of climate, of the order of a
century, have presumably always occurred, and are shown
in the thicknesses of the glacial varves, lake deposits and tree
rings, but they can be properly examined only when we can
assemble sufficient facts, especially about the prevailing winds,
to enable us to reconstruct the probable pressure distribution,
and this is not possible before the sixteenth century. Here we
have to add another factor to our list of causes, namely, the
variations of the atmospheric circulation referred to on
p. 66. The atmosphere, like the sea, is in a state of perpetual

oscillation, the " waves " varying in length from a few hours to many years, the result being highly complex changes in the distribution of pressure from day to day, month to month and year to year. These changes can be predicted for a day or two, and are the basis of modern weather forecasting. Much effort has been devoted to their analysis in the hope of forecasting for longer periods ahead, but they are only periodic to a slight extent and so far the attempts have been unsuccessful. It is highly probable, however, that the longer oscillations are due partly to variations of solar activity and partly to interactions between the circulations of the atmosphere and the oceans. They take the form of an alternate weakening and strengthening of the whole circulation of the atmosphere. In the periods of weak circulation the low pressure centres near Iceland and the Aleutians are smaller, shallower and less stable, and anticyclones readily develop over the western margins of the continents. The winds are variable and the climate is " continental," with cold dry winters and hot summers. In the periods of strong circulation the low pressure areas are enlarged and intensified and powerful south-westerly air streams invade the western parts of the continents. The climate becomes " oceanic " with mild rainy winters and cool summers. Even in the middle of an " oceanic " period, however, there are occasional " continental " years, such as 1921, and *vice versa*, and the change from one type to the other often seems to be abrupt.

1. The latter half of the sixteenth century appears to have been mainly " continental," rather dry on the whole in the north temperate zone. In Western Europe the winds were probably more easterly than now, and the winters were cold. This seems to have been a time of minimum solar activity. There are few records of storms, and a reasonable inference is that the floating ice-cap suffered little disturbance and was able to grow in extent and solidity. The result would be more frequent incursions of cold Arctic air over Russia and extensions of the Siberian anticyclone across Northern Europe. This would give Northern and Western Europe frequent easterly winds, cold in winter, hot in summer. Similar conditions probably occurred in North America. The main storm tracks were deflected southwards, and this period seems to have been rainy both in South-east Europe and in Yucatan.

2.　The great outburst of mountain glaciation which began at the end of the sixteenth or early in the seventeenth century was so remarkable that this period has been termed the "Little Ice-Age." In the Alps and Iceland it began about 1600 and reached a maximum about 1643. In both countries the advances exceeded those at any other period since late-glacial times. There was a retreat in the first half of the eighteenth century, followed by a readvance in the first half of the nineteenth century, which gave place to a rapid retreat after 1850. In Southern Norway and Alaska, on the other hand, the maximum advance did not occur until about 1750.

It is now generally agreed that the most favourable conditions for the growth of glaciers are snowy winters and cool damp summers. The snowfall, however, takes time to accumulate, and the maximum extension of a glacier lags behind the greatest accumulation of snow by a number of years, depending on the size and length of the glacier. Hence the "glacial period" which began about 1600 probably reflects the snowfall of the latter part of the sixteenth century. We have seen that this period was probably mainly anti-cyclonic over Northern Europe, and that depressions followed southerly tracks. The greatest snowfall occurs in the northern halves of depressions, consequently this was a time of heavy snowfall in the Alps, Pyrenees and Iceland. During this period the snowfall accumulated at high levels, but at first the glaciers were unable to extend into the valleys because of the low mean annual temperature, which decreased their viscosity and kept them frozen to the ground. As soon as this limitation was removed, the glaciers grew rapidly in extent. At this time, however, Norway and Alaska still remained under the influence of the northern anticyclone, and the glacial advance did not extend to those regions until later.

3.　The first half of the eighteenth century is the first period for which instrumental observations are available. This period was discussed by C. E. P. Brooks (9). The greater frequency of north-easterly winds and the light rainfall over Western Europe (see p. 309) point to a decreased intensity of the Icelandic low, and there is some suggestion that the Aleutian low was also weak. In summer, anticyclones tended to develop over Western Europe. The lack of snowfall and probable high summer temperatures caused a recession of the Alpine glaciers, but in Norway snowfall increased and the glaciers advanced. The weather type of 1921 was probably the norm instead of the exception.

Shortly after 1750 this continental type changed to a more oceanic type, with milder winters and cooler, rainier summers. In England the change seems to have taken place rather abruptly in 1752 and was attributed by the commonalty to the change of the calendar in that year. This oceanic type continued for about a century, culminating in 1850 with another maximum advance of the glaciers. There was a brief return of the continental type from 1794 to 1810, analysed by C. E. P. Brooks (10), which gave a famous period of severe winters in Western Europe.

4. Since 1850 winter temperatures have tended to rise over all the north temperate and Arctic regions and probably in corresponding latitudes of the Southern Hemisphere. The change was slow and irregular at first, but became very rapid after 1900. The rise in the mean temperature of the three winter months, from 1851-1900 to 1901-1930, amounted to 5° F. or more in Western and Central Europe. This change was associated with a marked strengthening of the atmospheric circulation and steady west-south-west winds in Western Europe. There was little change of summer temperature. Glaciers and ice-sheets receded very rapidly, and after 1918 little or no drift ice reached the shores of Iceland. The rise of winter temperature progressed from south to north, and Central Europe may have passed the crest as early as 1920 when the rise in the Arctic was in full swing. The magnitude of the change in the Arctic is shown by the mean winter temperatures of Spitsbergen, which rose by 16° F. between 1911-1920 and 1931-1935. The edge of the main area of Arctic ice also receded towards the pole by some hundreds of miles. Since January 1940 the winter climate of Europe has reverted abruptly to greater severity, but it is too soon to say whether this is the beginning of another long period of continental climate or only a temporary fluctuation.

This concludes the examination of historical changes of climate, and also the analysis of the causes of climatic variations. The problem has proved to be one of great complexity, but throughout the book I have tried to examine each suggested cause impartially. The results seem to me to point very strongly to the following conclusions :—

1. The major climatic oscillations, lasting millions of years, are due to the major cycles of mountain-building and degradation, and their geographical effects in the widest sense, which possibly include variations in the amount of carbon dioxide and volcanic dust in the atmosphere.

2. Climatic oscillations of the second order, lasting thousands or tens of thousands of years, are due to two or possibly three causes :

(a) Minor changes in the land and sea distribution, caused partly by the shifting of the load on the earth's crust by erosion and partly by the isostatic effects of the growth and decay of the ice-sheets themselves. These were mainly effective during periods of high orography.

(b) Astronomical changes—eccentricity of the earth's orbit, obliquity of the ecliptic, precession of the equinoxes and possibly other causes. These are continuously effective and can be traced in some of the warm periods. They may have caused the succession of glacial and interglacial periods.

(c) Possibly long period variations of solar activity. Climatic oscillations of a few hundred years appear to be related to solar changes and it is a reasonable inference that the range of solar activity in the course of tens of thousands of years has been greater than the range during the Christian era. If such changes did occur, they must have caused considerable changes of precipitation.

3. Climatic oscillations lasting a few hundred years. So far as the evidence goes, these seem to be due mainly to variations of solar activity.

4. Climatic oscillations lasting for shorter periods, up to a hundred years or so. These may be due in part to variations of solar activity but there is evidence that they are often due to changes in the general circulation of the atmosphere which may have no external cause. They result from the interaction of the winds, ocean currents and floating ice-fields which we know to occur, but which, in the present state of our knowledge, is incalculable. These changes must always

have occurred, but were probably on a smaller scale during the warm periods, when there were no polar ice-caps, than during the glacial periods.

Other factors, hitherto unsuspected, may be discovered and prove to be important, but as a result of this analysis there seems to be no necessity to introduce hypothetical agents such as clouds of cosmic dust, strange stars or great disturbances of the earth's axis of rotation. The known causes set out in this book suffice to account for the variations of climate during geological and historical time.

REFERENCES

(1) WALKER, SIR GILBERT. "Correlations in seasonal variation of weather. VIII. A preliminary study of world weather." Calcutta, *Indian Met. Mem.*, 24, Pt. 4, 1923.

(2) ——. *idem.* "VI. Sunspots and pressure." *idem.*, 21, Pt. 12, 1915.

(3) HUNTINGTON, E., and S. S. VISHER. "Climatic changes, their nature and causes." New Haven, 1922.

(4) BROOKS, C. E. P. "The variation of the annual frequency of thunderstorms in relation to sunspots." London, *Q. J. R. Meteor. Soc.*, 60, 1934, p. 153.

(5) Zurich, *Vierteljahrschrif*, 38, 1893, p. 77 ; and Washington, D.C., *Monthly Weather Rev.*, 30, 1902, p. 173.

(6) TURNER, H. H. "On a simple method of detecting discontinuities in a series of recorded observations, with an application to sunspots." London, *Mon. Not. R. astr. Soc.*, 74, 1913, p. 82.

(7) PETTERSSON, O. "Climatic variations in historic and prehistoric time." *Svenska Hydrogr.-Biol. Komm. Skriften*, 5, 1914.

(8) BROOKS, C. E. P., and J. GLASSPOOLE. "The drought of 1921." London, *Q. J. R. Meteor. Soc.*, 48, 1922, p. 139.

(9) BROOKS, C. E. P. "The climate of the first half of the eighteenth century." London, *Q. J. R. Meteor. Soc.*, 56, 1930, p. 389.

(10) BROOKS, C. E. P. "Winds in London during the early 19th Century." London, *Meteor. Mag.*, 67, 1932, p. 56.

APPENDIX I

The Geological Time-Scale

The Age of the Earth.—Various methods have been suggested from time to time by which we can determine the approximate period which has elapsed since the formation of the first solid crust of the earth (1). The older methods were based on the thickness of sedimentary rocks or the amount of salt in the ocean, divided by the present rate of accumulation ; they assumed that the present rate of geological processes is a fair average of their rate throughout geological times. This assumption we now know to be false ; the present is a period of abnormally high relief, and in addition the unconsolidated deposits of the last ice-age facilitate denudation. The maximum age of the oldest rocks calculated from the rate of denudation is 350 million $(3 \cdot 5 \times 10^8)$ years, and this is probably only about one-fourth of their real age.

Presumably the sun is older than the earth, and calculations of the age of the sun based on the supply of energy by the aggregation of hydrogen atoms gives 140,000 million $(1 \cdot 4 \times 10^{11})$ years as the age of the sun. The sun apparently existed alone for a long period before a passing star disrupted it to form the solar system, for a calculation of the time since Mercury first took shape as a planet indicates that the age of the solar system is probably not greater than 10,000 million (10^{10}) years.

The most reliable method of calculating the age of any particular portion of the earth's crust is based on the phenomena of radio-activity. As is now well known, the elements uranium and thorium are continually breaking up and passing through a series of changes, the end-products of which are lead and helium. So far as we know, in the natural state the rate at which each of these elements disintegrates is a peculiarity of the element itself, and is entirely independent of the physical changes, such as variations of pressure and temperature, which it undergoes. If a sample of rock contains a certain amount of uranium, at the end of about 5,000 million years it will contain half this amount, at the end of another 5,000 million years one-quarter, and so on, the amount remaining continually halving in 5,000 million years. The original mass of uranium will never quite disappear. Suppose, now, that a rock when it solidified contained a certain amount of uranium, but no lead or helium. To-day it contains uranium, lead, and helium, and from the ratio of the amount of uranium to the amount of lead, or to the amount of helium, we can calculate the number of millions of years which have elapsed since that particular rock was formed. The uranium-lead ratio gives the more reliable estimates, for helium, being a gas, is liable to escape, and the uranium-helium estimates are systematically too low. In this way the following ages (in millions of years) have been calculated for different geological periods :—

Oligocene	. . .	26	Devonian . .	310-340
Eocene	. . . .	60	Archæan . . .	560-1,340
Carboniferous	. .	260-300		

In 1947, A. Holmes (2) concluded that " on the evidence at present available, the most probable age of the earth is about 3,350 million years."

J. Joly (3) developed an interpretation of the geological history of the earth, which leads him to much smaller values for the ages of the rocks. Owing to the universal presence of radio-active material in the earth's crust, both sial and sima, there is a perpetual generation of heat. In the continental masses of sial this heat is able to escape, but in the deeper layer of sima it is unable to escape and so goes on accumulating until the sima reaches its melting point. Melting begins at a considerable depth and proceeds gradually upwards. Finally, melting extends to such a height that the accumulated heat is able to escape into the oceans, and the substratum again becomes solid. Owing to the changes of density and volume involved, the period of melting is one of continental subsidence, while the period of solidification is a time of mountain-building. Melting—escape of heat—solidification —constitute a cycle, and according to Joly's calculations a cycle requires from forty to sixty million years to consummate itself. The number of complete cycles recognised is quite small—four or five, according to different authors. The periods of mountain-building closing the cycles occur as follows :—Laurentian and Algoman revolutions in the Archæan, Huronian or Killarney closing the Archæan, Appalachian or Hercynian closing the Palæozoic, and Alpine in the Miocene or Pliocene. Other revolutions which are recognised by some geologists but not by others are the Caledonian, occurring in the Silurian, and the Laramide in the Cretaceous. Hence the total duration of time since the Laurentian can have been only 200 to 300 million years. This estimate is quite incompatible with the usually accepted data of radio-activity given by the uranium-lead ratio, and Joly seeks to explain the discrepancy by supposing that the speed of radio-active processes has not in fact been constant during geological time, but that part of the lead contained in the early rocks was formed from isotopes of uranium which disintegrated at a greater rate than the only form known at present. There are minute peculiarities in some of uranium effects in the early rocks which support this view, but it does not seem probable, since the longer periods fit in better with the mass of geophysical data. A. Holmes (4), reviewing Joly's work, states that the generally accepted ages of the rocks are not likely to be in error by more than 10 per cent. He accepts the period of about forty million years for one of the cycles, but thinks that there have been about five times as many cycles as Joly supposes, and suggests that the main revolutions recognised by Joly are the concluding stages of major cycles in which the melting and consolidation of the deeper seated magma is added to that of the more superficial magma which by itself formed only minor revolutions. Inspection of the later part of the geological record seems to support Holmes' view ; for instance, minor periods of orogenesis occurred at the end of the Jurassic, in the Upper Cretaceous, in the Oligocene-Miocene, and in the Quaternary. The age of the Eocene is given by the uranium-lead ratio as 60 million years, so that we may take 80 million years for the interval between the Upper Cretaceous mountain-building and the Quaternary, or a period of two minor cycles.

If we take the total thickness of the stratified rocks and calculate the time which would be required for their formation on the assumption that denudation has always proceeded at its present rate, we obtain for the age of the oldest rocks only

Era.	Formation.	Thickness of Deposits. 1,000 feet. T.	Mean Elevation. 1,000 feet. E.	Duration ratio. T/E.	Duration. 10^6 years.	Age of Base by Duration-ratio. 10^6 years.	Age by Radio-activity. 10^6 years.	Climate.
Quaternary	Recent	..	..	..	..	..	..	Glaciation in temperate latitudes.
	Pleistocene	..	..	..	1	..	1	Cool.
Tertiary	Pliocene	13	3·6	3·6	12	13	13	Moderate.
	Miocene	14	2·3	6·1	21	34	30	Moderate to warm.
	Oligocene	12	2·8	4·3	15	49	60	Moderate, becoming warm.
	Eocene	20	2·8	7·1	24	73	..	Moderate, becoming warm.
Mesozoic	Cretaceous	44	2·6	16·9	57	130	110	Moderate.
	Jurassic	8	1·2	6·7	23	153	155	Warm and equable.
	Trias	17	0·6	28·3	96	249	190	Warm and equable.
Palaeozoic	Permian	12	2·0	6·0	20	269	210-240	Glacial at first, becoming moderate.
	Carboniferous	29	3·5	8·3	28	297	260-300	Warm at first, becoming glacial.
	Devonian	22	1·1	20·0	68	365	310-340	Moderate, becoming warm.
	Silurian	15	3·1	4·8	16	381	340	Warm.
	Ordovician	17	0·8	21·3	72	453	400	Moderate to warm.
	Cambrian	26	2·6	10·0	34	487	510	Cold, becoming warm.
Proterozoic	Keweenawan	50	4	12·5	42	529	560	Glacial.
	Animikian	14	..	..	..	..	..	..
	Huronian	18	..	..	..	..	..	..

Table 31.—Ages of the various geological periods.

about 150 million years, which is much too low. A much better agreement is obtained if we assume that in each period the rate of denudation has been proportional to the average elevation. If for any geological period we divide the total thickness of the stratified rocks by the average elevation during that period (Table 31), we get a number which we may term the " duration-ratio " of that period. The sum of all the duration-ratios from the beginning of the Keweenawan series (Upper Proterozoic) to the close of the Pliocene is approximately 156. The age of some probably Upper Proterozoic uraninite from Morogoro, East Africa, is given by the uranium-lead ratio as 560 million years. Hence a " duration-ratio " of unity corresponds with an actual duration of 3·6 million years. Similarly, the sum of the duration-ratios since the beginning of the Devonian is 107, while the greatest age determined from the uranium-lead ratio for a Devonian rock is 340 million years, giving a value of 3·2 million years for a duration-ratio of unity. The corresponding value deduced from the Carboniferous radio-active rocks (duration-ratio 87, age 300 million years) is 3·4, and from the Eocene (duration-ratio 18, age 60 million years) is 3·3. These four values are in sufficiently good agreement, and we can adopt as the time-equivalent of a duration-ratio of unity the mean period of 3·4 million years. This gives us the ages and durations of the various geological periods shown in Table 31, which are in sufficiently good agreement with the determinations by radio-active ratios.

REFERENCES

(1) JEFFREYS, H. " The earth, its origin, history, and physical constitution." 2 ed. Cambridge, 1929.
(2) HOLMES, A. " A revised estimate of the age of the earth." London, *Nature*, 159, 1947, p. 127.
(3) JOLY, J. " The surface-history of the earth." Oxford, 1925.
(4) HOLMES, A. " Radio-activity and geology." *Nature*, 116, 1925, p. 891.

APPENDIX II

Theories of Climatic Change

The various theories of the causes of climatic change are set out below, classified into types. The intention is not to provide a complete bibliography, but to give some indication as to where a description of the theory can be most readily found.

I. Cosmical.

CALVERWELL, E. P. *Geol. Mag.*, 32, 1895, p. 64. [Gas-filled regions in space.]

GUILLEMIN, —. *Arch. sci. phys.*, (3) 22, p. 585. [Cosmical dust in space.]

HOYLE, F., and R. A. LYTTLETON. See Chapter IV.

IVES, R. See Chapter IV.

NOELKE, F. "Das Problem der Entwicklungsgeschichte unseres Planetensystems." Berlin, 1908. [Cosmical dust in space.]

II. Solar Radiation.

DUBOIS, E. See Chapter IV.

FAYE, —. "Concordance des époques géologiques avec les époques cosmogoniques." Paris, *C. R. Acad. Sci.*, 100, 1885, p. 926.

FISCHER, E. "Eiszeittheorie." Heidelberg, 1902. [The sun moves in an elliptical orbit ; when far from focus velocity falls off and temperature decreases, giving an ice-age.]

HUNTINGTON, E., and S. S. VISHER. See Chapter IV.

JAEKEL, O. *Zs. D. Geol. Ges.*, 57, 1905, Monatsber., p. 223. [The separation of each of the inner planets from the sun was accompanied by a decrease of radiation and an ice-age on the earth ; obsolete.]

SIMPSON, SIR GEORGE. See Chapter IV.

III. Astronomical.

ADHÉMAR, J. "Les révolutions de la mer. Déluges périodiques." Paris, 1842. [Glaciation with winter in aphelion at maximum eccentricity of earth's orbit.]

BALL, SIR R. "The cause of an ice-age." London, 1891. [Precession of the equinoxes.]

CROLL, J. "Climate and time in their geological relations." London, 1875. [Eccentricity, glaciation with winter in aphelion.]

EKHOLM, N. London, *Q. J. R. Meteor. Soc.*, 27, 1901, p. 27. [Obliquity of ecliptic.]

HILDEBRANDT, M. "Die Eiszeiten der Erde." Berlin, 1901 [Glaciation with small eccentricity.]

MILANKOVITCH, M. See Chapter V.

MURPHY, J. J. London, *Q. J. Geol. Soc.*, 32, 1876, p. 400. [Glaciation with summer in aphelion at maximum eccentricity.]

PETTERSSON, O. (Tidal variations.) See Chapter XXII.

SPITALER, R. See Chapter V.

IV. EARTH HEAT.

HOFFMANN, J. F. *Beitr. Geophys.*, 9, 1908, p. 405. [Warm periods due to heat set free by decomposition of organisms in strata.]

MANSON, MARSDEN. See Chapter VII.

PROBST, J. "Klima und Gestaltung der Erde in ihren Wechselwirkungen." Stuttgart, 1887. [Warm springs.]

WAGNER, A. (Radio-active heat.) See Chapter X.

We may include here :—

FRANKLIN, A. V. Toronto, *J. R. Astr. Soc. Canada*, 12, 1918, p. 450. [Radiation from a warm moon.]

V. POLE-MOVEMENTS AND DRIFT OF CRUST.

KREICHGAUER, D. "Die Äquatorfrage in der Geologie." Steyl, 1902. [Polar movements.]

SIMROTH, H. "Die Pendulationstheorie." Leipzig, 1907.

OLDHAM, R. D. *Geol. Mag.*, 23, 1886, p. 300.

WEGENER, A. See Chapter XIII.

VI. ELEVATION.

ENQUIST, F. *Bull. Geol. Inst. Upsala*, 12, 1915, p. 35. [Deepening of ocean basins.]

GRÉGOIRE, A. *Bull. Soc. Belge Géol.*, 23, 1909, p. 154. [The sea floor is colder than the land, hence when there is a reversal the sea is warmed, increasing evaporation, which causes greater snowfall on cold land.]

LE CONTE, J. "The Ozarkian and its significance." *J. Geol.*, Chicago, 7, 1899, p. 525.

RAMSAY, W. *Ofversigt af Finska Vetenskaps Soc. Förh.*, 52, 1910. Afd. H. See also Chapter X.

SCHUCHERT, CH. Carnegie Inst., Washington, Publ. 192, 1914, p. 263.

UPHAM, W. *Amer. Geol.*, 6, 1890, p. 327.

VII. Land and Sea Distribution.

> Brooks, C. E. P.   See Chapter VIII.
>
> Harmer, the late F. W.   London, *Q. J. R. Meteor. Soc.*, 51, 1925, p. 247.
>
> Kerner, F. v.   Wien, *Sitzungsber. K. Akad. Wiss., Math.-nat. Kl.*, 122, Abt. 2*a*, 1913, p. 233.
>
> Lyell, Ch.   "Principles of geology." 11th ed.   London, 1892.
>
> Semper, M.   *Zs. D. Geol. Ges.*, 48, 1896, p. 261.

VIII. Ocean Currents.

> Chamberlin, T. C.   See Chapter III.   [Reversal of deep sea circulation.]
>
> Hull, E.   London, *Q. J. Geol. Soc.*, 53, 1897, p. 107. [Deflection of Gulf Stream by Antillean continent.]
>
> Klein, H. J.   *Gaea*, 41, 1905, p. 449.   [Deflection of Gulf Stream by land projection from Newfoundland towards Cape Verde Islands.]

IX. Changes in Composition of Atmosphere.

> Arrhenius, S.   *Phil. Mag.*, 41, 1896, p. 237.   [Carbon dioxide.]
>
> Callendar, G. S.   (Carbon dioxide.)   See Chapter VI.
>
> Chamberlin, T. C.   *J. Geol.*, Chicago, 7, 1899, pp. 545, 667, 752.   [Carbon dioxide.]
>
> Frech, F.   *Zs. Ges. Erdk.*, Berlin, 37, 1902, p. 611.   [Carbon dioxide.]
>
> Harboe, E. G.   *Zs. D. Geol. Ges.*, 50, 1898, p. 441.   [Water vapour from volcanoes.]
>
> Harlé, E. and A.   *Bull. Soc. Geol. France*, 11, 1911, p. 118. [Probable greater pressure of air in geological times.   See Chapter II.]
>
> Manson, Marsden.   See Chapter VII.

X. Volcanic Dust.

> Humphreys, W. H.   See Chapter VI.
>
> Sarasin, P. and F.   Basle, *Verh. Naturf. Ges.*, 13, 1901, p. 603.

XI. Changes of Atmospheric Circulation.

> Abbe, C.   Washington, U.S. Weather Bureau.   *Monthly Weather Review*, 34, 1906, p. 559.   [Slight changes of circulation.]
>
> Deeley, R. M.   *Geol. Mag.*, (6) 2, 1915, p. 450.   [Effect of great polar water areas on stratosphere.]
>
> Dines, W. H.   See Chapter II.
>
> Harmer, F. W.   See Chapter II.

INDEX

SOME DOVER SCIENCE BOOKS

SOME DOVER SCIENCE BOOKS

WHAT IS SCIENCE?,
Norman Campbell
This excellent introduction explains scientific method, role of mathematics, types of scientific laws. Contents: 2 aspects of science, science & nature, laws of science, discovery of laws, explanation of laws, measurement & numerical laws, applications of science. 192pp. 5⅜ x 8.   60043-2 Paperbound $1.25

FADS AND FALLACIES IN THE NAME OF SCIENCE,
Martin Gardner
Examines various cults, quack systems, frauds, delusions which at various times have masqueraded as science. Accounts of hollow-earth fanatics like Symmes; Velikovsky and wandering planets; Hoerbiger; Bellamy and the theory of multiple moons; Charles Fort; dowsing, pseudoscientific methods for finding water, ores, oil. Sections on naturopathy, iridiagnosis, zone therapy, food fads, etc. Analytical accounts of Wilhelm Reich and orgone sex energy; L. Ron Hubbard and Dianetics; A. Korzybski and General Semantics; many others. Brought up to date to include Bridey Murphy, others. Not just a collection of anecdotes, but a fair, reasoned appraisal of eccentric theory. Formerly titled *In the Name of Science*. Preface. Index. x + 384pp. 5⅜ x 8.
20394-8 Paperbound $2.00

PHYSICS, THE PIONEER SCIENCE,
L. W. Taylor
First thorough text to place all important physical phenomena in cultural-historical framework; remains best work of its kind. Exposition of physical laws, theories developed chronologically, with great historical, illustrative experiments diagrammed, described, worked out mathematically. Excellent physics text for self-study as well as class work. Vol. 1: Heat, Sound: motion, acceleration, gravitation, conservation of energy, heat engines, rotation, heat, mechanical energy, etc. 211 illus. 407pp. 5⅜ x 8. Vol. 2: Light, Electricity: images, lenses, prisms, magnetism, Ohm's law, dynamos, telegraph, quantum theory, decline of mechanical view of nature, etc. Bibliography. 13 table appendix. Index. 551 illus. 2 color plates. 508pp. 5⅜ x 8.
60565-5, 60566-3 Two volume set, paperbound $5.50

THE EVOLUTION OF SCIENTIFIC THOUGHT FROM NEWTON TO EINSTEIN,
A. d'Abro
Einstein's special and general theories of relativity, with their historical implications, are analyzed in non-technical terms. Excellent accounts of the contributions of Newton, Riemann, Weyl, Planck, Eddington, Maxwell, Lorentz and others are treated in terms of space and time, equations of electromagnetics, finiteness of the universe, methodology of science. 21 diagrams. 482pp. 5⅜ x 8.
20002-7 Paperbound $2.50

CHANCE, LUCK AND STATISTICS: THE SCIENCE OF CHANCE,
Horace C. Levinson
Theory of probability and science of statistics in simple, non-technical language.
Part I deals with theory of probability, covering odd superstitions in regard to
"luck," the meaning of betting odds, the law of mathematical expectation,
gambling, and applications in poker, roulette, lotteries, dice, bridge, and other
games of chance. Part II discusses the misuse of statistics, the concept of statistical probabilities, normal and skew frequency distributions, and statistics applied to various fields—birth rates, stock speculation, insurance rates, advertis-
ing, etc. "Presented in an easy humorous style which I consider the best kind of
expository writing," Prof. A. C. Cohen, Industry Quality Control. Enlarged
revised edition. Formerly titled *The Science of Chance*. Preface and two new
appendices by the author. xiv + 365pp. 5⅜ x 8. 21007-3 Paperbound $2.00

BASIC ELECTRONICS,
prepared by the U.S. Navy Training Publications Center
A thorough and comprehensive manual on the fundamentals of electronics.
Written clearly, it is equally useful for self-study or course work for those with
a knowledge of the principles of basic electricity. Partial contents: Operating
Principles of the Electron Tube; Introduction to Transistors; Power Supplies
for Electronic Equipment; Tuned Circuits; Electron-Tube Amplifiers; Audio
Power Amplifiers; Oscillators; Transmitters; Transmission Lines; Antennas and
Propagation; Introduction to Computers; and related topics. Appendix. Index.
Hundreds of illustrations and diagrams. vi + 471pp. 6½ x 9¼.
61076-4 Paperbound $2.95

BASIC THEORY AND APPLICATION OF TRANSISTORS,
prepared by the U.S. Department of the Army
An introductory manual prepared for an army training program. One of the
finest available surveys of theory and application of transistor design and
operation. Minimal knowledge of physics and theory of electron tubes required.
Suitable for textbook use, course supplement, or home study. Chapters: Introduction; fundamental theory of transistors; transistor amplifier fundamentals;
parameters, equivalent circuits, and characteristic curves; bias stabilization;
transistor analysis and comparison using characteristic curves and charts; audio
amplifiers; tuned amplifiers; wide-band amplifiers; oscillators; pulse and switch-
ing circuits; modulation, mixing, and demodulation; and additional semi-
conductor devices. Unabridged, corrected edition. 240 schematic drawings,
photographs, wiring diagrams, etc. 2 Appendices. Glossary. Index. 263pp.
6½ x 9¼. 60380-6 Paperbound $1.75

GUIDE TO THE LITERATURE OF MATHEMATICS AND PHYSICS,
N. G. Parke III
Over 5000 entries included under approximately 120 major subject headings of
selected most important books, monographs, periodicals, articles in English,
plus important works in German, French, Italian, Spanish, Russian (many
recently available works). Covers every branch of physics, math, related engineering. Includes author, title, edition, publisher, place, date, number of
volumes, number of pages. A 40-page introduction on the basic problems of
research and study provides useful information on the organization and use of
libraries, the psychology of learning, etc. This reference work will save you
hours of time. 2nd revised edition. Indices of authors, subjects, 464pp. 5⅜ x 8.
60447-0 Paperbound $2.75

THE RISE OF THE NEW PHYSICS (formerly THE DECLINE OF MECHANISM),
A. d'Abro
This authoritative and comprehensive 2-volume exposition is unique in scientific publishing. Written for intelligent readers not familiar with higher mathematics, it is the only thorough explanation in non-technical language of modern mathematical-physical theory. Combining both history and exposition, it ranges from classical Newtonian concepts up through the electronic theories of Dirac and Heisenberg, the statistical mechanics of Fermi, and Einstein's relativity theories. "A must for anyone doing serious study in the physical sciences," *J. of Franklin Inst.* 97 illustrations. 991pp. 2 volumes.
20003-5, 20004-3 Two volume set, paperbound $5.50

THE STRANGE STORY OF THE QUANTUM, AN ACCOUNT FOR THE GENERAL READER OF THE GROWTH OF IDEAS UNDERLYING OUR PRESENT ATOMIC KNOWLEDGE, *B. Hoffmann*
Presents lucidly and expertly, with barest amount of mathematics, the problems and theories which led to modern quantum physics. Dr. Hoffmann begins with the closing years of the 19th century, when certain trifling discrepancies were noticed, and with illuminating analogies and examples takes you through the brilliant concepts of Planck, Einstein, Pauli, de Broglie, Bohr, Schroedinger, Heisenberg, Dirac, Sommerfeld, Feynman, etc. This edition includes a new, long postscript carrying the story through 1958. "Of the books attempting an account of the history and contents of our modern atomic physics which have come to my attention, this is the best," H. Margenau, Yale University, in *American Journal of Physics*. 32 tables and line illustrations. Index. 275pp. 5⅜ x 8.
20518-5 Paperbound $2.00

GREAT IDEAS AND THEORIES OF MODERN COSMOLOGY,
Jagjit Singh
The theories of Jeans, Eddington, Milne, Kant, Bondi, Gold, Newton, Einstein, Gamow, Hoyle, Dirac, Kuiper, Hubble, Weizsäcker and many others on such cosmological questions as the origin of the universe, space and time, planet formation, "continuous creation," the birth, life, and death of the stars, the origin of the galaxies, etc. By the author of the popular *Great Ideas of Modern Mathematics*. A gifted popularizer of science, he makes the most difficult abstractions crystal-clear even to the most non-mathematical reader. Index. xii + 276pp. 5⅜ x 8½.
20925-3 Paperbound $2.50

GREAT IDEAS OF MODERN MATHEMATICS: THEIR NATURE AND USE,
Jagjit Singh
Reader with only high school math will understand main mathematical ideas of modern physics, astronomy, genetics, psychology, evolution, etc., better than many who use them as tools, but comprehend little of their basic structure. Author uses his wide knowledge of non-mathematical fields in brilliant exposition of differential equations, matrices, group theory, logic, statistics, problems of mathematical foundations, imaginary numbers, vectors, etc. Original publications, appendices. indexes. 65 illustr. 322pp. 5⅜ x 8. 20587-8 Paperbound $2.25

THE MATHEMATICS OF GREAT AMATEURS, *Julian L. Coolidge*
Great discoveries made by poets, theologians, philosophers, artists and other non-mathematicians: Omar Khayyam, Leonardo da Vinci, Albrecht Dürer, John Napier, Pascal, Diderot, Bolzano, etc. Surprising accounts of what can result from a non-professional preoccupation with the oldest of sciences. 56 figures. viii + 211pp. 5⅜ x 8½.
61009-8 Paperbound $2.00

COLLEGE ALGEBRA, *H. B. Fine*

Standard college text that gives a systematic and deductive structure to algebra; comprehensive, connected, with emphasis on theory. Discusses the commutative, associative, and distributive laws of number in unusual detail, and goes on with undetermined coefficients, quadratic equations, progressions, logarithms, permutations, probability, power series, and much more. Still most valuable elementary-intermediate text on the science and structure of algebra. Index. 1560 problems, all with answers. x + 631pp. 5⅜ x 8. 60211-7 Paperbound $2.75

HIGHER MATHEMATICS FOR STUDENTS OF CHEMISTRY AND PHYSICS, *J. W. Mellor*

Not abstract, but practical, building its problems out of familiar laboratory material, this covers differential calculus, coordinate, analytical geometry, functions, integral calculus, infinite series, numerical equations, differential equations, Fourier's theorem, probability, theory of errors, calculus of variations, determinants. "If the reader is not familiar with this book, it will repay him to examine it," *Chem. & Engineering News.* 800 problems. 189 figures. Bibliography. xxi + 641pp. 5⅜ x 8.   60193-5 Paperbound $3.50

TRIGONOMETRY REFRESHER FOR TECHNICAL MEN, *A. A. Klaf*

A modern question and answer text on plane and spherical trigonometry. Part I covers plane trigonometry: angles, quadrants, trigonometrical functions, graphical representation, interpolation, equations, logarithms, solution of triangles, slide rules, etc. Part II discusses applications to navigation, surveying, elasticity, architecture, and engineering. Small angles, periodic functions, vectors, polar coordinates, De Moivre's theorem, fully covered. Part III is devoted to spherical trigonometry and the solution of spherical triangles, with applications to terrestrial and astronomical problems. Special time-savers for numerical calculation. 913 questions answered for you! 1738 problems; answers to odd numbers. 494 figures. 14 pages of functions, formulae. Index. x + 629pp. 5⅜ x 8.

20371-9 Paperbound $3.00

CALCULUS REFRESHER FOR TECHNICAL MEN, *A. A. Klaf*

Not an ordinary textbook but a unique refresher for engineers, technicians, and students. An examination of the most important aspects of differential and integral calculus by means of 756 key questions. Part I covers simple differential calculus: constants, variables, functions, increments, derivatives, logarithms, curvature, etc. Part II treats fundamental concepts of integration: inspection, substitution, transformation, reduction, areas and volumes, mean value, successive and partial integration, double and triple integration. Stresses practical aspects! A 50 page section gives applications to civil and nautical engineering, electricity, stress and strain, elasticity, industrial engineering, and similar fields. 756 questions answered. 556 problems; solutions to odd numbers. 36 pages of constants, formulae. Index. v + 431pp. 5⅜ x 8.   20370-0 Paperbound $2.25

INTRODUCTION TO THE THEORY OF GROUPS OF FINITE ORDER, *R. Carmichael*

Examines fundamental theorems and their application. Beginning with sets, systems, permutations, etc., it progresses in easy stages through important types of groups: Abelian, prime power, permutation, etc. Except 1 chapter where matrices are desirable, no higher math needed. 783 exercises, problems. Index. xvi + 447pp. 5⅜ x 8.   60300-8 Paperbound $3.00

FIVE VOLUME "THEORY OF FUNCTIONS" SET BY KONRAD KNOPP

This five-volume set, prepared by Konrad Knopp, provides a complete and readily followed account of theory of functions. Proofs are given concisely, yet without sacrifice of completeness or rigor. These volumes are used as texts by such universities as M.I.T., University of Chicago, N. Y. City College, and many others. "Excellent introduction . . . remarkably readable, concise, clear, rigorous," *Journal of the American Statistical Association.*

ELEMENTS OF THE THEORY OF FUNCTIONS,
Konrad Knopp
This book provides the student with background for further volumes in this set, or texts on a similar level. Partial contents: foundations, system of complex numbers and the Gaussian plane of numbers, Riemann sphere of numbers, mapping by linear functions, normal forms, the logarithm, the cyclometric functions and binomial series. "Not only for the young student, but also for the student who knows all about what is in it," *Mathematical Journal.* Bibliography. Index. 140pp. 5⅜ x 8.   60154-4 Paperbound $1.50

THEORY OF FUNCTIONS, PART I,
Konrad Knopp
With volume II, this book provides coverage of basic concepts and theorems. Partial contents: numbers and points, functions of a complex variable, integral of a continuous function, Cauchy's integral theorem, Cauchy's integral formulae, series with variable terms, expansion of analytic functions in power series, analytic continuation and complete definition of analytic functions, entire transcendental functions, Laurent expansion, types of singularities. Bibliography. Index. vii + 146pp. 5⅜ x 8.   60156-0 Paperbound $1.50

THEORY OF FUNCTIONS, PART II,
Konrad Knopp
Application and further development of general theory, special topics. Single valued functions. Entire, Weierstrass, Meromorphic functions. Riemann surfaces. Algebraic functions. Analytical configuration, Riemann surface. Bibliography. Index. x + 150pp. 5⅜ x 8.   60157-9 Paperbound $1.50 ·

PROBLEM BOOK IN THE THEORY OF FUNCTIONS, VOLUME 1.
Konrad Knopp
Problems in elementary theory, for use with Knopp's *Theory of Functions,* or any other text, arranged according to increasing difficulty. Fundamental concepts, sequences of numbers and infinite series, complex variable, integral theorems, development in series, conformal mapping. 182 problems. Answers. viii + 126pp. 5⅜ x 8.   60158-7 Paperbound $1.50

PROBLEM BOOK IN THE THEORY OF FUNCTIONS, VOLUME 2,
Konrad Knopp
Advanced theory of functions, to be used either with Knopp's *Theory of Functions,* or any other comparable text. Singularities, entire & meromorphic functions, periodic, analytic, continuation, multiple-valued functions, Riemann surfaces, conformal mapping. Includes a section of additional elementary problems. "The difficult task of selecting from the immense material of the modern theory of functions the problems just within the reach of the beginner is here masterfully accomplished," *Am. Math. Soc.* Answers. 138pp. 5⅜ x 8.
60159-5 Paperbound $1.50

NUMERICAL SOLUTIONS OF DIFFERENTIAL EQUATIONS,
H. Levy & E. A. Baggott
Comprehensive collection of methods for solving ordinary differential equations of first and higher order. All must pass 2 requirements: easy to grasp and practical, more rapid than school methods. Partial contents: graphical integration of differential equations, graphical methods for detailed solution. Numerical solution. Simultaneous equations and equations of 2nd and higher orders. "Should be in the hands of all in research in applied mathematics, teaching," *Nature.* 21 figures. viii + 238pp. 5⅜ x 8.   60168-4 Paperbound $1.85

ELEMENTARY STATISTICS, WITH APPLICATIONS IN MEDICINE AND THE BIOLOGICAL SCIENCES, *F. E. Croxton*
A sound introduction to statistics for anyone in the physical sciences, assuming no prior acquaintance and requiring only a modest knowledge of math. All basic formulas carefully explained and illustrated; all necessary reference tables included. From basic terms and concepts, the study proceeds to frequency distribution, linear, non-linear, and multiple correlation, skewness, kurtosis, etc. A large section deals with reliability and significance of statistical methods. Containing concrete examples from medicine and biology, this book will prove unusually helpful to workers in those fields who increasingly must evaluate, check, and interpret statistics. Formerly titled "Elementary Statistics with Applications in Medicine." 101 charts. 57 tables. 14 appendices. Index. vi + 376pp. 5⅜ x 8.   60506-X Paperbound $2.25

INTRODUCTION TO SYMBOLIC LOGIC,
S. Langer
No special knowledge of math required — probably the clearest book ever written on symbolic logic, suitable for the layman, general scientist, and philosopher. You start with simple symbols and advance to a knowledge of the Boole-Schroeder and Russell-Whitehead systems. Forms, logical structure, classes, the calculus of propositions, logic of the syllogism, etc. are all covered. "One of the clearest and simplest introductions," *Mathematics Gazette.* Second enlarged, revised edition. 368pp. 5⅜ x 8.   60164-1 Paperbound $2.25

A SHORT ACCOUNT OF THE HISTORY OF MATHEMATICS,
W. W. R. Ball
Most readable non-technical history of mathematics treats lives, discoveries of every important figure from Egyptian, Phoenician, mathematicians to late 19th century. Discusses schools of Ionia, Pythagoras, Athens, Cyzicus, Alexandria, Byzantium, systems of numeration; primitive arithmetic; Middle Ages, Renaissance, including Arabs, Bacon, Regiomontanus, Tartaglia, Cardan, Stevinus, Galileo, Kepler; modern mathematics of Descartes, Pascal, Wallis, Huygens, Newton, Leibnitz, d'Alembert, Euler, Lambert, Laplace, Legendre, Gauss, Hermite, Weierstrass, scores more. Index. 25 figures. 546pp. 5⅜ x 8.
20630-0 Paperbound $2.75

INTRODUCTION TO NONLINEAR DIFFERENTIAL AND INTEGRAL EQUATIONS,
Harold T. Davis
Aspects of the problem of nonlinear equations, transformations that lead to equations solvable by classical means, results in special cases, and useful generalizations. Thorough, but easily followed by mathematically sophisticated reader who knows little about non-linear equations. 137 problems for student to solve. xv + 566pp. 5⅜ x 8½.   60971-5 Paperbound $2.75

AN INTRODUCTION TO THE GEOMETRY OF N DIMENSIONS,
D. H. Y. Sommerville
An introduction presupposing no prior knowledge of the field, the only book in English devoted exclusively to higher dimensional geometry. Discusses fundamental ideas of incidence, parallelism, perpendicularity, angles between linear space; enumerative geometry; analytical geometry from projective and metric points of view; polytopes; elementary ideas in analysis situs; content of hyper-spacial figures. Bibliography. Index. 60 diagrams. 196pp. 5⅜ x 8.
60494-2 Paperbound $1.50

ELEMENTARY CONCEPTS OF TOPOLOGY, *P. Alexandroff*
First English translation of the famous brief introduction to topology for the beginner or for the mathematician not undertaking extensive study. This unusually useful intuitive approach deals primarily with the concepts of complex, cycle, and homology, and is wholly consistent with current investigations. Ranges from basic concepts of set-theoretic topology to the concept of Betti groups. "Glowing example of harmony between intuition and thought," David Hilbert. Translated by A. E. Farley. Introduction by D. Hilbert. Index. 25 figures. 73pp. 5⅜ x 8. 60747-X Paperbound $1.25

ELEMENTS OF NON-EUCLIDEAN GEOMETRY,
D. M. Y. Sommerville
Unique in proceeding step-by-step, in the manner of traditional geometry. Enables the student with only a good knowledge of high school algebra and geometry to grasp elementary hyperbolic, elliptic, analytic non-Euclidean geometries; space curvature and its philosophical implications; theory of radical axes; homothetic centres and systems of circles; parataxy and parallelism; absolute measure; Gauss' proof of the defect area theorem; geodesic representation; much more, all with exceptional clarity. 126 problems at chapter endings provide progressive practice and familiarity. 133 figures. Index. xvi + 274pp. 5⅜ x 8. 60460-8 Paperbound $2.00

INTRODUCTION TO THE THEORY OF NUMBERS, *L. E. Dickson*
Thorough, comprehensive approach with adequate coverage of classical literature, an introductory volume beginners can follow. Chapters on divisibility, congruences, quadratic residues & reciprocity. Diophantine equations, etc. Full treatment of binary quadratic forms without usual restriction to integral coefficients. Covers infinitude of primes, least residues. Fermat's theorem. Euler's phi function, Legendre's symbol, Gauss's lemma, automorphs, reduced forms, recent theorems of Thue & Siegel, many more. Much material not readily available elsewhere. 239 problems. Index. I figure. viii + 183pp. 5⅜ x 8.
60342-3 Paperbound $1.75

MATHEMATICAL TABLES AND FORMULAS,
compiled by Robert D. Carmichael and Edwin R. Smith
Valuable collection for students, etc. Contains all tables necessary in college algebra and trigonometry, such as five-place common logarithms, logarithmic sines and tangents of small angles, logarithmic trigonometric functions, natural trigonometric functions, four-place antilogarithms, tables for changing from sexagesimal to circular and from circular to sexagesimal measure of angles, etc. Also many tables and formulas not ordinarily accessible, including powers, roots, and reciprocals, exponential and hyperbolic functions, ten-place logarithms of prime numbers, and formulas and theorems from analytical and elementary geometry and from calculus. Explanatory introduction. viii + 269pp. 5⅜ x 8½. 60111-0 Paperbound $1.50

A SOURCE BOOK IN MATHEMATICS,
D. E. Smith
Great discoveries in math, from Renaissance to end of 19th century, in English translation. Read announcements by Dedekind, Gauss, Delamain, Pascal, Fermat, Newton, Abel, Lobachevsky, Bolyai, Riemann, De Moivre, Legendre, Laplace, others of discoveries about imaginary numbers, number congruence, slide rule, equations, symbolism, cubic algebraic equations, non-Euclidean forms of geometry, calculus, function theory, quaternions, etc. Succinct selections from 125 different treatises, articles, most unavailable elsewhere in English. Each article preceded by biographical introduction. Vol. I: Fields of Number, Algebra. Index. 32 illus. 338pp. 5⅜ x 8. Vol. II: Fields of Geometry, Probability, Calculus, Functions, Quaternions. 83 illus. 432pp. 5⅜ x 8.
60552-3, 60553-1 Two volume set, paperbound $5.00

FOUNDATIONS OF PHYSICS,
R. B. Lindsay & H. Margenau
Excellent bridge between semi-popular works & technical treatises. A discussion of methods of physical description, construction of theory; valuable for physicist with elementary calculus who is interested in ideas that give meaning to data, tools of modern physics. Contents include symbolism; mathematical equations; space & time foundations of mechanics; probability; physics & continua; electron theory; special & general relativity; quantum mechanics; causality. "Thorough and yet not overdetailed. Unreservedly recommended," *Nature* (London). Unabridged, corrected edition. List of recommended readings. 35 illustrations. xi + 537pp. 5⅜ x 8.
60377-6 Paperbound $3.50

FUNDAMENTAL FORMULAS OF PHYSICS,
ed. by D. H. Menzel
High useful, full, inexpensive reference and study text, ranging from simple to highly sophisticated operations. Mathematics integrated into text—each chapter stands as short textbook of field represented. Vol. 1: Statistics, Physical Constants, Special Theory of Relativity, Hydrodynamics, Aerodynamics, Boundary Value Problems in Math, Physics, Viscosity, Electromagnetic Theory, etc. Vol. 2: Sound, Acoustics, Geometrical Optics, Electron Optics, High-Energy Phenomena, Magnetism, Biophysics, much more. Index. Total of 800pp. 5⅜ x 8.
60595-7, 60596-5 Two volume set, paperbound $4.75

THEORETICAL PHYSICS,
A. S. Kompaneyets
One of the very few thorough studies of the subject in this price range. Provides advanced students with a comprehensive theoretical background. Especially strong on recent experimentation and developments in quantum theory. Contents: Mechanics (Generalized Coordinates, Lagrange's Equation, Collision of Particles, etc.), Electrodynamics (Vector Analysis, Maxwell's equations, Transmission of Signals, Theory of Relativity, etc.), Quantum Mechanics (the Inadequacy of Classical Mechanics, the Wave Equation, Motion in a Central Field, Quantum Theory of Radiation, Quantum Theories of Dispersion and Scattering, etc.), and Statistical Physics (Equilibrium Distribution of Molecules in an Ideal Gas, Boltzmann Statistics, Bose and Fermi Distribution. Thermodynamic Quantities, etc.). Revised to 1961. Translated by George Yankovsky, authorized by Kompaneyets. 137 exercises. 56 figures. 529pp. 5⅜ x 8½.
60972-3 Paperbound $3.50

MATHEMATICAL PHYSICS, *D. H. Menzel*
Thorough one-volume treatment of the mathematical techniques vital for classical mechanics, electromagnetic theory, quantum theory, and relativity. Written by the Harvard Professor of Astrophysics for junior, senior, and graduate courses, it gives clear explanations of all those aspects of function theory, vectors, matrices, dyadics, tensors, partial differential equations, etc., necessary for the understanding of the various physical theories. Electron theory, relativity, and other topics seldom presented appear here in considerable detail. Scores of definition, conversion factors, dimensional constants, etc. "More detailed than normal for an advanced text . . . excellent set of sections on Dyadics, Matrices, and Tensors," *Journal of the Franklin Institute.* Index. 193 problems, with answers. x + 412pp. 5⅜ x 8.   60056-4 Paperbound $2.50

THE THEORY OF SOUND, *Lord Rayleigh*
Most vibrating systems likely to be encountered in practice can be tackled successfully by the methods set forth by the great Nobel laureate, Lord Rayleigh. Complete coverage of experimental, mathematical aspects of sound theory. Partial contents: Harmonic motions, vibrating systems in general, lateral vibrations of bars, curved plates or shells, applications of Laplace's functions to acoustical problems, fluid friction, plane vortex-sheet, vibrations of solid bodies, etc. This is the first inexpensive edition of this great reference and study work. Bibliography, Historical introduction by R. B. Lindsay. Total of 1040pp. 97 figures. 5⅜ x 8.   60292-3, 60293-1 Two volume set, paperbound $6.00

HYDRODYNAMICS, *Horace Lamb*
Internationally famous complete coverage of standard reference work on dynamics of liquids & gases. Fundamental theorems, equations, methods, solutions, background, for classical hydrodynamics. Chapters include Equations of Motion, Integration of Equations in Special Gases, Irrotational Motion, Motion of Liquid in 2 Dimensions, Motion of Solids through Liquid-Dynamical Theory, Vortex Motion, Tidal Waves, Surface Waves, Waves of Expansion, Viscosity, Rotating Masses of Liquids. Excellently planned, arranged; clear, lucid presentation. 6th enlarged, revised edition. Index. Over 900 footnotes, mostly bibliographical. 119 figures. xv + 738pp. 6⅛ x 9¼.   60256-7 Paperbound $4.00

DYNAMICAL THEORY OF GASES, *James Jeans*
Divided into mathematical and physical chapters for the convenience of those not expert in mathematics, this volume discusses the mathematical theory of gas in a steady state, thermodynamics, Boltzmann and Maxwell, kinetic theory, quantum theory, exponentials, etc. 4th enlarged edition, with new material on quantum theory, quantum dynamics, etc. Indexes. 28 figures. 444pp. 6⅛ x 9¼.
60136-6 Paperbound $2.75

THERMODYNAMICS, *Enrico Fermi*
Unabridged reproduction of 1937 edition. Elementary in treatment; remarkable for clarity, organization. Requires no knowledge of advanced math beyond calculus, only familiarity with fundamentals of thermometry, calorimetry. Partial Contents: Thermodynamic systems; First & Second laws of thermodynamics; Entropy; Thermodynamic potentials: phase rule, reversible electric cell; Gaseous reactions: van't Hoff reaction box, principle of LeChatelier; Thermodynamics of dilute solutions: osmotic & vapor pressures, boiling & freezing points; Entropy constant. Index. 25 problems. 24 illustrations. x + 160pp. 5⅜ x 8.   60361-X Paperbound $2.00

CELESTIAL OBJECTS FOR COMMON TELESCOPES,
Rev. T. W. Webb
Classic handbook for the use and pleasure of the amateur astronomer. Of inestimable aid in locating and identifying thousands of celestial objects. Vol I, The Solar System: discussions of the principle and operation of the telescope, procedures of observations and telescope-photography, spectroscopy, etc., precise location information of sun, moon, planets, meteors. Vol. II, The Stars: alphabetical listing of constellations, information on double stars, clusters, stars with unusual spectra, variables, and nebulae, etc. Nearly 4,000 objects noted. Edited and extensively revised by Margaret W. Mayall, director of the American Assn. of Variable Star Observers. New Index by Mrs. Mayall giving the location of all objects mentioned in the text for Epoch 2000. New Precession Table added. New appendices on the planetary satellites, constellation names and abbreviations, and solar system data. Total of 46 illustrations. Total of xxxix + 606pp. 5⅜ x 8.   20917-2, 20918-0 Two volume set, paperbound $5.00

PLANETARY THEORY,
E. W. Brown and C. A. Shook
Provides a clear presentation of basic methods for calculating planetary orbits for today's astronomer. Begins with a careful exposition of specialized mathematical topics essential for handling perturbation theory and then goes on to indicate how most of the previous methods reduce ultimately to two general calculation methods: obtaining expressions either for the coordinates of planetary positions or for the elements which determine the perturbed paths. An example of each is given and worked in detail. Corrected edition. Preface. Appendix. Index. xii + 302pp. 5⅜ x 8½.   61133-7 Paperbound $2.25

STAR NAMES AND THEIR MEANINGS,
Richard Hinckley Allen
An unusual book documenting the various attributions of names to the individual stars over the centuries. Here is a treasure-house of information on a topic not normally delved into even by professional astronomers; provides a fascinating background to the stars in folk-lore, literary references, ancient writings, star catalogs and maps over the centuries. Constellation-by-constellation analysis covers hundreds of stars and other asterisms, including the Pleiades, Hyades, Andromedan Nebula, etc. Introduction. Indices. List of authors and authorities. xx + 563pp. 5⅜ x 8½.   21079-0 Paperbound $3.00

A SHORT HISTORY OF ASTRONOMY, *A. Berry*
Popular standard work for over 50 years, this thorough and accurate volume covers the science from primitive times to the end of the 19th century. After the Greeks and the Middle Ages, individual chapters analyze Copernicus, Brahe, Galileo, Kepler, and Newton, and the mixed reception of their discoveries. Post-Newtonian achievements are then discussed in unusual detail: Halley, Bradley, Lagrange, Laplace, Herschel, Bessel, etc. 2 Indexes. 104 illustrations, 9 portraits. xxxi + 440pp. 5⅜ x 8.   20210-0 Paperbound $2.75

SOME THEORY OF SAMPLING, *W. E. Deming*
The purpose of this book is to make sampling techniques understandable to and useable by social scientists, industrial managers, and natural scientists who are finding statistics increasingly part of their work. Over 200 exercises, plus dozens of actual applications. 61 tables. 90 figs. xix + 602pp. 5⅜ x 8½.   61755-6 Paperbound $3.50

PRINCIPLES OF STRATIGRAPHY,
A. W. Grabau

Classic of 20th century geology, unmatched in scope and comprehensiveness. Nearly 600 pages cover the structure and origins of every kind of sedimentary, hydrogenic, oceanic, pyroclastic, atmoclastic, hydroclastic, marine hydroclastic, and bioclastic rock; metamorphism; erosion; etc. Includes also the constitution of the atmosphere; morphology of oceans, rivers, glaciers; volcanic activities; faults and earthquakes; and fundamental principles of paleontology (nearly 200 pages). New introduction by Prof. M. Kay, Columbia U. 1277 bibliographical entries. 264 diagrams. Tables, maps, etc. Two volume set. Total of xxxii + 1185pp. 5⅜ x 8. 60686-4, 60687-2 Two volume set, paperbound $6.25

SNOW CRYSTALS, *W. A. Bentley and W. J. Humphreys*

Over 200 pages of Bentley's famous microphotographs of snow flakes—the product of painstaking, methodical work at his Jericho, Vermont studio. The pictures, which also include plates of frost, glaze and dew on vegetation, spider webs, windowpanes; sleet; graupel or soft hail, were chosen both for their scientific interest and their aesthetic qualities. The wonder of nature's diversity is exhibited in the intricate, beautiful patterns of the snow flakes. Introductory text by W. J. Humphreys. Selected bibliography. 2,453 illustrations. 224pp. 8 x 10¼. 20287-9 Paperbound $3.25

THE BIRTH AND DEVELOPMENT OF THE GEOLOGICAL SCIENCES,
F. D. Adams

Most thorough history of the earth sciences ever written. Geological thought from earliest times to the end of the 19th century, covering over 300 early thinkers & systems: fossils & their explanation, vulcanists vs. neptunists, figured stones & paleontology, generation of stones, dozens of similar topics. 91 illustrations, including medieval, renaissance woodcuts, etc. Index. 632 footnotes, mostly bibliographical. 511pp. 5⅜ x 8. 20005-1 Paperbound $2.75

ORGANIC CHEMISTRY, *F. C. Whitmore*

The entire subject of organic chemistry for the practicing chemist and the advanced student. Storehouse of facts, theories, processes found elsewhere only in specialized journals. Covers aliphatic compounds (500 pages on the properties and synthetic preparation of hydrocarbons, halides, proteins, ketones, etc.), alicyclic compounds, aromatic compounds, heterocyclic compounds, organophosphorus and organometallic compounds. Methods of synthetic preparation analyzed critically throughout. Includes much of biochemical interest. "The scope of this volume is astonishing," *Industrial and Engineering Chemistry*. 12,000-reference index. 2387-item bibliography. Total of x + 1005pp. 5⅜ x 8. 60700-3, 60701-1 Two volume set, paperbound $4.50

THE PHASE RULE AND ITS APPLICATION,
Alexander Findlay

Covering chemical phenomena of 1, 2, 3, 4, and multiple component systems, this "standard work on the subject" (*Nature,* London), has been completely revised and brought up to date by A. N. Campbell and N. O. Smith. Brand new material has been added on such matters as binary, tertiary liquid equilibria, solid solutions in ternary systems, quinary systems of salts and water. Completely revised to triangular coordinates in ternary systems, clarified graphic representation, solid models, etc. 9th revised edition. Author, subject indexes. 236 figures. 505 footnotes, mostly bibliographic. xii + 494pp. 5⅜ x 8. 60091-2 Paperbound $2.75

CATALOGUE OF DOVER BOOKS

The Principles of Electrochemistry,
D. A. MacInnes
Basic equations for almost every subfield of electrochemistry from first principles, referring at all times to the soundest and most recent theories and results; unusually useful as text or as reference. Covers coulometers and Faraday's Law, electrolytic conductance, the Debye-Hueckel method for the theoretical calculation of activity coefficients, concentration cells, standard electrode potentials, thermodynamic ionization constants, pH, potentiometric titrations, irreversible phenomena. Planck's equation, and much more. 2 indices. Appendix. 585-item bibliography. 137 figures. 94 tables. ii + 478pp. 5⅜ x 8⅜.
60052-1 Paperbound $3.00

Mathematics of Modern Engineering,
E. G. Keller and R. E. Doherty
Written for the Advanced Course in Engineering of the General Electric Corporation, deals with the engineering use of determinants, tensors, the Heaviside operational calculus, dyadics, the calculus of variations, etc. Presents underlying principles fully, but emphasis is on the perennial engineering attack of set-up and solve. Indexes. Over 185 figures and tables. Hundreds of exercises, problems, and worked-out examples. References. Total of xxxiii + 623pp. 5⅜ x 8.   60734-8, 60735-6 Two volume set, paperbound $3.70

Aerodynamic Theory: A General Review of Progress,
William F. Durand, editor-in-chief
A monumental joint effort by the world's leading authorities prepared under a grant of the Guggenheim Fund for the Promotion of Aeronautics. Never equalled for breadth, depth, reliability. Contains discussions of special mathematical topics not usually taught in the engineering or technical courses. Also: an extended two-part treatise on Fluid Mechanics, discussions of aerodynamics of perfect fluids, analyses of experiments with wind tunnels, applied airfoil theory, the nonlifting system of the airplane, the air propeller, hydrodynamics of boats and floats, the aerodynamics of cooling, etc. Contributing experts include Munk, Giacomelli, Prandtl, Toussaint, Von Karman, Klemperer, among others. Unabridged republication. 6 volumes. Total of 1,012 figures, 12 plates, 2,186pp. Bibliographies. Notes. Indices. 5⅜ x 8½.   61709-2, 61710-6, 61711-4, 61712-2, 61713-0, 61715-9   Six volume set, paperbound $13.50

Fundamentals of Hydro- and Aeromechanics,
L. Prandtl and O. G. Tietjens
The well-known standard work based upon Prandtl's lectures at Goettingen. Wherever possible hydrodynamics theory is referred to practical considerations in hydraulics, with the view of unifying theory and experience. Presentation is extremely clear and though primarily physical, mathematical proofs are rigorous and use vector analysis to a considerable extent. An Engineering Society Monograph, 1934. 186 figures. Index. xvi + 270pp. 5⅜ x 8.
60374-1 Paperbound $2.25

Applied Hydro- and Aeromechanics,
L. Prandtl and O. G. Tietjens
Presents for the most part methods which will be valuable to engineers. Covers flow in pipes, boundary layers, airfoil theory, entry conditions, turbulent flow in pipes, and the boundary layer, determining drag from measurements of pressure and velocity, etc. Unabridged, unaltered. An Engineering Society Monograph. 1934. Index. 226 figures, 28 photographic plates illustrating flow patterns. xvi + 311pp. 5⅜ x 8.   60375-X Paperbound $2.50

A COURSE IN MATHEMATICAL ANALYSIS,
Edouard Goursat

Trans. by E. R. Hedrick, O. Dunkel, H. G. Bergmann. Classic study of fundamental material thoroughly treated. Extremely lucid exposition of wide range of subject matter for student with one year of calculus. Vol. 1: Derivatives and differentials,· definite integrals, expansions in series, applications to geometry. 52 figures, 556pp. 60554-X Paperbound $3.00. Vol. 2, Part I: Functions of a complex variable, conformal representations, doubly periodic functions, natural boundaries, etc. 38 figures, 269pp. 60555-8 Paperbound $2.25. Vol. 2, Part II: Differential equations, Cauchy-Lipschitz method, nonlinear differential equations, simultaneous equations, etc. 308pp. 60556-6 Paperbound $2.50. Vol. 3, Part I: Variation of solutions, partial differential equations of the second order. 15 figures, 339pp. 61176-0 Paperbound $3.00. Vol. 3, Part II: Integral equations, calculus of variations. 13 figures, 389pp. 61177-9 Paperbound $3.00 60554-X, 60555-8, 60556-6 61176-0, 61177-9 Six volume set,
paperbound $13.75

PLANETS, STARS AND GALAXIES,
A. E. Fanning

Descriptive astronomy for beginners: the solar system; neighboring galaxies; seasons; quasars; fly-by results from Mars, Venus, Moon; radio astronomy; etc. all simply explained. Revised up to 1966 by author and Prof. D. H. Menzel, former Director, Harvard College Observatory. 29 photos, 16 figures. 189pp. 5⅜ x 8½.   21680-2 Paperbound $1.50

GREAT IDEAS IN INFORMATION THEORY, LANGUAGE AND CYBERNETICS,
Jagjit Singh

Winner of Unesco's Kalinga Prize covers language, metalanguages, analog and digital computers, neural systems, work of McCulloch, Pitts, von Neumann, Turing, other important topics. No advanced mathematics needed, yet a full discussion without compromise or distortion. 118 figures. ix + 338pp. 5⅜ x 8½.
21694-2 Paperbound $2.25

GEOMETRIC EXERCISES IN PAPER FOLDING,
T. Sundara Row

Regular polygons, circles and other curves can be folded or pricked on paper, then used to demonstrate geometric propositions, work out proofs, set up well-known problems. 89 illustrations, photographs of actually folded sheets. xii + 148pp. 5⅜ x 8½.   21594-6 Paperbound $1.00

VISUAL ILLUSIONS, THEIR CAUSES, CHARACTERISTICS AND APPLICATIONS,
M. Luckiesh

The visual process, the structure of the eye, geometric, perspective illusions, influence of angles, illusions of depth and distance, color illusions, lighting effects, illusions in nature, special uses in painting, decoration, architecture, magic, camouflage. New introduction by W. H. Ittleson covers modern developments in this area. 100 illustrations. xxi + 252pp. 5⅜ x 8.
21530-X Paperbound $1.50

ATOMS AND MOLECULES SIMPLY EXPLAINED,
B. C. Saunders and R. E. D. Clark

Introduction to chemical phenomena and their applications: cohesion, particles, crystals, tailoring big molecules, chemist as architect, with applications in radioactivity, color photography, synthetics, biochemistry, polymers, and many other important areas. Non technical. 95 figures. x + 299pp. 5⅜ x 8½.
21282-3 Paperbound $1.50

APPLIED OPTICS AND OPTICAL DESIGN,
A. E. Conrady

With publication of vol. 2, standard work for designers in optics is now complete for first time. Only work of its kind in English; only detailed work for practical designer and self-taught. Requires, for bulk of work, no math above trig. Step-by-step exposition, from fundamental concepts of geometrical, physical optics, to systematic study, design, of almost all types of optical systems. Vol. 1: all ordinary ray-tracing methods; primary aberrations; necessary higher aberration for design of telescopes, low-power microscopes, photographic equipment. Vol. 2: (Completed from author's notes by R. Kingslake, Dir. Optical Design, Eastman Kodak.) Special attention to high-power microscope, anastigmatic photographic objectives. "An indispensable work," *J., Optical Soc. of Amer.* Index. Bibliography. 193 diagrams. 852pp. 6⅛ x 9¼.

60611-2, 60612-0 Two volume set, paperbound $8.00

MECHANICS OF THE GYROSCOPE, THE DYNAMICS OF ROTATION,
R. F. Deimel, Professor of Mechanical Engineering at Stevens Institute of Technology

Elementary general treatment of dynamics of rotation, with special application of gyroscopic phenomena. No knowledge of vectors needed. Velocity of a moving curve, acceleration to a point, general equations of motion, gyroscopic horizon, free gyro, motion of discs, the damped gyro, 103 similar topics. Exercises. 75 figures. 208pp. 5⅜ x 8.

60066-1 Paperbound $1.75

STRENGTH OF MATERIALS,
J. P. Den Hartog

Full, clear treatment of elementary material (tension, torsion, bending, compound stresses, deflection of beams, etc.), plus much advanced material on engineering methods of great practical value: full treatment of the Mohr circle, lucid elementary discussions of the theory of the center of shear and the "Myosotis" method of calculating beam deflections, reinforced concrete, plastic deformations, photoelasticity, etc. In all sections, both general principles and concrete applications are given. Index. 186 figures (160 others in problem section). 350 problems, all with answers. List of formulas. viii + 323pp. 5⅜ x 8.

60755-0 Paperbound $2.50

HYDRAULIC TRANSIENTS,
G. R. Rich

The best text in hydraulics ever printed in English . . . by former Chief Design Engineer for T.V.A. Provides a transition from the basic differential equations of hydraulic transient theory to the arithmetic integration computation required by practicing engineers. Sections cover Water Hammer, Turbine Speed Regulation, Stability of Governing, Water-Hammer Pressures in Pump Discharge Lines, The Differential and Restricted Orifice Surge Tanks, The Normalized Surge Tank Charts of Calame and Gaden, Navigation Locks, Surges in Power Canals—Tidal Harmonics, etc. Revised and enlarged. Author's prefaces. Index. xiv + 409pp. 5⅜ x 8½.

60116-1 Paperbound $2.50

Prices subject to change without notice.

Available at your book dealer or write for free catalogue to Dept. Adsci, Dover Publications, Inc., 180 Varick St., N.Y., N.Y. 10014. Dover publishes more than 150 books each year on science, elementary and advanced mathematics, biology, music, art, literary history, social sciences and other areas.